An Evolutionary Pathway for
Coping with Emerging Infectious Disease

To Harry –
incomparable friend and host –
with warmest regards
Ralph Audy april 1968

Sek. Zenian Berge
Cairo, 1964

Image of J. Ralph Audy made while visiting Harry Hoogstraal at the US Naval Medical Research Unit No. 3 (NAMRU-3). Dr. Audy was former Professor of Tropical Medicine and Human Ecology in the Department of Medicine and Professor of International Health and Human Ecology in the Department of International Health and director of the Hooper Foundation, University of California, San Francisco.

An Evolutionary Pathway
for Coping with
Emerging Infectious Disease

Edited by

Scott L. Gardner, Daniel R. Brooks, Walter A. Boeger, and Eric P. Hoberg

Zea Books
Lincoln, Nebraska
2023

ISBN 978-1-60962-299-2 paperback

ISBN 978-1-60962-300-5 ebook

doi: 10.32873/unl.dc.zea.1504

Composed in Cambria types.

Zea Books are published by the University of Nebraska–Lincoln Libraries.

Nebraska
UNIVERSITY OF
Lincoln

This book is dedicated to the biodiversitists of the world
who are working to keep the place habitable.

Contents

Illustrations

Figures

Tables

Contributors

Salvatore J. Agosta
Center for Environmental Studies, VCU Life Sciences, Virginia Commonwealth University, Richmond, Virginia, 23284, USA, and Stellenbosch Institute for Advanced Study, Stellenbosch, South Africa
ORCID https://orcid.org/0000-0003-1874-5572
Email sagosta@vcu.edu

Sabrina B. L. Araujo
Biological Interactions, Programa de Pós-graduação em Ecologia e Conservação, and Departamento de Física, Universidade Federal do Paraná, Curitiba, Paraná 81531-980, Brazil
Email araujosbl@gmail.com

Walter A. Boeger
Biological Interactions and Programa de Pós-graduação em Ecologia e Conservação, Universidade Federal do Paraná, Curitiba, Paraná, 81531-980, Brazil
ORCID https://orcid.org/0000-0002-6004-2822
Email wboeger@gmail.com

Daniel R. Brooks
Centre for Ecological Research, Institute of Evolution, 1121 Budapest, Konkoly-Thege Miklós út 29-33, Hungary; Stellenbosch Institute for Advanced Study, Stellenbosch, South Africa; Harold W. Manter Laboratory of Parasitology, University of Nebraska State Museum, University of Nebraska–Lincoln, Lincoln, Nebraska, USA, 68588-0514; and Department of Ecology and Evolutionary Biology, University of Toronto, Toronto, Ontario, M5S3B2 Canada
ORCID https://orcid.org/0000-0002-7891-9821
Email dnlbrooks@gmail.com

Jocelyn P. Colella
Biodiversity Institute, University of Kansas, Lawrence, Kansas, USA, 66045
ORCID https://orcid.org/0000-0003-2463-1029
Email jcollela.jc@gmail.com

Joseph A. Cook
Museum of Southwestern Biology, Department of Biology, University of New Mexico, MSC03 2020, Albuquerque, New Mexico, USA, 87138
ORCID https://orcid.org/0000-0003-3985-0670
Email tucojoe@gmail.com

Jonathan L. Dunnum
Museum of Southwestern Biology, Department of Biology, University of New Mexico, MSC03 2020, Albuquerque, New Mexico, USA, 87138
ORCID https://orcid.org/0000-0001-7868-3719
Email jldunnum@gmail.com

Gábor Földvári
Institute of Evolution, Centre for Ecological Research, 1121 Budapest, Konkoly-Thege Miklós út 29-33, Hungary; Centre for Eco-Epidemiology, National Laboratory for Health Security, 1121 Budapest, Konkoly-Thege Miklós út 29-33, Hungary ORCID https://orcid.org/0000-0001-5297-9036
Email foldvarigabor@gmx.de

Scott L. Gardner
Harold W. Manter Laboratory of Parasitology, W-529 Nebraska Hall, University of Nebraska State Museum and School of Biological Sciences, University of Nebraska–Lincoln, Lincoln, Nebraska, USA, 68588-0514
ORCID https://orcid.org/0000-0003-3133-740X
Email slg@unl.edu

Eric P. Hoberg
Museum of Southwestern Biology, Department of Biology, University of New Mexico, MSC03 2020, Albuquerque, New Mexico, USA, 87138, and Department of Pathobiological Sciences, School of Veterinary Medicine, University of Wisconsin–Madison, Wisconsin, USA, 53716
ORCID https://orcid.org/0000-0003-0819-7437
Email geocolonizer@gmail.com

Alicia Juarrero
VectorAnalytica, Inc., Washington, DC, USA, 20007
Email aliciajuarrero@gmail.com

Vitaliy Kharchenko
I.I. Schmalhausen Institute of Zoology, vul. B. Khmelnyts'kogo, 15 Kyiv, 01054, Ukraine
ORCID https://orcid.org/0000-0002-3824-2078
Email vit.khark@gmail.com

Marina Knickel
Konrad Lorenz Institute for Evolution and Cognition
Research, Klosterneuburg, Austria, and the Department
of Agriculture, Food and Environment, University of Pisa,
Italy
ORCID https://orcid.org/0000-0003-1496-7864
Email knickel.marina@gmail.com

Christine Marizzi
BioB\us, New York City, New York, USA
ORCID https://orcid.org/0000-0001-9065-3352
Email christine@biobus.org

Orsolya Molnár
Medical University of Vienna, Institute for Hygiene
and Applied Immunology, Center for Pathophysiology,
Infectiology and Immunology, Kinderspitalgasse 15, Vienna,
1090, Austria
ORCID https://orcid.org/0000-0002-2458-4659
Email orsolya.bajer-molnar@meduniwien.ac.at

Eloy Ortíz
VectorAnalytica, Inc., 6904 N. Kendall Drive, Apt. F302,
Miami, Florida, 33156, USA
Email eortiz@vectoranalytica.com

Bernd Panassiti
Independent researcher, Munich 81543, Germany
ORCID https://orcid.org/0000-0003-1415-4097
Email bernd.panassiti@gmail.com

Wolfgang Preiser
Department of Pathology, Faculty of Medicine and Health
Sciences, Stellenbosch University, and NHLS Tygerberg
Francie van Zijl Rylaan Drive, Tygerberg, Cape Town,
South Africa
ORCID https://orcid.org/0000-0002-0254-7910
Email preiser@sun.ac.za

Angie T. C. Souza
Biological Interactions, Universidade Federal do Paraná,
Curitiba, Paraná 81531-980, Brazil, and Programa de
Pós-graduação em Ecologia e Conservação, Universidade
Federal do Paraná, Curitiba, Paraná 81531-980, Brazil.
ORCID https://orcid.org/0000-0003-0592-8070
Email angietcsouza@gmail.com

Éva Szabó
Institute of Evolution, Centre for Ecological Research,
1121 Budapest, Konkoly-Thege Miklós út 29-33, Hungary;
Centre for Eco-Epidemiology, National Laboratory for
Health Security, 1121 Budapest, Konkoly-Thege Miklós út
29-33, Hungary; Doctoral School of Biology, Institute of
Biology, Eötvös Loránd University, Budapest, Hungary
Email gagarinalba@gmail.com

Valeria Trivellone
Illinois Natural History Survey, Prairie Research Institute,
University of Illinois at Urbana-Champaign, Champaign,
Illinois, USA, 61821
ORCID https://orcid.org/0000-0003-1415-4097
Email valeria3@illinois.edu

Acknowledgments

Salvatore J. Agosta (Chapters 1, 7, Conclusion)
SJA thanks the editors of this volume for inviting him to write this paper. Thanks also to Virginia Commonwealth University, VCU Life Sciences, Center for Environmental Studies, for providing a short sabbatical and to the Stellenbosch Institute for Advanced Study for a visiting fellowship that provided the ideal setting to develop many of the ideas in this paper.

Sabrina B. L. Araujo (Chapter 4)
SBLA thanks the support provided by the Conselho Nacional de Pesquisa e Desenvolvimento, Brazil, grant number 311284/2021-3.

Walter A. Boeger (Introduction, chapters 2, 4, 6, 7, 8)
WAB acknowledges the support provided by the Conselho Nacional de Pesquisa e Desenvolvimento, Brazil, through the grant "O Paradigma de Estocolmo: explorando as previsões do paradigma sobre os padrões e processos em sistemas parasitos e hospedeiros" (No. 302708/2020-0). WAB expresses profound personal thanks for the personal and scientific interactions and opportunities provided by many people, among them Daniel Brooks, Eric Hoberg, Salvatore Agosta, Scott Gardner, Sabrina Araujo, Valeria Trivellone, and many recent graduate and undergraduate students—it was these interactions that stimulated many of the ideas and studies on the Stockholm Paradigm. Daniel Brooks, especially, supported most of his daydreams in evolution; some made sense and resulted in actual publications.

Daniel R. Brooks (Introduction, chapters 2, 5, 6, 7, 8, Conclusion)
DRB expresses personal thanks to the long-time collaboration with Deborah McLennan, most notably in cowriting *Phylogeny, Ecology, and Behavior: A Research Program in Comparative Biology* (1990), *Parascript: Parasites and the Language of Evolution* (1993), and *The Nature of Diversity: An Evolutionary Voyage of Discovery* (2002); to Sal Agosta, whose proposals of sloppy fitness space and of ecological fitting in sloppy fitness space established the dynamical basis for the Stockholm Paradigm (SP) for evolutionary studies in general and for emerging infectious diseases in particular; to Scott Gardner, who organized, hosted, and funded a 2013 workshop at the Cedar Point Biological Station of the University of Nebraska–Lincoln, where the DAMA protocol was first formulated; to Sven Jakobsson whose invitation to speak in 2013 at the 20th anniversary of the Blod Bad conference in Tovetorp, Sweden, provided the opportunity to first use the term "Stockholm Paradigm" to honor the essential contributions by researchers at Stockholm University; to Walter Boeger, who obtained funding through the Ciencias sem Fronteras program that allowed DRB to work with WB in 2014–16 and to establish the collaboration with Sabrina Araujo that led to the modeling framework for the SP; to Soren Nylin and Niklas Janz, who organized, hosted, and obtained funding from the Wallenberg Foundation for a workshop on ecological associations at the Tovetorp research station; to the Stellenbosch Institute for Advanced Study for a 2017 fellowship, where most of the writing of *The Stockholm Paradigm: Climate Change and Emerging Disease* was done, and a 2021 fellowship with Sal Agosta; to the Institute for Advanced Studies, Köszeg, Hungary, for fellowships during the period of 2016–23; to Eörs Szathmáry, for organizing, hosting, and funding a 2016 workshop on the SP and DAMA protocol at Balatonfured, Hungary, and for organizing and hosting a 2017 workshop exploring the potential for a COST proposal to establish a DAMA network in EU at the Parmenides Foundation in Pullach, Germany, as well as for his support for a Hungarian Academy of Sciences Distinguished Guest Scientist Fellowship during March–May 2022; to Gabor Földvári, for organizing and leading the successful effort to obtain the COST Action CA21170 Prevention, Anticipation, and Mitigation of Tick-Borne Disease Risk Applying the DAMA Protocol (PRAGMATICK) funded by the European Union and for organizing and hosting the 2023 PRAGMATICK workshop—Stockholm Paradigm, DAMA Protocol, and Citizen Science: Applications for the PRAGMATICK COST-Action; to colleagues of the Para Limes Foundation and Chatham House, who have provided opportunities for DRB to present these ideas to diverse audiences and to learn from them; and to the colleagues who have participated as coauthors of this volume for showing the value of convivial collaboration in elaborating a new paradigm to help cope effectively with the dual threats of climate change and emerging infectious disease; and to Jan Vasbinder, director of the Para Limes Foundation, for asking DRB to participate in the 2021 virtual conference Buying Time; to Tim Benton of Chatham House for inviting DRB to speak in the 2021 Systemic Risk workshop of Chatham House.

Jocelyn P. Colella (Chapter 7)
JPC is Assistant Curator of Mammals in the Biodiversity Institute, University of Kansas, and assistant professor of ecology and evolutionary biology. JPC is central to

the development biorepository protocols within PICANTE and DAMA with funding from the National Science Foundation through Collections in Support of Biological Research (CSBR), DBI 2100955.

Joseph A. Cook (Chapter 7)
JAC is Distinguished Professor of Biology and the Curator of Mammals in the Museum of Southwestern Biology at the University of New Mexico. Concepts explored in DAMA were facilitated through development of new methods and protocols for the holistic specimen in multiple integrated field inventories across three continents and by building extensive biorepositories and databases that connect pathogens and hosts. Funding from the National Science Foundation was provided to JAC and EPH through Collaborative Research: Integrated Inventory of Biomes of the Arctic (NSF DEB 125801) to JAC, EPH, JPC, and others in the PICANTE Consortium (Pathogen Informatics Center: Analyses, Networking, Translation, & Education) through the Predictive Intelligence for Pandemic Prevention Program (NSF 2155222); to JAC and SLG through Mongolia Vertebrate Parasite Project (NSF 0717214); and to JAC through NSF 2026377 Steppe Parasite Networks and NSF 1901920 Terrestrial Parasite Tracker.

Jonathan L. Dunnum (Chapter 7)
JLD is collections manager for the Division of Mammalogy in the Museum of Southwestern Biology at the University of New Mexico. JLD has been central to creating methodologies for field inventories, databasing, and development of biorepositories for hosts and pathogens as a fundamental foundation of PICANTE and the DAMA protocol.

Gábor Földvári (Chapters 5, 6, 11)
GF was supported through grants from the COST Action CA21170 Prevention, Anticipation, and Mitigation of Tick-Borne Disease Risk Applying the DAMA Protocol (PRAGMATICK) and the National Research, Development, and Innovation Office (RRF-2.3.1-21-2022-00006 and K143622) .

Scott L. Gardner (Introduction, chapter 6)
SLG is curator and director of the Harold W. Manter Laboratory of Parasitology (HWML), Division of Parasitology, University of Nebraska State Museum, University of Nebraska–Lincoln (UNL). He is also full professor in the School of Biological Sciences at UNL and adjunct professor in the Department of Environmental, Agricultural, and Occupational Health, College of Public Health, University

of Nebraska Medical Center, Omaha. The concept for the DAMA protocol was developed during a workshop at UNL's Cedar Point Biological Station with sponsorship by the HWML and the National Science Foundation. All work on development of this book and ideas of parasite-mammal collecting was funded by both the HWML and the National Science Foundation, including grants DEB-0717214 (SLG and JAC), DBI-0646356, DBI-0646356, DBI-1458139, DBI-1756397, and DBI-1901911 to SLG and collaborators.

Eric P. Hoberg (Introduction, chapters 2, 5, 6, 7, 8)
EPH is former curator of the US National Parasite Collection, Agricultural Research Service, USDA, and holds current appointments in the Museum of Southwestern Biology, University of New Mexico in Albuquerque, and in the Department of Pathobiological Sciences in the School of Veterinary Medicine at University of Wisconsin–Madison. EPH is coauthor of *The Stockholm Paradigm: Climate Change and Emerging Disease* with DRB and WB. EPH is appreciative of more than three decades of illuminating collaborations with DRB, WAB, JAC, SLG, and colleagues near and far that have culminated here in our current explorations of the SP and DAMA protocol. Gratitude to Daniel Brooks, Walter Boeger, Salvatore Agosta, and Valeria Trivellone for the continuing and expanding synthesis that is the Stockholm Paradigm. Concepts explored in the SP and DAMA for integrative inventories and archival specimen biorepositories for pathogens and hosts were in part developed with support from the National Science Foundation (NSF). Funding was provided to JAC and EPH and the Museum of Southwestern Biology, University of New Mexico, for Collaborative Research: Integrated Inventory of Biomes of the Arctic through NSF Division of Environmental Biology, DEB 1258010; and to JAC, JPC, EPH, and others in the PICANTE Consortium (Pathogen Informatics Center: Analyses, Networking, Translation, & Education) through the NSF PIPP (Predictive Intelligence for Pandemic Prevention Program), NSF 2155222. The foundations for "knowing the biosphere" and an integrative DAMA protocol presented here were refined by EPH, DRB, WAB, OM, VT, and colleagues during the PRAGMATICK workshop Stockholm Paradigm, DAMA Protocol, and Citizen Science: Applications for the PRAGMATICK COST-Action of the National Research, Development and Innovation Office in Hungary (RRF-2.3.1-21-2022-00006) and the COST Action CA21170 Prevention, Anticipation, and Mitigation of Tick-Borne Disease Risk Applying the DAMA Protocol (PRAGMATICK), European Union, Budapest, Hungary, 16–17 February 2023.

Alicia Juarrero (Chapters 6, 9)
This research was funded by VectorAnalytica, Inc.

Marina Knickel (Chapter 10)
MK thanks Adél Marx for her help in collecting and reviewing literature, Mátyás Massár for assistance with visualization tools, and Dr. Valeria Trivellone for her valuable insights on crop diseases, as well as three anonymous reviewers for their insightful comments.

Christine Marizzi (Chapter 10)
CM thanks Adél Marx for her help in collecting and reviewing literature, Mátyás Massár for assistance with visualization tools, and Dr. Valeria Trivellone for her valuable insights on crop diseases, as well as three anonymous reviewers for their insightful comments.

Orsolya Molnár (Chapters 5, 6, 10)
OM thanks Adél Marx for her help in collecting and reviewing literature, Mátyás Massár for assistance with visualization tools, and Dr. Valeria Trivellone for her valuable insights on crop diseases, as well as three anonymous reviewers for their insightful comments.

Eloy Ortíz (Chapters 6, 9)
This research was funded by VectorAnalytica, Inc.

Bernd Panassiti (Chapter 3)
BP thanks Daniel R. Brooks for his valuable insights on cophylogenetic analyses and for helping us in developing critical thinking skills by integrating complex aspects of the biological nature, Walter A. Boeger for driving us through a better understanding of the PACT algorithm, and all the editors and reviewers who provided precious feedback to improve the structure, grammar, and presentation of our chapter. Also, BP is very grateful to Valeria Trivellone for introducing him to the fascinating world of the Stockholm Paradigm and all the people involved in pushing this concept forward. BP thanks from the bottom of his heart his wife, Marisa Gloria Panassiti, and his family for never losing patience with his scientific endeavors.

Angie T. C. Souza (Chapter 4)
ATCS was supported by a scholarship from CAPES—Coordenação de Aperfeiçoamento de Pessoal de Nível Superior (Finance Code 001: ATCS).

Éva Szabó (Chapter 11)
ÉS was supported through grants from the COST Action CA21170 "Prevention, anticipation and mitigation of tick-borne disease risk applying the DAMA protocol (PRAGMATICK)" and the National Research, Development and Innovation Office (RRF-2.3.1-21-2022-00006 and K143622) as well as an Eötvös Loránd Research Network Young Researcher's Grant.

Valeria Trivellone (Chapters 3, 5, 6, 7, 8)
VT thanks Daniel R. Brooks for his valuable insights on cophylogenetic analyses and for helping me develop critical thinking skills by integrating complex aspects of the biological nature, Walter A. Boeger for driving me through a better understanding of the PACT algorithm, and Eric P. Hoberg for opening my eyes to broad concepts such as the biosphere and interfaces between habitats and for guiding me expertly to a better interpretation of the interactions among the living entities on earth. A special thanks to Chris H. Dietrich for relentlessly mentoring me on the difficult journey of integrating ecological and evolutionary studies. Without their experiences, support, and the gift of their time, my professional path would have been much longer and more winding. Thank you to other coauthors, Bernd Panassiti, Salvatore J. Agosta, Jocelyn P. Colella, Joseph A. Cook, Jonathan L. Dunnum, Gabor Földvári, Scott L. Gardner, Alicia Juarrero, Vitaliy Kharchenko, Orsolya Molnár, Eloy Ortíz, and Wolfgang Preiser, who provided inspiration and compelling feedback, and to all the editors and reviewers who provided precious feedback to improve the structure and presentation of the chapters.

Introduction

The Emerging Infectious Disease Crisis: Changing a Losing Game

Daniel R. Brooks, Eric P. Hoberg, Walter A. Boeger, and Scott L. Gardner

Abstract

Emerging infectious disease (EID) represents an existential threat to humanity. EIDs are increasing in frequency and impact because of climate change and other human activities. We are losing the battle against EIDs because of improper assessment of the risk of EID. This stems from adherence to a failed paradigm of pathogen-host associations that suggests EIDs ought to be both unpredictable and rare. That, in turn, leads to policies suggesting that crisis response is the best we can do. Real-time and phylogenetic assessments show EIDs to be neither rare nor unpredictable—this is the parasite paradox that shows the failures of the traditional paradigm. The Stockholm Paradigm (SP) resolves the parasite paradox, based on the notion that EIDs are expressions of preexisting capacities of pathogens that colonize susceptible but previously unexposed hosts when environmental perturbations create new opportunities. This makes risk space much larger than thought; moreover, climate change and anthropogenic activities increase the risk of EID. The policy extension of the SP is the DAMA protocol (Document, Assess, Monitor, Act). Preexisting capacities for colonizing new hosts given the opportunity are both specific and phylogenetically conservative, hence, highly predictable. This provides hope that we can prevent at least some EIDs and mitigate the impacts of those we cannot prevent. Novel variants arise only after new hosts are colonized and are thus both likely and unpredictable. This makes the DAMA protocol the essential starting point for a clear pathway for coping effectively with the EID crisis. This volume explores the state of the art with respect to the SP and the DAMA protocol.

Keywords: emerging infectious diseases, climate change, Stockholm Paradigm, parasite paradox, DAMA protocol

Risky Business

Humanity is inundated daily with reports about human impacts on climate and climate impacts on humans. By now there ought to be significant and accurate assessment of the risks associated with climate change. However, this does not appear to be the case. In his recent book *Five Times Faster: Rethinking the Science, Economics, and Diplomacy of Climate Change*, Simon Sharpe (2023) emphasizes how poor risk assessment of various threat multipliers associated with climate change—in particular emissions technology—has jeopardized humanity's future. The contributors to this volume agree with Sharpe's assertion and believe that efforts to cope with emerging infectious diseases (EIDs) have also suffered from poor risk assessment. We believe the risk space has been grossly underestimated and that risk assessment for the role of climate change and global trade and travel in catalyzing EIDs in the wildlands and managed landscapes have been uniformly poor.

Humans are prone to poor risk assessment because they have a strong need for drama and heroism, a strong attraction to magic, and a strong aversion to bad news, especially news that involves taking personal responsibility. In the case of EID, inadequate risk assessment is associated with ecologists, clinicians, and policy makers having convinced themselves that what they are already doing is the best that can be done. The people who have gotten us into this situation are continually asked for solutions, and their reply is to demand more money while repeating the activities that are not working. In many ways, they see EID as a great mystery, much as did Soper (1919) when editorializing about the Spanish Influenza pandemic.

We think inquiring people may be asking the right questions about the EID crisis (e.g., Astbury et al., 2022) but not asking them of the right people. We know this because

14

the people who are being asked give answers that don't work. The contributions in this volume, based on critical insights offered by people more than half a century ago, in particular J. Ralph Audy (1958) and Charles S. Elton (1958), suggest that the people being asked the questions are doing the best they can, following the dictates of a failed scientific paradigm. At the same time, they present an alternative paradigm that explains why risk assessment has been so poor, why the risk is much greater than suspected, and what we can do to mitigate the risks associated with the EID crisis.

According to the historical background provided by Rosenthal et al. (2015) and Ndow et al. (2019), the term "emerging infectious disease" first appeared in the literature in the title of an article by Maurer (1962) about equine piroplasmosis being introduced to the United States, but it seems to have been in use among disease specialists as early as the mid-1950s. The term became popular in the 1980s, stimulated by the genital herpes and HIV/AIDS outbreaks, and was given impetus in major statements by the Institute of Medicine (1992) and the Centers for Disease Control and Prevention (1994). EIDs are caused by pathogens that affect humans and every species upon which humans depend. They are the result of pathogens that may be quite common (and not necessarily known to cause disease) in one host colonizing one or more new hosts as a result of changing geographic distribution and ecological interactions. They include diseases never seen before as well as diseases thought to have been contained or eradicated. EIDs are maintained and transmitted in water, food, soil, and air and through casual contact between hosts or with contaminated surfaces, sex, wildlife, domestic animals and plants (from companion animals to livestock and from garden plants to crops), and anything that bites or feeds on us and the species upon which we rely for survival. They occur in rural, urban, and wildland settings, from tropical to arctic regions, in developed and developing countries.

EIDs are literally beyond belief—they do not favor national borders; ethnic, social, or cultural groups; or economic, political, or religious belief systems. Affecting water, food, public health, and sociocultural and economic stability, they are a security concern for every country. EIDs are increasingly common, and they are expensive (Sokolow et al., 2022). Even if each one is only slightly detrimental, their accumulation (*pathogen pollution*: Daszak et al., 2000; Cunningham et al., 2003) increases the total economic burden of EID. Prior to the COVID pandemic, annual treatment costs and production losses caused by high probability/low

impact EID cost the world an estimated $1.3 trillion per year. This figure is greater than the gross domestic product (GDP) of all but 15 countries and weakens humanity's capacity to cope with low-probability/high-impact EIDs like SARS-CoV-2. A recent analysis by Trivellone et al. (2022; see also Brooks et al., 2022) indicated that the cost of EID in agriculture and livestock production in the United States (US) will exceed the US GDP by 2080, and if those costs are transferred to consumers, food will be unsustainably expensive. The longer a problem persists, the more expensive fixing it becomes.

Something Is Wrong

. . . you would think that the public health
response would be a well-oiled machine by now.
But yet again we have been blindsided.
—*New Scientist*, 30 January 2016

Why is there an EID crisis? And why is humanity losing the battle to cope with EIDs? The perspective of the contributors to this volume is that our scientific beliefs about the nature of pathogen-host systems, and the clinical and public policy actions stemming from those beliefs, do not accord with what is happening in the world, nor do they accord with our fundamental understanding of the nature of the biosphere. Pathogens are not enemies, or threats, or agents of illness and death. They are natural components of the biosphere, and their effects on various hosts need to be understood evolutionarily. Disease is not a natural purpose; it is an indication of marginal existence. Pathogens do not magically appear and disappear; they are present even when they are not causing disease or making headlines, living in hosts that are not diseased.

The Traditional Paradigm

The traditional assumption is that pathogens should rarely "change allegiances" because they are so strongly co-adapted to their hosts. Under this still-prevailing paradigm, a pathogen must evolve novel genetic capacity to colonize a new host, and we cannot predict when, where, or how that will happen, so we have no option except to wait for an infectious disease to emerge and then try to cope with it after the fact. Furthermore, the assumption that particular mutations must arise in order for a pathogen to move into a new host gives us a false sense of security. The risk space is assumed to be quite small, and EIDs are expected to be rare.

The notion of rarity comes into play because the traditional view holds that hosts and parasites become so specialized with each other that the parasite is not capable of colonizing a new host. If this were true, then parasites could colonize new hosts only when the right random genetic mutations just happen to produce the ability to live in a new host at the right place and time. So, the expectation is that EID will be rare. And if infectious diseases are both rare and unpredictable, responding in crisis mode after the fact may not be just all we can do, it might actually be cost-effective. This would be analogous to saying that because tornadoes are rare events, there is no reason to take precautions before tornado season. Just clean up after the fact. However, if something like climate change drives an increase in tornadoes to the point they are common, insurance companies begin to require tornado insurance as a precaution. At that point, trying to minimize the potential effects of tornadoes becomes more cost-effective than waiting until one happens. We suggest that the world is in a directly analogous state of affairs with respect to EID.

The Parasite Paradox

The traditional view of pathogen-host associations led to what is called the "parasite paradox" (Agosta et al., 2010). EIDs ought to be rare because strong coevolutionary ties between pathogens and hosts ought to greatly limit the capacity of pathogens to colonize new hosts, and colonizing new hosts requires the evolution of specific novel genetic capacities, which ought to be rare. Pathogens are indeed ecological specialists closely connected to their hosts. EIDs, however, occur often and rapidly, and phylogenetic studies show that pathogens changing hosts has been a common theme throughout evolutionary history. In fact, there is a recurring pattern in evolutionary history of pathogen host ranges increasing and decreasing, and then increasing again. This pattern has been called the "oscillation hypothesis" (Janz and Nylin, 2008; Janz, 2011; Nylin et al., 2018), and it highlights the enormous capacity pathogens have for survival and diversification. The disconnect between the theoretical framework of the prevailing paradigm and the accumulated empirical data on EID points clearly to the need for a new paradigm that resolves the parasite paradox.

Resolving the Parasite Paradox: The Stockholm Paradigm

The new alternative paradigm, called the Stockholm Paradigm (SP), has been explained in detail in multiple publications (e.g., Brooks et al., 2014; Hoberg and Brooks, 2015; Brooks and Boeger, 2019; Brooks et al., 2019; Agosta and Brooks, 2020; Agosta, 2022; Brooks et al., 2022) and in the next two chapters in this volume (Agosta, 2023, this volume; Brooks et al., 2023, this volume). The SP makes assertions that solve the parasite paradox and make clear the connection between general environmental perturbations, including climate change, and EID in that:

(1) Pathogens do not need new genetic capacities to colonize new hosts. All that is needed is a change in circumstances, so pathogens have opportunities for contact with susceptible but previously unexposed hosts. Because the triggering mechanism for EID no longer needs to be thought as one of capacity but can be viewed rather as one of opportunity, the arguments that EID ought to be rare and unpredictable are cast into doubt.

(2) Because EIDs are expressions of *preexisting capacities* of actual or potential pathogens, given *new opportunities*, infectious agents can spread quickly, and subsequent disease can emerge rapidly. When new genetic capacities of the pathogen arise in a new host, these new capacities emerge *after* the host is colonized.

(3) Perturbations in environmental conditions are associated with geographic expansion and colonization of additional hosts. Host ranges (the number of species used as hosts by the pathogen) expand and contract as geographic ranges of the host populations expand and contract, and this dynamic creates new opportunity space, giving rise to the evolutionary patterns of host ranges described by the oscillation hypothesis.

(4) Pathogens are the ultimate survivors. During periods of environmental stability, pathogens specialize, often in isolation, inhabiting a small proportion of the population of all susceptible hosts, increasing potential opportunities for colonizing new hosts when conditions change; conversely, during periods of environmental perturbations, pathogens generalize, colonizing increasing numbers of susceptible but previously unexposed hosts, increasing opportunities for *additional specializing*.

The Stockholm Paradigm and EID Risk Assessment: Bad News and Good

Evolutionary risk space is much larger than imagined
The SP suggests that the risk space for EID is far larger than many recent estimates (e.g., Carlson et al., 2022; Harvey

and Holmes, 2022). The SP also explains how environmental perturbations, such as climate change or anthropogenic impacts, play a critical role in setting the stage for EID.

Studies during the past 30 years of the here and now and of evolutionary history show us that (a) the traditional model is not correct—there is real-time and historical evidence that host colonizations (i.e., emergent diseases) happen often, but (b) those colonizations do not occur at random in evolutionary history (summarized in Brooks et al., 2019). They are clumped, they come in pulses, and those pulses correspond to episodes of large-scale environmental perturbations, including climate change (Hoberg et al., 2008). So, what is happening now is not a new phenomenon; it is simply new for contemporary humanity. Furthermore, because a newly colonized host has never been exposed to a given pathogen, there has never been a chance for resistance to evolve (natural selection cannot act on something before it happens). Each host colonization will thus produce acute disease until resistance evolves, which it usually quickly does. After that, the "new" parasite becomes a chronic rather than an acute problem *because it never disappears*. We assume that all EIDs "cost" something when they are acute, and cost much less when they are chronic. But they are never free of cost so long as they continue to persist as "pathogen pollution." The costs of acute disease may fluctuate, but the overall cost of EID will continue to rise.

Climate change directly increases EID risk

Biodiversity specialists, who recognized the role of climate change in altering host ranges in the wildlands, were initially slow to connect their findings with the EID phenomenon (e.g., Gardner and Campbell, 1992; Hoberg, 1997; Brooks and Hoberg, 2000, 2001; Hoberg and Adams, 2000). Initial suggestions of a link between climate change and biodiversity dynamics and EID emerged early in the new millennium (Brooks and Ferrao, 2005; Kutz et al., 2005; Brooks and Hoberg, 2006, 2007; Brooks, León-Règagnon, et al., 2006; Brooks, McLennan, et al., 2006; Cook et al., 2006; Zarlenga et al., 2006; Brooks and Hoberg, 2007; Waltari et al., 2007; Hoberg et al., 2008; Agosta et al., 2010; Hoberg, 2010; Hoberg and Brooks, 2010; Brooks and Hoberg, 2013; Hoberg and Brooks, 2013) but were not widely recognized by the EID community (but see, e.g., Mas-Coma et al., 2008; Altizer et al., 2011, 2013). The first published versions of the SP—resolving the parasite paradox and explaining how climate change catalyzed changes in host range, geographic distribution, and trophic interactions that initiate EID—appeared in 2014 and 2015 (Brooks et al., 2014; Hoberg and Brooks, 2015).

Alternatively, EID specialists were slow to accept a connection between climate change and EID. Early recognition of a relationship between biodiversity and EID did not make a specific connection to climate change as a driving force in producing EID (e.g., Morse and Schluederberg, 1990; Dobson and Carper 1996; Daszak et al., 2000; Cunningham et al., 2003; Dobson, 2005). Then, as now, the primary focus was on anthropogenic impacts on landscapes, primarily managed landscapes (e.g., Gottdenker et al., 2014; Lajaunie et al., 2015). At about the time the SP was first published as a cohesive paradigm, an edited volume on climate change and global health (Butler, 2014) contained four chapters discussing how known vector-borne diseases might be spread by climate change, and an edited volume about climate change and health (Levy and Patz, 2015) included just two chapters discussing how known vector-borne as well as waterborne and foodborne diseases might be spread by climate change.

Between the first publications (Brooks et al., 2014; Hoberg and Brooks, 2015) and the first book-length treatment of the SP and its relevance to climate change and EID (Brooks et al., 2019), the pendulum had begun to swing in the direction of emerging recognition of some kind of correlation between climate change and EID. Recent articles suggest that a link between climate change and EID is now well established (Short et al., 2017; Blum and Hotez, 2018; Semenza and Suk, 2018; Waits et al., 2018; Wilkinson et al., 2018; Carlson et al., 2019, 2020, 2022; Barbier, 2021; Garcia-Peña et al., 2021; Baker et al., 2022; Lorenz et al., 2022; McDermott, 2022; Mora et al., 2022; Sun et al., 2022; Vinson et al., 2022; Wang et al., 2022; Wieler, 2022; Zhang et al., 2022; Adepoju et al., 2023; Petrone et al., 2023; Yeh et al., 2023), without explaining what is causing that link, other than a general sense that climate change is accelerating, anthropogenic impacts affecting climate change continue, and the number of EID is increasing annually.

Human activities directly increase EID risk

The past 11,000 years of human evolution has created massive new risk space for EID. Agriculture and domestication, global trade and travel, conflict, and migration all create new opportunities for EID (Maurelli et al., 2022; Prati et al., 2022; Tazerji et al., 2022). Humans have become the world's most significant ecological super-spreaders of EID (Boeger et al., 2022; Hoberg et al., 2022). Ironically, one of the mainstays of modern human civilization, the large urban center, is an example of humanity's fragility when it comes to EID. Large urban centers are density and connectivity traps (Brooks et al., 2019). They are not self-sufficient, requiring constant flows of water and material goods, many of which carry pathogens. Urban centers are warm, the largest of them being 2–3°C warmer than their surroundings; in this regard, they may be thought of as large petri dishes for cultivating pathogens. Within these cities, urban green spaces, zoos, and botanical gardens (e.g., Borba da Silva et al., 2022; Romero-Salas, 2022), large numbers and densities of companion animals, and large quantities of waste provide excellent environments for pathogens and their reservoir hosts (e.g., Földvári et al., 2011, 2014, 2022; Cunningham et al., 2014; Szekeres et al., 2016, 2019; Rothenburger et al., 2017; Ortiz et al., 2021; Albery et al., 2022; Dobigny and Morand, 2022; Ecke et al., 2022; Francisco et al., 2022; Paris et al., 2023; Upton et al., 2023).

Large cities are characterized by high population density, meaning that the risk of EID transmitted through myriad means of casual contact is high, with low kinship ties, making it difficult to know who you can count on in an emergency. Not knowing who to cooperate with puts people at risk of not being able to mobilize effectively. A lack of trust is generally compensated for by government services, assumed to be neutral and objective with respect to all citizens. In the largest urban centers as well, services seem to be delivered in a nonuniform manner. Cities owe their economic power and presumptive efficiency to extreme division of labor and extreme interdependency, but this means even a small disease outbreak with minimal deaths can inflict serious damage on infrastructure. And finally, modern urban centers support a lifestyle for many at the expense of people so poor they must go to work even when they are sick. And those people, often undocumented, are largely invisible to public health services.

More than half the human population now lives in cities. As the world becomes increasingly urbanized, the challenge of EID will grow, exacerbated by climate displacement, food and water insecurity, and inequality (e.g., Whitmee et al., 2015; Eskew and Olival, 2018; Brooks et al., 2019; Brown and Brooks, 2021; Garcia-Peña et al., 2021; Ortiz et al., 2021; Trivellone et al., 2022).

Minimizing failure: We can be proactive about coping with EID

The SP treats all environmental perturbations as opportunities for pathogens to explore new fitness space, whether or not those perturbations are anthropogenic. On this point, that anthropogenic change increases the risk for EID, the traditional EID specialists and the SP specialists agree. The potential for finding common ground exists, but collaborations have not yet been effectively developed. If we are to cope effectively with the EID crisis, everyone involved must understand first that *the direct connection between climate change and the EID crisis is part of a larger phenomenon.* The EID crisis is an indication that the biosphere is beginning to cope with climate change without asking our permission or waiting for us to choose survival. That sets the stage for different groups of specialists to collaborate and find effective ways to mitigate the risk.

The EID crisis will continue so long as global climate change persists. Humanity shows every indication of continuing to augment extensive environmental perturbations. EID constitute a minefield of evolutionary accidents waiting to happen (Brooks and Ferrao, 2005). Until the formulation of the SP, infectious diseases were not thought to be preventable. The traditional assumption was that we must wait for a pathogen to evolve the capacity to get into a new host, and we cannot predict when, where, or with which pathogen that will happen, so we have no option except to wait for an EID to emerge and then try to cope with it after the fact. Furthermore, the assumption that particular mutations must arise in order for a pathogen to move into a new host gave us a false sense of security. The risk space was assumed to be quite small. The SP, in contrast, suggests that the risk space is quite large. This is consistent with Audy's (1958) findings that by the time a pathogen announces itself in the form of a disease outbreak, it has been emerging for quite some time, spreading via misdiagnosis, asymptomatic hosts, and hosts to which we do not pay attention. But there is a glimmer of hope.

The SP suggests that colonizing a new host is based on preexisting capacities, especially those related to transmission dynamics and microhabitat preferences. Both the host resources needed for colonization and the modes of transmission from host to host are generally highly specific but phylogenetically conservative. That means that we can largely predict how a given known pathogen, or a previously unknown close relative of a known pathogen, might behave when it enters a new ecosystem or encounters a novel susceptible host (to be forewarned is to be forearmed). If that is true, the ways in which EID occur can largely be anticipated, so while we can do little to control *inherited pathogen capacity*, which largely determines where and in association with which hosts a pathogen could survive, we can reduce *pathogen opportunity*, which largely determines where and in association with which hosts a pathogen actually resides.

Once this hopeful implication of the SP was recognized, it became clear that putting the SP to work in the EID crisis would require a comprehensive policy extension of the

SP. The extension that emerged from a workshop held in 2013 at Cedar Point Biological Station of the University of Nebraska–Lincoln is the DAMA protocol (Document, Assess, Monitor, Act) (Brooks et al., 2014). DAMA aims to prevent at least some outbreaks from occurring and to prevent at least some outbreaks that do occur from becoming pandemics (an ounce of prevention is worth a pound of cure). Attempting to *anticipate to mitigate* EID requires information about pathogens that might cause but are not currently causing an outbreak and about occurrence in hosts that are infected but not diseased, so we can better assess the risk of outbreaks before they produce a crisis. Increasing the geographic or host range of a pathogen may not always cause disease (Audy, 1958). Some host populations are less tolerant than others, some pathogen variants may be inherently more pathogenic than others, and some pathogen variants may be more or less pathogenic in different hosts. This variation means a lag may occur between the arrival of a pathogen in a host population or a geographic area and disease outbreak. If we can *find them before they find us*, we might be able to prevent or at least mitigate disease outbreaks.

The recent report that monkeypox infections in humans are decreasing (Khan et al., 2023) is not a reason to celebrate. It is a reason to discover if the virus has found welcome homes in new nonhuman hosts, especially rodents. We need to know what pathogens are doing when they are not infecting humans or species of economic importance to us. That requires extensive and intensive accounting of the geographic distribution and fundamental fitness space—including actual and potential host range and full range of genetic variation, especially the rare genotypes—for all pathogens deemed to pose an EID risk. This work has proven to be easier said than done.

Professional Impediments to Progress

> *It turned that we didn't have much time after all . . .*
> *But little as it was, we threw it away unused.*
> *. . . What do we do now?*
> —Isaac Asimov, *Foundation* (1951)

Until now, humanity's failure to cope effectively with the EID crisis has been no one's fault because most specialists in the scientific, clinical, and policy communities have been convinced that their activities associated with the EID crisis were the best that anyone could possibly do. The advent of the SP and the DAMA protocol cast doubt on

that conviction. We now know what is happening and why, thanks to the SP. We know that we can more effectively cope with the EID crisis, especially in the arena of prevention and mitigation, thanks to the DAMA protocol. We have sufficient technology and know-how to make it work, and there is no shortage of well-intentioned people who want to succeed. In principle, few doubt that an ounce of prevention is worth a pound of cure, but we face a major stumbling block in convincing various parties interested in EID to alter their fundamental paradigm—the persistent belief that EID cannot be prevented, so spending money on prevention will be wasted—and embrace a new policy platform. The economic impact of EID in treatment costs and production losses is already unsustainably high and rising. What we are doing clearly is not working, so why not try something new? What we lack is the professional will to change our behavior.

With few exceptions (e.g., Nylin et al., 2018), the *scientific community* has ignored the existence, much less the implications, of the parasite paradox. Nor has much attention been paid to the implications stemming from observations that a minority of hosts are infected and a majority of those are not diseased. Anderson and May (1982, 1985) showed that under such circumstances, even highly pathogenic pathogens could exist indefinitely in association with any given host (rabies is an excellent example). This has allowed them to avoid dealing with the internal contradictions of the traditional model of strong coevolution between pathogens and hosts, despite extended discussions of the problem for more than twenty years (Brooks and McLennan, 2002; Brooks et al., 2019; Agosta and Brooks, 2020, and references therein). That, in turn, has fueled the false narrative that the current EID crisis is something novel, related solely to recent anthropogenic influences.

Nothing in the SP conflicts with the basic principles of Darwinian evolution. The SP is the most general statement of the dynamic Darwin envisioned in the final paragraph of *Origin of Species*, the so-called entangled bank statement (Agosta and Brooks, 2020; Agosta, 2022; Brooks and Agosta, 2023). The problem has been the insistence on simple, singular solutions to complex problems. The various components of the SP are each well-known research topics, and each element has been put forward by at least one research group as the singular explanation for interspecific associations, including pathogens and their hosts. This is the focus of the impediment. It's all about who gets credit, and it's all about wanting a singular explanation, a silver bullet, for what Darwin recognized as a complex problem. It is well to remember David Hull's observation that successful scientists are those who are able to build bridges and cooperate with each other (Hull, 1988). In a sense, the SP requires that everyone give a little of their personal vested interests for the common goal of achieving a truly unified evolutionary perspective on the diversification of interspecific associations. If we can see the SP as a case of "something for everyone," perhaps that will help overcome this understandable but unacceptable impediment.

The *clinical health community* pays attention to the consensus of the scientific community to the extent that it conforms to their traditional understanding of and their general approach to dealing with infectious disease. This community comprises mostly people without training in evolutionary principles and with limited experience in scientific research. Its members have largely accepted the traditional view not because it fits what is happening but because it fits the traditional *raison d'etre* for their existence, and they do not know anything different, reinforcing the notion that EID cannot be predicted.

As recently as 2016, the World Health Organization (2016) concluded that "having the ability to anticipate epidemic-prone emerging infectious diseases will give us the necessary edge to battle outbreaks which are becoming more frequent. This foresight, if reliable, is central to global health security and provides the tools and strategies to reduce avoidable loss of life, minimize illness and suffering, and reduce harm to national and global economies." Paradoxically, another conclusion from the same event was that "there's no crystal ball" and, hence, our only option is to be "better prepared for what we do know." Their focus is mainly on reactive response to reemergences of known pathogens and not emergences of previously unknown pathogens, despite recognition of the manifest risks (e.g., Romanello et al., 2021).

The US made global health a federal priority because of national security concerns in 2009 (Global Preparedness Monitoring Board, 2023). During the past two decades, there has been explosive growth in Global Health programs, departments, and institutes (Guzman and Potter, 2021). Today, there are more than 100 schools and approximately 60 undergraduate, master of public health (MPH), and doctor of public health (DrPH) degrees programs with more than 60,000 students enrolled nationwide. Recognizing a great lack of basic knowledge, PREDICT, a collaboration between the United States Agency for International Development (USAID) and an array of international participants and partners (United States Agency for International Development, 2014, 2016) was founded in 2009, its fundamental reason for existence being:

Despite intensive, high-quality research efforts globally, we are still not able to predict which viruses will become pathogenic to people; which will cause new epidemics in animals; nor where and under what circumstances disease will emerge.

The One Health movement, led mostly by veterinarians, is motivated to break down barriers that separate veterinary clinicians working with pathogens shared among wildlife and livestock from public health clinicians who work with pathogens infecting humans, encouraging them to work cooperatively (e.g., Decaro et al., 2021).

The perspectives espoused by proponents of the Global Health and One Health movements, with their emphasis on cooperation and inclusion, are clearly appropriate for coping with a global problem like the EID crisis. Their vision thus far has not included plant pathogens, even though EID in crops and other commercially important plants are a significant global economic burden and expanding threat for food security (Trivellone et al., 2022; Brooks et al., 2022). Even within their acknowledged purview, Global Health and One Health are largely aspirational rather than operational (Costello et al., 2011; Baum et al., 2017; Cunningham et al., 2017; Buttke et al., 2021; Decaro et al., 2021; Fisher and Murray, 2021; Sonne, 2022; Lefrançois et al., 2023), not for lack of interest or desire and not for lack of funds. Rather, they are limited because their approach is based on a failed paradigm that leads to the parasite paradox, and from that to economically unsustainable crisis response to EID. All the current iterations of PREDICT and One Health are incomplete, because they do not translate data into actionable information, or actionable information into effective plans based on *prevention* rather than crisis response. As Molnár et al. have discussed (Molnár, Hoberg et al., 2022; Molnár, Knickel et al., 2022; Molnár, Hoberg et al., 2023, this volume; Molnár, Knickel et al., 2023, this volume), policy documents routinely use the term "preventing pandemics" as a synonym for "mitigating the impact of outbreaks to reduce the chances that they become pandemics," which devolves to preparation for more rapid crisis response to the emergence or reemergence of known pathogens.

Implementing and embracing the SP requires a massive sea change in the ways health specialists are trained. It is essential to avoid training this growing number of people in a failed paradigm. Current clinical training provides too narrow a perspective, leading clinicians to be susceptible to "anchoring bias" (Aguirre et al., 2019), which is making diagnoses based on correlating signs and symptoms with previously known diseases and pathogens. Most are not trained to look for something unexpected or to search for asymptomatic patients during the earliest stages of an outbreak, something Audy (1958) warned about. The clinical community is guided by the principle of "first, do no harm," which leads to an emphasis on preparing for a return of what is already known, followed by crisis response. In a world of climate change, "do no harm" is unacceptably negligent. This must be replaced by something like the *precautionary principle* used in biodiversity studies, which states that incomplete knowledge is not a justification for inaction. A modification of "first, do no harm" could be something like "first, engage in as much prevention as possible."

Finally, we have the *health policy community*, those who control public policy, institutional activities, funding agencies and foundations, highly visible scientific journals, and media outlets. We suggest to this community that it is time to stop promoting a paradigm that has served us so poorly for so long. We understand this will be difficult to achieve. Clinicians tell this community that this is the best that can be done, that their activities are following the best science of the day, and if they could only get more funding to keep doing business-as-usual activities, eventually they will prevail. The members of the health policy community also live in a world in which there is more profit in suffering than in well-being. This situation creates a nightmare for the public policy community. If we begin with crisis response, the economic costs are unsustainably high. It is, however, difficult to imagine policy priorities that conflict with a global economy largely driven by a perceived need for growing profits.

Crisis response, however unsustainably expensive and increasingly ineffective it is, lends itself to heroic appearances (people in hazmat suits wandering about spraying bleach on sidewalks, impassioned officials at press conferences assuring the public that everyone is doing the best they can do, "following the science") and a reliance on magical palliation, medicines and vaccines that themselves are increasingly expensive and ineffective. Such crisis response contrasts with empirical evidence; throwing money at palliative measures is the best we can do once a pandemic has occurred, *but it is not the best we can do* (Molnár, Hoberg et al., 2022; Molnár, Knickel et al., 2022; Molnár, Hoberg et al., 2023, this volume; Molnár, Knickel et al., 2023, this volume). We are encouraged by a recent policy statement from the Global Preparedness Monitoring Board (GPMB, 2023) recognizing that business-as-usual approaches failed with respect to the COVID pandemic, which indicates strategic flaws in policy approaches to EID. The board's stated

goals will be more effectively achieved by recognizing the conceptual framework and policy extension provided by the SP and the DAMA protocol.

Our Job Going Forward Is to Save as Many as We Can

We are not here to curse the darkness, but to light a candle that can guide us through the darkness to a safe and sure future.
For the world is changing. The old era is ending. The old ways will not do.
—John F. Kennedy

This volume is not focused on condemning the failures of the past and present but on minimizing failure in the future. For that, we need a new paradigm of the evolution of pathogen-host associations. If you follow the wrong paradigm, at worst you throw money away and achieve nothing; at best, you achieve less than you could. So, it comes down to this question: are we dealing with something that involves only current fixed ecologies being pushed around by climate change and human activities or are we are dealing with an evolutionary phenomenon in which the ecologies of pathogens are historically conservative but flexible, so their behaviors in the face of being pushed around by climate change and human activities reflect primarily their evolutionary history?

If we are to contain the risk of continuous failure while not falling prey to the "illusion" that we can "control" the workings of complex systems, we need a deeper understanding of the thesis that "prevention" must supplement "preparation." We must view every disease emergence as a failure to anticipate and prevent and place far more emphasis on prevention. At the same time, we would be foolish to think we could ever fully prevent disease outbreaks. Results of DAMA protocol programs would therefore provide information about outbreaks we can prevent as well as outbreaks we will not be able to prevent everywhere. Preparation, therefore, is also essential—not trying to prepare for all possible pandemics, not focusing on the most recently concluded pandemic (never make the mistake of fighting the last war) but focusing on the next likely one(s) as indicated by the result of DAMA activities. In preparing for those, we can mitigate our failures to prevent. When our preparations fail and an outbreak becomes a pandemic, palliation is our last line of defense. Antibiotics, vaccines, respirators, and dialysis machines are evidence that we have failed, first to prevent and second to mitigate. With each failure, preserving society becomes increasingly costly.

Global climate change, global trade and travel, poverty, urbanization, conflict, and migration all increase the risk, so we are unlikely to see a decrease in EID in our lifetimes. The longer we let this solvable problem persist, the more likely it is that we will be overcome by the costs. The more success we have with prevention, the less costly the EID crisis becomes.

We *know* why the EID crisis is happening, and we know how to mitigate its impacts; *what we lack so far is the will to change our behavior*.

Literature Cited

Adepoju, O.A.; Afinowi, O.A.; Tauheed, A.M.; Danazumi, A.U.; Dibba, L.B.S.; Balogun, J.B.; et al. 2023. Multisectoral perspectives on global warming and vector-borne diseases: a focus on southern Europe. Current Tropical Medicine Reports 10: 47–70. https://doi.org/10.1007/s40475-023-00283-y

Agosta, S.J. 2022. The Stockholm paradigm explains the dynamics of Darwin's entangled bank, including emerging infectious disease. MANTER: Journal of Parasite Diversity 27. https://doi.org/10.32873/unl.dc.manter27

Agosta, S.J. 2023. The Stockholm Paradigm explains the eco-evolutionary dynamics of the biosphere in a changing world, including emerging infectious disease. In: An Evolutionary Pathway for Coping with Emerging Infectious Disease. S.L. Gardner, D.R. Brooks, W.A. Boeger, E.P. Hoberg (eds.). Zea Books, Lincoln, NE.

Agosta, S.J.; Brooks, D.R. 2020. The Major Metaphors of Evolution: Darwinism Then and Now. Springer International, New York. https://doi.org/10.1007/978-3-030-52086-1

Agosta, S.J.; Janz, N.; Brooks, D.R. 2010. How specialists can be generalists: resolving the "parasite paradox" and implications for emerging infectious disease. Zoologia 27: 151–162. https://doi.org/10.1590/S1984-46702010000200001

Aguirre, L.E.; Chueng, T.; Lorio, M.; Mueller, M. 2019. Anchoring bias, Lyme disease, and the diagnosis conundrum. Cureus 11: e4300. https://doi.org/10.7759/cureus.4300

Albery, G.F.; Carlson, C.J.; Cohen, L.E.; Eskew, E.A.; Gibb, R.; Ryan, S.J., et al. 2022. Urban-adapted mammal species have more known pathogens. Nature Ecology & Evolution 6: 794–801. https://doi.org/10.1038/s41559-022-01723-0

Altizer, S.; Bartel, R.; Han, B.A. 2011. Animal migration and infectious disease risk. Science 331: 296–302. https://doi.org/10.1126/science.1194694

Altizer, S.; Ostfeld, R.S.; Johnson, P.T.J.; Kutz, S.; Harvell, C.D. 2013. Climate change and infectious diseases: from evidence to a predictive framework. Science 341: 514–519. https://doi.org/10.1126/science.1239401

Anderson, R.M.; May, R.M. 1982. Coevolution of hosts and parasites. Parasitology 85: 411–426.

Anderson, R.M.; May, R.M. 1985. Epidemiology and genetics in the coevolution of parasites and hosts. Proceedings of the Royal Society B 219: 281–283.

Astbury, C.; Lee, K.M.; Aguiar, R.; Atique, A.; Balolong, M.; Clarke, J.; et al. 2022. Policies to prevent zoonotic spillover: protocol for a systematic scoping review of evaluative evidence. BMJ Open 12: e058437. https://doi.org/10.1136/bmjopen-2021-058437

Audy, J.R. 1958. The localization of disease with special reference to the zoonoses. Transactions of the Royal Society of Tropical Medicine & Hygiene 52: 308–334. https://doi.org/10.1016/0035-9203(58)90045-2

Baker, R.E.; Mahmud, A.S.; Miller, I.F.; Rajeev, M.; Rasambainarivo, F.; Rice, B.L.; et al. 2022. Infectious disease in an era of global change. Nature Reviews Microbiology 20: 193–205. https://doi.org/10.1038/s41579-021-00639-z

Barbier, E.B. 2021. Habitat loss and the risk of disease outbreak. Journal of Environmental Economics and Management 108: 102451. https://doi.org/10.1016/j.jeem.2021.102451

Baum, S.; Machalaba, C.; Daszak, P.; Salerno, R.H.; Karesh, W.B. 2017. Evaluating One Health: are we demonstrating effectiveness? One Health 3: 5–10. https://doi.org/10.1016%2Fj.onehlt.2016.10.004

Blum, A.J.; Hotez, P.J. 2018. Global "worming": climate change and its projected general impact on human helminth infections. PLOS Neglected Tropical Diseases 12: e0006370. https://doi.org/10.1371/journal.pntd.0006370

Boeger, W.A.; Brooks, D.R.; Trivellone, V.; Agosta, S.J.; Hoberg, E.P. 2022. Ecological super-spreaders drive host-range oscillations: Omicron and risk space for emerging infectious disease. Transboundary and Emerging Diseases 69: e1280–e1288. https://doi.org/10.1111/tbed.14557

Borba da Silva, M.; Froes de Oliveira, D.; Santos, F.V.; Aguiar, C.D.S.; Prado, I.S.; Brandão, D.A.; et al. 2022. Gastrointestinal parasites in wild and exotic animals from a zoo in the State of Bahia, Brazil—first record. Research, Society and Development 11: e19111334959. https://doi.org/10.33448/rsd-v11i13.34959

Brooks, D.R.; Agosta, S.J. 2023. A Darwinian Survival Guide: Hope for the Twenty-first Century. MIT Press, Boston.

Brooks, D.R.; Boeger W.A. 2019. Climate change and emerging infectious diseases: evolutionary complexity in action. Current Opinion in Systems Biology 13: 75–81. https://doi.org/10.1016/j.coisb.2018.11.001

Brooks, D.R.; Boeger, W.A.; Hoberg, E.P. 2023. The Stockholm Paradigm: the conceptual platform for coping with the emerging infectious disease crisis. In: An Evolutionary Pathway for Coping with Emerging Infectious Disease. S.L. Gardner, D.R. Brooks, W.A. Boeger, E.P. Hoberg (eds.). Zea Books, Lincoln, NE.

Brooks, D.R.; Ferrao, A. 2005. The historical biogeography of co-evolution: emerging infectious diseases are evolutionary accidents waiting to happen. Journal of Biogeography. 32: 1291–1299. https://doi:10.1111/j.1365-2699.2005.01315.x

Brooks, D.R.; Hoberg, E.P. 2000. Triage for the biosphere: the need and rationale for taxonomic inventories and phylogenetic studies of parasites. Comparative Parasitology 67: 1–25.

Brooks, D.R.; Hoberg, E.P. 2001. Parasite systematics in the 21st century: opportunities and obstacles. Trends in Parasitology 17: 273–275.

Brooks, D.R.; Hoberg, E.P. 2006. Systematics and emerging infectious diseases: from management to solution. Journal of Parasitology 92: 426–429.

Brooks, D.R.; Hoberg, E.P. 2007. How will global climate change affect parasites? Trends in Parasitology 23: 571–574. https://doi.org/10.1016/j.pt.2007.08.016

Brooks, D.R.; Hoberg, E.P. 2013. The emerging infectious diseases crisis and pathogen pollution. In: The Balance of Nature and Human Impact. K. Rhode (ed.). Cambridge University Press, Cambridge, UK. 215–229 p. https://doi.org/10.1017/CBO9781139095075

Brooks, D.R.; Hoberg, E.P; Boeger, W.A. 2019. The Stockholm Paradigm: Climate Change and Emerging Disease. University of Chicago Press, Chicago.

Brooks, D.R.; Hoberg, E.P.; Boeger, W.A.; Gardner, S.L.; Galbraith, K.E.; Herczeg, D.; et al. 2014. Finding them before they find us: informatics, parasites, and environments in accelerating climate change. Comparative Parasitology 81: 155–164.

Brooks, D.R.; Hoberg, E.P.; Boeger W.A.; Trivellone, V. 2022. Emerging infectious disease: an underappreciated area of strategic concern for food security. Transboundary and Emerging Diseases 69: 254–267. https://doi.org/10.1111/tbed.14009

Brooks, D.R.; León-Règagnon, V.; McLennan, D.A.; Zelmer, D. 2006. Ecological fitting as a determinant of the community structure of platyhelminth parasites of anurans. Ecology 87: S76–S85.

Brooks, D.R.; McLennan, D. 2002. The Nature of Diversity: An Evolutionary Voyage of Discovery. University of Chicago Press, Chicago.

Brooks, D.R.; McLennan, D.A.; León-Règagnon, V.; Hoberg, E. 2006. Phylogeny, ecological fitting and lung flukes: helping solve the problem of emerging infectious diseases. Revista Mexicana de Biodiversidad 77: 225–234.

Brown, H.A.; Brooks, D.R. 2021. How 'managed retreat' from climate change could revitalize rural America: revisiting the Homestead Act. The Conversation. https://theconversation.com/how-managed-retreat-from-climate-change-could-revitalize-rural-america-revisiting-the-homestead-act-169007

Butler, C.D. (ed.). 2014. Climate Change and Global Health. CABI, Wallingford; Boston.

Buttke, D.; Wild, M.; Monello, R.; Schuurman, G.; Hahn, M.; Jackson, K. 2021. Managing wildlife disease under climate change. EcoHealth 18: 406–410. https://doi.org/10.1007/s10393-021-01542-y

Carlson, C.J.; Albery, G.F.; Merow, C.; Trisos, C.H.; Zipfel, C.M.; et al. 2022. Climate change increases cross-species viral transmission risk. Nature 607: 555–562. https://doi.org/10.1038/s41586-022-04788-w

Carlson, C.J.; Dallas, T.A.; Alexander, L.W.; Phelan, A.L.; Phillips, A.J. 2020. What would it take to describe the global diversity of parasites? Proceedings of the Royal Society B 287: 20201841. http://dx.doi.org/10.1098/rspb.2020.1841

Carlson, C.J.; Zipfel, C.M.; Garnier, R.; Bansal, S. 2019. Global estimates of mammalian viral diversity accounting for host sharing. Nature Ecology Evolution 3: 1070–1075. https://doi.org/10.1038/s41559-019-0910-6

Centers for Disease Control and Prevention. 1994. Addressing emerging infectious disease threats: a prevention strategy for the United States. Morbidity and Mortality Weekly Report 43 (RR-5). 18 pp.

Cook, J.A.; Hoberg, E.P.; Koehler, A.; Henttonen, H.; Wickström, L.; et al. 2006. Beringia: intercontinental exchange and diversification of high latitude mammals and their parasites during the Pliocene and Quaternary. Mammal Study 30: S33–S44. https://doi.org/10.3106/1348-6160(2005)30[33:BIEADO]2.0.CO;2

Costello, A.; Maslin, M.; Montgomery, H.; Johnson, A.M.; Ekins, P. 2011. Global health and climate change: moving from denial and catastrophic fatalism to positive action. Philosophical Transactions of the Royal Society A 369: 1866–1882. https://doi.org/10.1098/rsta.2011.0007

Cunningham, A.A.; Daszak, P.; Rodríguez, J.P. 2003. Pathogen pollution: defining a parasitological threat to biodiversity conservation. Journal of Parasitology 89: S78–S83.

Cunningham, A.A.; Daszak, P.; Wood, J.L.N. 2017. One Health, emerging infectious diseases, and wildlife: two decades of progress? Philosophical Transactions of the Royal Society B 372: 20160167. http://dx.doi.org/10.1098/rstb.2016.0167

Cunningham, A.A.; Lawson, B.; Hopkins, T.; Toms, M.; Wormald, K.; Peck, K. 2014. Monitoring diseases in garden wildlife. Veterinary Record 174: 126. https://doi.org/10.1136/vr.g1295

Daszak, P.; Cunningham, A.A.; Hyatt, A.D. 2000. Emerging infectious diseases of wildlife—threats to biodiversity and human health. Science 287: 443–449.

Decaro, N.; Lorusso, A.; Capua, I. 2021. Erasing the invisible line to empower the pandemic response. Viruses 13: 348. https://doi.org/10.3390/v13020348

Dobigny, G.; Morand, S. 2022. Zoonotic emergence at the animal-environment-human interface: the forgotten urban socio-ecosystems. Peer Community Journal 2: e79. https://doi.org/10.24072/pcjournal.206

Dobson, A.P. 2005. What links bats to emerging infectious diseases? Science 310: 628–629.

Dobson, A.P.; Carper, E.R. 1996. Infectious diseases and human population history. Bioscience 46: 115–126.

Ecke, F.; Han, B.A.; Hörnfeldt, B.; Khalil, H.; Magnusson, M.; Singh, N.J.; Ostfeld, R.S. 2022. Population fluctuations and synanthropy explain transmission risk in rodent-borne zoonoses. Nature Communications 13: 7532. https://doi.org/10.1038/s41467-022-35273-7

Elton, C.S. 1958. The Ecology of Invasions by Animals and Plants. Methuen, London.

Eskew, E.A.; Olival, K.J. 2018. De-urbanization and zoonotic disease risk. EcoHealth 15: 707–712. https://doi.org/10.1007/s10393-018-1359-9

Fisher, M.C.; Murray, K.A. 2021. Emerging infections and the integrative environment-health sciences: the road ahead. Nature Reviews Microbiology 19: 133–135. https://doi.org/10.1038/s41579-021-00510-1

Földvári, G.; Jahfari, S.; Rigó, K.; Jablonszky, M.; Szekeres, S.; Majoros, G.; et al. 2014. *Candidatus* Neoehrlichia mikurensis and *Anaplasma phagocytophilum* in urban hedgehogs. Emerging Infectious Diseases 20: 496–498. https://doi.org/10.3201/eid2003.130935

Földvári, G.; Rigó, K.; Jablonszky, M.; Biró, N.; Majoros, G.; Molnár, V.; Tóth, M. 2011. Ticks and the city: ectoparasites of the northern white-breasted hedgehog (*Erinaceus roumanicus*) in an urban park. Ticks and Tick-Borne Diseases 2: 231–234. https://doi.org/10.1016/j.ttbdis.2011.09.001

Földvári, G.; Szabó, E.; Tóth, G.E.; Lanszki, Z.; Zana, B.; Varga, Z.; Kemenesi, G. 2022. Emergence of *Hyalomma marginatum* and *Hyalomma rufipes* adults revealed by citizen science tick monitoring in Hungary. Transboundary and Emerging Diseases 69: e2240–e2248. https://doi.org/10.1111/tbed.14563

Francisco, I.; Bailey, S.; Bautista, T.; Diallo, D.; Gonzalez, J.; Gonzalez, J.; et al. 2022. Detection of velogenic avian paramyxoviruses in rock doves in New York City, New York. Microbiology Spectrum 10: e02061-21. https://doi.org/10.1128/spectrum.02061-21

García-Peña, G.E.; Rubio, A.V.; Mendoza, H.; Fernández, M.; Milholland, M.T.; Aguirre, A.A.; et al. 2021. Land-use change and rodent-borne diseases: hazards on the shared socioeconomic pathways. Philosophical Transactions of the Royal Society B 376: 20200362. https://doi.org/10.1098/rstb.2020.0362

Gardner, S.L.; Campbell, M.L. 1992. Parasites as probes for biodiversity. Journal of Parasitology 78: 596–600.

Global Preparedness Monitoring Board. 2023. GPMB Monitoring Framework for Preparedness: Technical Framework and Methodology. World Health Organization, Geneva, Switzerland. https://www.gpmb.org/annual-reports/overview/item/gpmb-monitoring-framework-full

Gottdenker, N.L.; Streicker, D.G.; Faust, C.L.; Carroll, C.R. 2014. Anthropogenic land use change and infectious diseases: a review of the evidence. EcoHealth 11: 619–632. https://doi.org/10.1007/s10393-014-0941-z

Guzman, C.A.F.; Potter, T. (eds.). 2021. The Planetary Health Education Framework. Planetary Health Alliance. 61 p. http://dx.doi.org/10.13140/RG.2.2.27505.20320

Harvey, E.; Holmes, E.C. 2022. Diversity and evolution of the animal virome. Nature Reviews Microbiology 20: 321–334. https://doi.org/10.1038/s41579-021-00665-x

Hoberg, E.P. 1997. Phylogeny and historical reconstruction: host-parasite systems as keystones in biogeography and ecology. In: Biodiversity II: Understanding and Protecting Our Biological Resources. M.L. Reaka-Kudla, D.E. Wilson, E. Wilson (eds.). Joseph Henry Press, Washington. 243–261 p.

Hoberg, E.P. 2010. Invasive processes, mosaics and the structure of helminth parasite faunas. Revue Scientifique et Technique (Office International des Épizooties) 29: 255–272.

Hoberg, E.P.; Adams, A. 2000. Phylogeny, history and biodiversity: understanding faunal structure and biogeography in the marine realm. Bulletin of the Scandinavian Society for Parasitology 10: 19–37.

Hoberg, E.P., Boeger, W.A.; Brooks, D.R.; Trivellone, V.; Agosta, S.J. 2022. Stepping-stones and mediators of pandemic expansion—a context for humans as ecological super-spreaders. MANTER: Journal of Parasite Biodiversity 18. https://doi.10.32873/unl.dc.manter18

Hoberg, E.P.; Brooks, D.R. 2008. A macroevolutionary mosaic: episodic host-switching, geographic colonization, and diversification in complex host-parasite systems. Journal of Biogeography 35: 1533–1550.

Hoberg, E.P.; Brooks, D.R. 2010. Beyond vicariance: integrating taxon pulses, ecological fitting, and oscillation in historical biogeography and evolution. In: The Geography of Host-Parasite Interactions. S. Morand, B. Krasnov (eds.). Oxford University Press, Oxford, UK. 7–20 p.

Hoberg, E.P.; Brooks, D.R. 2013. Episodic processes, invasion, and faunal mosaics in evolutionary and ecological time. In: The Balance of Nature and Human Impact. K. Rohde (ed.). Cambridge University Press, Cambridge, UK. 199–214 p.

Hoberg, E.P.; Brooks, D.R. 2015. Evolution in action: climate change, biodiversity dynamics and emerging infectious disease. Philosophical Transactions of the Royal Society B 370: 20130553. http://dx.doi.org/10.1098/rstb.2013.0553

Hoberg, E.P.; Polley, L.; Jenkins, E.J.; Kutz, S.J. 2008. Pathogens of domestic and free-ranging ungulates: global climate change in temperate to boreal latitudes across North America. Revue Scientifique et Technique (Office International des Épizooties) 27: 511–528.

Hull, D. 1988. Science as a Process. University of Chicago Press, Chicago.

Institute of Medicine. 1992. Emerging Infections: Microbial Threats to Health in the United States. National Academies Press, Washington, D.C.

Janz, N. 2011. Ehrlich and Raven revisited: mechanisms underlying codiversification of plants and enemies. Annual Review of Ecology, Evolution, and Systematics 42: 71–89. https://doi.org/10.1146/annurev-ecolsys-102710-145024

Janz, N.; Nylin, S. 2008. The oscillation hypothesis of host-plant range and speciation. In: Specialization, Speciation, and Radiation: The Evolutionary Biology of Herbivorous Insects. K.J. Tilmon (ed.). University of California Press, Berkeley. 203–215 p.

Khan, M.R.; Hossain, M.J.; Roy, A.; Islam, M.R. 2023. Decreasing trend of monkeypox cases in Europe and America shows hope for the world: evidence from the latest epidemiological data. Health Science Reports 6: e1030. https://doi.org/10.1002/hsr2.1030

Kutz, S.J.; Hoberg, E.P.; Polley, L.; Jenkins, E.J. 2005. Global warming is changing the dynamics of arctic host-parasite systems. Proceedings of the Royal Society of London 272: 2571–2576.

Lajaunie, C.; Morand, S.; Binot, A. 2015. The link between health and biodiversity in Southeast Asia through the example of infectious diseases. Environmental Justice 8: 26–31. https://doi.org/10.1089/env.2014.0017

Lefrançois, T.; Malvy, D.; Atlani-Duault, L.; Benamouzig, D.; Druais, P.-L.; et al. 2023. After 2 years of the COVID-19 pandemic, translating One Health into action is urgent. The Lancet 401: 789–794. https://doi.org/10.1016/s0140-6736(22)01840-2

Levy, B.S.; Patz, J. (eds.). 2015. Climate Change and Public Health. Oxford University Press, New York.

Lorenz, C.; De Azevedo, T.S.; Chiaravalloti-Neto, F. 2022. Impact of climate change on West Nile virus distribution in South America. Transactions of the Royal Society of Tropical Medicine & Hygiene 116: 1043–1053. https://doi.org/10.1093/trstmh/trac044

Mas-Coma, S.; Valero, M.A.; Bargues, M.D. 2008. Effects of climate change on animal and zoonotic helminthiases. Revue Scientifique et Technique 27: 443–457. http://dx.doi.org/10.20506/rst.27.2.1822

Maurelli, M.P.; Pepe, P.; Gualdieri, L.; Bosco, A.; Cringoli, G.; Rinaldi, L. 2022. Improving diagnosis of intestinal parasites towards a migrant-friendly health system. Current Tropical Medicine Reports 10: 17–25. https://doi.org/10.1007/s40475-022-00280-7

Maurer, F.D. 1962. Equine piroplasmosis—another emerging disease. Journal of the American Veterinary Association 141: 699–702.

McDermott, A., 2022. Climate change hastens disease spread across the globe. Proceedings of the National Academy of Sciences 119: e2200481119. https://doi.org/10.1073/pnas.2200481119

Molnár, O.; Hoberg, E.; Trivellone, V.; Földvári, G.; Brooks, D.R. 2022. The 3P framework: a comprehensive approach to coping with the emerging infectious disease crisis. MANTER: Journal of Parasite Biodiversity 23. http://dx.doi.org/10.32873/unl.dc.manter23

Molnár, O.; Hoberg, E.P.; Trivellone, V.; Földvári, G.; Brooks, D.R. 2023. Prevent-Prepare-Palliate: the 3P framework—integrating the DAMA protocol into global public health systems. In: An Evolutionary Pathway for Coping with Emerging Infectious Disease. S.L. Gardner, D.R. Brooks, W.A. Boeger, E.P. Hoberg (eds.). Zea Books, Lincoln, NE.

Molnár, O.; Knickel, M.; Marizzi, C. 2022. Taking action: turning evolutionary theory into preventive policies. MANTER: Journal of Parasite Biodiversity 28. http://dx.doi.org/10.32873/unl.dc.manter28

Molnár, O.; Knickel, M.; Marizzi, C. 2023. All hands on deck: turning evolutionary theory into preventive policies. In: An Evolutionary Pathway for Coping with Emerging Infectious Disease. S.L. Gardner, D.R. Brooks, W.A. Boeger, E.P. Hoberg (eds.). Zea Books, Lincoln, NE.

Mora, C.; McKenzie, T.; Gaw, I.M.; Dean, J.M.; von Hammerstein, H.; Knudson, T.A.; et al. 2022. Over half of known human pathogenic diseases can be aggravated by climate change. Nature Climate Change 12: 869–875. https://doi.org/10.1038/s41558-022-01426-1

Morse, S.S.; Schluederberg, A. 1990. Emerging viruses: the evolution of viruses and viral diseases. The Journal of Infectious Diseases 162: 1–7.

Ndow, G.; Ambe, J.R.; Tomori, O. 2019. Emerging infectious diseases: a historical and scientific review. In: Socio-cultural Dimensions of Emerging Infectious Diseases in Africa. G.B. Tangwa, A. Abayomi, S.J. Ujewe, N.S. Munung (eds.). Springer Nature Switzerland AG, Basel. 31–40 p. https://link.springer.com/chapter/10.1007/978-3-030-17474-3_3

Nylin, S.; Agosta, S.; Bensch, S.; Boeger, W.; Braga, M.P.; Brooks, D.R.; et al. 2018. Embracing colonizations: a new paradigm for species association dynamics. Trends in Ecology & Evolution 33: 4–14. https://doi.org/10.1016/j.tree.2017.10.005

Ortiz, D.I.; Piche-Ovares, M.; Romero-Vega, L.M.; Wagman, J.; Troyo, A. 2021. The impact of deforestation, urbanization, and changing land use patterns on the ecology of mosquito and tick-borne diseases in Central America. Insects 13: 20. https://doi.org/10.3390/insects13010020

Paris, V.; Rane, R.V.; Mee, P.T.; Lynch, S.E.; Hoffmann, A.A.; Schmidt, T.L. 2023. Urban population structure and dispersal of an Australian mosquito (Aedes notoscriptus) involved in disease transmission. Heredity 130: 99–108. https://doi.org/10.1038/s41437-022-00584-4

Petrone, M.E.; Holmes, E.C.; Harvey, E. 2023. Through an ecological lens: an ecosystem-based approach to zoonotic risk assessment. EMBO Reports 24: e56578. https://doi.org/10.15252/embr.202256578

Prati, S.; Grabner, D.S.; Pfeifer, S.M.; Lorenz, A.W.; Sures, B. 2022. Generalist parasites persist in degraded environments: a lesson learned from microsporidian diversity in amphipods. Parasitology 149: 973–982. https://doi.org/10.1017/s0031182022000452

Romanello, M.; McGushin, A.; Di Napoli, C.; Drummond, P.; Hughes, N.; Jamart, L.; et al. 2021. The 2021 report of the Lancet Countdown on health and climate change: code red for a healthy future. The Lancet 398: 1619–1662. https://doi.org/10.1016/S0140-6736(21)01787-6

Romero-Salas, D.; Sánchez-Montes, S.; Bravo-Ramos, J.L.; Sánchez-Otero, M.G.; Diaz-Lopez, C.G.; Salguero-Romero, J.L.; Cruz-Romero, A. 2022. First report of Babesia bigemina in Lama glama in a zoological garden of Veracruz, Mexico. Veterinary Parasitology: Regional Studies and Reports 33: 100756. https://doi.org/10.1016/j.vprsr.2022.100756

Rosenthal, S.R.; Ostfeld, R.S.; McGarvey, S.T.; Lurie, M.N.; Smith, K.F. 2015: Redefining disease emergence to improve prioritization and macro-ecological analyses. One Health 1: 17–23. https://doi.org/10.1016/j.onehlt.2015.08.001

Rothenburger, J.L.; Himsworth, C.H.; Nemeth, N.M.; Pearl, D.L.; Jardine, C.M. 2017. Environmental factors and zoonotic pathogen ecology in urban exploiter species. EcoHealth 14: 630–641. https://doi.org/10.1007/s10393-017-1258-5

Semenza, J.C.; Suk, J.E. 2018. Vector-borne diseases and climate change: a European perspective. FEMS Microbiology Letters 365: fnx244. https://doi.org/10.1093/femsle/fnx244

Sharpe, S. 2023. Five Times Faster: Rethinking the Science, Economics, and Diplomacy of Climate Change. Cambridge University Press, Cambridge, UK.

Short, E.E.; Caminade, C.; Thomas, B.N. 2017. Climate change contribution to the emergence or re-emergence of parasitic diseases. Infectious Diseases: Research and Treatment 10: 117863361773229. https://doi.org/10.1177/1178633617732296

Sokolow, S.H.; Nova, N.; Jones, I.J.; Wood, C.L.; Lafferty, K.D.; Garchitorena, A.; et al. 2022. Ecological and socioeconomic factors associated with the human burden of environmentally mediated pathogens: a global analysis. The Lancet Planetary Health 6: e870–e879. https://doi.org/10.1016/S2542-5196(22)00248-0

Sonne, M. 2022. A review on potential influence of climate change on vector born and zoonotic diseases: prevalence and recommended action for earlier disease detection in humans and animal. International Journal of Research and Analytical Reviews 9: 684–708.

Soper, G.A. 1919. The lessons of the pandemic. Science 59: 501–506.

Sun, Y.; Sun, L.; Sun, S.; Tu, Z.; Liu, Y.; Yi, L.; et al. 2023. Virome profiling of an eastern roe deer reveals spillover of viruses from domestic animals to wildlife. Pathogens 12: 156. https://doi.org/10.3390/pathogens12020156

Szekeres, S.; van Leeuwen, A.D.; Rigó, K.; Jablonszky, M.; Majoros, G., et al. 2016. Prevalence and diversity of human pathogenic rickettsiae in urban versus rural habitats, Hungary. Experimental and Applied Acarology 68: 223–226. https://doi.org/10.1007/s10493-015-9989-x

Szekeres, S.; van Leeuwen, A.D.; Tóth, E.; Majoros, G.; Sprong, H.; Földvári, G. 2019. Road-killed mammals provide insight into tick-borne bacterial pathogen communities within urban habitats. Transboundary and Emerging Diseases 66: 277–286. https://doi.org/10.1111/tbed.13019

Tazerji, S.S.; Nardini, R.; Safdar, M.; Shehata, A.A.; Duarte, P.M. 2022. An overview of anthropogenic actions as drivers for emerging and re-emerging zoonotic diseases. Pathogens 11: 1376. https://doi.org/10.3390/pathogens11111376

Trivellone, V.; Hoberg, E.P.; Boeger, W.A.; Brooks, D.R. 2022. Food security and emerging infectious disease: risk assessment and risk management. Royal Society Open Science 9: 211687. https://doi.org/10.1098/rsos.211687

United States Agency for International Development. 2014. Reducing Pandemic Risk, Promoting Global Health. PREDICT 1 (2009–2014) Final Report.

United States Agency for International Development. 2016. Reducing Pandemic Risk, Promoting Global Health, Supporting the Global Health Security Agenda. PREDICT 2016 Annual Report, Washington D.C.

Upton, K.E.; Budke, C.M.; Verocai, G.G. 2023. Heartworm, Dirofilaria immitis, in carnivores kept in zoos located in Texas, USA: risk perception, practices, and antigen detection. Parasites & Vectors 16: 150. https://doi.org/10.1186/s13071-023-05750-z

Vinson, J.E.; Gottdenker, N.L.; Chaves, L.F.; Kaul, R.B.; Kramer, A.M.; Drake, J.M.; Hall, R.J. 2022. Land reversion and zoonotic spillover risk. Royal Society Open Science 9: 220582. https://doi.org/10.1098/rsos.220582

Waits, A.; Emelyanova, A.; Oksanen, A.; Abass, K.; Rautio, A. 2018. Human infectious diseases and the changing climate in the Arctic. Environment International 121: 703–713. https://doi.org/10.1016/j.envint.2018.09.042

Waltari, E.; Hoberg, E.P.; Lessa, E.P.; Cook, J.A. 2007. Eastward ho: phylogeographical perspectives on colonization of hosts and parasites across the Beringian nexus. Journal of Biogeography 34: 561–574. https://doi.org/10.1111/j.1365-2699.2007.01705.x

Wang, G.; Zhang, D.; Khan, J.; Guo, J.; Feng, Q.; et al. 2022. Predicting the impact of climate change on the distribution of a neglected arboviruses vector (Armigeres subalbatus) in China. Tropical Medicine and Infectious Disease 7: 431. https://doi.org/10.3390/tropicalmed7120431

Whitmee, S.; Haines, A.; Beyrer, C.; Boltz, F.; Capon, A.G.; de Souza Dias, B.F.; et al. 2015. Safeguarding human health in the Anthropocene epoch: report of the Rockefeller Foundation–Lancet Commission on planetary health. The Lancet 386: 1973–2028. https://doi.org/10.1016/S0140-6736(15)60901-1

Wieler, L.H. 2022. Climate change—a burning topic for public health. Journal of Health Monitoring 7: 3–5. https://doi.org/10.25646/10387

Wilkinson, D.A.; Marshall, J.C.; French, N.P.; Hayman, D.T.S. 2018: Habitat fragmentation, biodiversity loss and the risk of novel infectious disease emergence. Journal of the Royal Society Interface 15: 20180403. https://doi.org/10.1098/rsif.2018.0403

World Health Organization. 2016. Anticipating Emerging Infectious Disease Epidemics (Meeting Report, WHO Informal Consultation). World Health Organization, Geneva, Switzerland.

Yeh, K.B.; Parekh, F.K.; Mombo, I.; Leimer, J.; Hewson, R.; Olinger, G.; et al. 2023. Climate change and infectious disease: a prologue on multidisciplinary cooperation and predictive analytics. Frontiers in Public Health 11: 1018293. https://doi.org/10.3389/fpubh.2023.1018293

Zarlenga, D. S.; Rosenthal, B.M.; La Rosa, G.; Pozio, E.; Hoberg, E.P. 2006. Post-Miocene expansion, colonization, and host switching drove speciation among extant nematodes of the archaic genus Trichinella. Proceedings of the National Academy of Sciences USA 103: 7354–7359. https://doi.org/10.1073/pnas.0602466103

Zhang, L.; Rohr, J.; Cui, R.; Xin, Y.; Han, L.; Yang, X.; et al. 2022. Biological invasions facilitate zoonotic disease emergences. Nature Communications 13: 1762. https://doi.org/10.1038/s41467-022-29378-2

Section I:

The Conceptual Framework

1

The Stockholm Paradigm Explains the Eco-Evolutionary Dynamics of the Biosphere in a Changing World, Including Emerging Infectious Disease

Salvatore J. Agosta

Abstract

Pathogens and their hosts are embedded within the larger biosphere. Emerging infectious disease occurs when a parasite "switches" to a new host. Understanding the dynamics of emerging infectious disease requires understanding the dynamics of host-switching, which requires a more general understanding of how the biosphere and its constituent members cope when conditions change. The Stockholm Paradigm is an integrative eco-evolutionary framework that describes how living systems cope with change by oscillating between exploiting and exploring the geographical and functional dimensions of their environments. It combines organismal capacity, ecological opportunity, and the repeated external perturbations to the conditions that drive the interaction between capacity and opportunity, catalyzing the evolutionary dynamics of species, their interactions, and the ecosystems they form. The Stockholm Paradigm makes clear that emerging infectious disease is an expected outcome of the expression of the same evolutionary potential that governs the response of the rest of the biosphere when conditions change.

Keywords: capacity, Darwinism, ecological fitting, ecosystem, evolvability, fitness, host switch, inheritance, opportunity, parasite, pathogen, phylogenetic conservatism

Introduction

> *... when a parasite arrives in a new habitat, it will feed on those species whose defense traits it can circumvent because of the abilities it carries at the time.*
>
> Janzen (1980)

Parasitism may be the most common mode of life on the planet (Price, 1980; Brooks and McLennan, 1993). An enormous diversity of organisms lives and feeds on or inside other organisms, sapping energy and nutrients, and sometimes becoming pathogenic. A broad definition includes viruses, bacteria, fungi, protozoans, worms, plant-feeding insects, and plants that parasitize other plants (Nylin et al., 2018). It is safe to say that most species, including parasites, play host to at least one other species that parasitizes it and may cause disease.

Emerging infectious diseases (EIDs) caused by parasites cost the world an estimated \$1.3 trillion per year in treatment and production losses prior to the COVID-19 pandemic, with costs in the USA predicted to exceed the gross domestic product (GDP) of the entire country within 80 years (Brooks et al., 2019, 2022; Trivellone, Hoberg, et al., 2022). An EID occurs when a parasite colonizes an evolutionarily novel host or an old host that it has not been in contact with for some time. Commonly referred to as *host-switching*, it is fundamentally a result of the interaction between the *capacity* of a given parasite to infect new organisms beyond their current hosts and the *opportunity* to meet these new organisms (Agosta et al., 2010; Araujo et al., 2015; Braga et al., 2018; Nylin et al., 2018; Brooks et al., 2019; Feronato et al., 2021; Brooks et al., 2022). Traditionally, parasites have been viewed as exemplars of the evolution of specialization—they have highly intimate physiological, morphological, and ecological relationships with their hosts. Parasites *need* their hosts to survive. Combine this with the observation that most parasites are restricted to just a few closely related host species both ecologically (i.e., at a given time or place) and historically (i.e., through evolutionary time) and it is not surprising that many have assumed they are the classic evolutionary "dead end" (Moran, 1988; Wiegmann et al., 1993; Kelley and Farrell, 1998): when the host goes extinct, so does the parasite. The long-standing assumption has been that parasites are in a constant coevolutionary "arms" race" with their hosts in which natural selection favors increasing specialization, leaving them with little or no capacity to colonize new hosts (for critical reviews of traditional coevolutionary theory, see Brooks and McLennan, 2002; Janz, 2011; Trivellone and Panassiti, 2022).

Yet, the world is replete with examples of parasites colonizing new hosts (Agosta, 2006; de Vienne et al., 2013; Nylin et al., 2018; Trivellone and Panassiti, 2022), often closely related to their old hosts and often as a response to a change in conditions that catalyzes the movement of species (Brooks and McLennan, 2002; Hoberg and Brooks, 2008; Agosta et al., 2010; Hoberg et al., 2017; Brooks et al., 2019; Carlson et al., 2022). EID is an emergent property of this larger phenomena of host-switching by parasites. The current EID crisis is the recognition that emerging diseases are increasing in frequency, implying that the capacity for host-switching is large and that the opportunities to do so are common (Brooks and Ferrao, 2005; Brooks et al., 2019, 2022).

The Parasite Paradox

If parasites are so highly specialized to their hosts that they lack the capacity to use new hosts, then host-switching and EID should be rare to nonexistent. This is the parasite paradox (Agosta et al., 2010): how do otherwise highly specialized parasites switch to new hosts? The answer is that parasites, like all organisms, maintain the capacity to respond in novel ways when conditions change, using what they have inherited from their ancestors to survive as best they can, including switching to new hosts (Agosta et al., 2010; Brooks et al., 2019). The assumption that narrow host range is synonymous with a lack of capacity to colonize new hosts fails to recognize that (1) parasites are specialized on *specific resources* not specific species, (2) these resources may be phylogenetically conserved across a wider array of species than the ancestral hosts, and (3) parasites have the capacity to use *any* species that contain the specific resource if given the opportunity (Brooks, 1979; Janzen, 1979, 1980; Brooks and McLennan, 1993, 2002; Agosta, 2006; Brooks et al., 2006, 2019, 2022; Agosta et al., 2010; Araujo et al., 2015; Malcicka et al., 2015; Nylin et al., 2018; Boeger et al., 2022; Lajoie and Parfrey, 2022). In the case of SARS-CoV-2 and COVID-19, all mammals possess

the angiotensin-converting enzyme 2 (ACE2) receptor that is the entry point for infection, and it is not surprising that it moved from its ancestral bat hosts to other mammals, including humans, when given the opportunity (Brooks et al., 2020; Low-Gan et al., 2021; Boeger et al., 2022; Ruiz-Aravena et al., 2022).

The key to understanding EID lies not in understanding how the evolution of host specialization restricts host use but in understanding how and under what conditions parasites originally thought to have a very narrow host range can colonize new hosts (Agosta et al., 2010; Araujo et al., 2015; Braga et al., 2018; Brooks et al., 2019, 2022). The issue is part of a more general evolutionary question: how can an organism that evolves the capacity to function under one set of conditions keep functioning when those conditions change? The short answer was provided by Darwin more than 160 years ago: it uses the information it inherited from its ancestors, which contains the potential to do something new, like switch to a new host. If this were not true, EID would not occur.

Parasites and their hosts are embedded within the larger biosphere. Understanding the dynamics of EID amounts to understanding the more general evolutionary problem of how organisms use the limited information they inherit to cope with unpredictable changes in their surroundings (Brooks et al., 2019; Agosta and Brooks, 2020). The core mechanism is "ecological fitting in sloppy fitness space" (Agosta and Klemens, 2008), a key piece of a larger theoretical framework—the Stockholm Paradigm (SP)—that explains the overall eco-evolutionary dynamics of the biosphere and how it responds to change (Brooks et al., 2014; Hoberg et al., 2015; Hoberg and Brooks, 2015; Hoberg et al., 2017; Brooks et al., 2019; Agosta and Brooks, 2020). From the perspective of the SP, it becomes clear that *EID is an expected outcome of the same Darwinian dynamics and expression of the same evolutionary potential that governs the response of the rest of the biosphere during periods of environmental change.*

The Stockholm Paradigm: How the Biosphere Copes with Change by Changing

The SP describes the integration of the capacities for organisms to engage functionally with the environment, the ecological opportunities to use those capacities, and the repeated perturbations to the conditions of life (e.g., climate change) that drive the interaction between capacity and opportunity and catalyze evolutionary diversification and complexity in living systems (Brooks et al., 2014; Hoberg et al., 2015; Hoberg and Brooks, 2015; Hoberg et al., 2017; Brooks et al., 2019). Although only recently proposed as a synthetic framework, the SP has been under construction since 1859 when Darwin published the first edition of *On the Origin of Species* (see Agosta and Brooks, 2020). Darwin presented two major metaphors for his grand theory of "how nature works": the tree of life, depicting the selective accumulation of biodiversity and its evolutionary history of common descent, and the entangled bank, portraying the interactions among biodiversity that give rise to our modern notion of complex ecosystems composing the biosphere. The entangled bank is the "interaction arena" where the members of the tree of life coexist, competing, cooperating, predating, and parasitizing each other.

At any given moment, the entangled bank is a snapshot of a dynamic evolutionary system with the central question being how such systems and their constituent members persist in the face of constant change (Agosta and Brooks, 2020). Coupled with the tree of life, it represents the interplay between evolutionary history and the current ecological conditions, from which emerges natural selection and the interacting web of biodiversity that we observe. This web comprises individual organisms, each with inherited capacities to interact with and engage functionally with the surroundings, including parasitizing other organisms. Understanding "how nature works" means understanding how the members of the tree of life interact to form the persistent entangled bank that characterizes the biosphere. This requires building an explanatory framework from the level of *inheritance systems* (individual organisms, populations, and species) to the emergent level of ecosystems.

Part 1. How Individual Inheritance Systems Cope with Change

For much of human history, species were largely thought of as static—the species we see today have always been here, and their characteristics that seemingly match their environments so well are evidence that they are "perfectly fit" to their surroundings. Lamarck (1809) was the first person to seriously challenge the notion of immutable species. In essence, he argued that organisms were able to evolve directly and instantaneously to changes in their surroundings, with the environment somehow driving them to get better by producing "the right adaptation at the right time." Darwin showed that this notion of evolution was wrong. He summarized his views succinctly in the second paragraph in the final (6th) edition of *On the Origin of Species* (1872; my boldface, [my insert]):

*. . . there are two factors: [in evolution] namely, the **nature of the organism**, and the **nature of the conditions**. The former seems to be much more important; for nearly similar variations sometimes arise under, as far as we can judge, dissimilar conditions; and, on the other hand, dissimilar variations arise under conditions which appear to be nearly uniform*

Contrary to Lamarck, Darwin rejected the notion that the environment, or "nature of the conditions," was the driver of evolution. Instead of viewing the surroundings as a creative force that "pushes" or "pulls" organisms to "get better," he viewed it as the arena into which organisms imposed themselves, using the informational capacities that they inherited from their ancestors to survive and reproduce as best they can, given the current conditions. Thus, he recognized the primacy of what he called the "nature of the organism" in evolution, and that evolution was the result of the outcome of the interaction between organisms, with their inherited capacities, and their environments, which were always changing. From this emerged his idea of *natural selection*.

Nature of the organism: metabolism plus inheritance

Organisms are cohesive functional wholes with the informational capacity to impose themselves on their surroundings in ways that facilitate survival and reproduction (Collier, 1988, 1998, 2003; Collier and Hooker, 1999). Most fundamentally, they are *combined metabolic-inheritance systems* (Gánti, 1979, 2003; Maynard Smith and Szathmáry, 1995) with the ability to both *exploit* and *explore* their environments (Brooks et al., 2019; Agosta and Brooks, 2020). Organisms use metabolism to stay alive, exploiting their surroundings as best they can with their inherited capacities and "buying" the time for reproduction to occur. Inheritance is how organisms extend themselves through time, exploring their surroundings through the production of highly similar but variable offspring. The combined metabolic-inheritance system is a functional whole with the information encoded in inheritance specifying the metabolic system, and with metabolism fueling inheritance.

Organisms are also historical entities. They form "communities of descent" with shared evolutionary history, represented by the tree of life. Organisms retain so much of their history in inheritance that common descent is always the predominant explanation for their current form and function, not the surroundings. If fruit flies are reared in the wild or in a glass bottle in a laboratory, the result is still fruit flies. The primacy of the *nature of the organism* over the *nature of the conditions* stems from four fundamental aspects of inheritance:

(1) Inheritance is highly conservative. While the inheritance system is open to change through genetic mutation, duplication, recombination, and so forth, it is highly constrained by the requirement for functional integration with the rest of the system.
(2) Inheritance produces indefinite variation. Despite its conservative nature, because of genetic mutation, imperfect copying, duplication, and recombination, all offspring are highly similar but unique. Even clones are not identical (Cepelewicz, 2020).
(3) Inheritance is highly historical. While each organism is unique, each bears a strong resemblance to its relatives. Some of these resemblances are truly ancient, like the Hox genes that specify development of metazoans.
(4) Inheritance is superfluous. Organisms produce as many offspring as possible *without regard for environmental conditions*. Therefore, there is frequent reproductive overrun, with the production of many more offspring than the environment can support.

Conservative inheritance means that history will always be the dominant causal explanation for the present in biological systems, and that a "perfect fit" between organisms and their constantly changing environments can never be achieved. Because change in the inheritance system is highly constrained, there is an inherited *evolutionary lag-load* (Maynard Smith, 1976) or *phylogenetic constraint* (Brooks and McLennan, 2002) that makes it impossible for organisms to simply evolve new capacities in the moment change occurs. Instead, as Darwin recognized, evolution requires a constant supply of preexisting variation. When the first tetrapods transitioned from water to land, for example, all the necessary traits needed for surviving on land, including lungs for breathing air, limbs for walking, and eggs capable of surviving buried in soil, had already evolved in the aquatic environment (Skulan, 2000; McLennan, 2008).

The superfluous reproduction of variable but highly similar offspring without regard for the conditions produces Darwin's "necessary misfit" (Brooks and Hoberg, 2008). The combination of reproductive overrun and conservative inheritance guarantees an imperfect fit between organisms and their environments, but within this imperfect fit lies the potential for coping with future change (Agosta and Brooks, 2020). If all organisms were "perfectly

fit" to the current conditions, there would be little capacity to respond when those conditions change. For Darwin, "adaptation" was the process of coping with change by using preexisting capacities to survive as best as possible, reinforced by natural selection in the new conditions. Natural selection emerges from Darwin's necessary misfit and is proportional to the amount of mismatch between organisms and their surroundings (Brooks and Hoberg, 2008).

Life's only discernable "goal" is continued survival. This is fueled by metabolism but achieved through reproduction and inheritance. The information contained within the inheritance system specifies the development of a new organism including the metabolic system and other capacities to engage functionally with the environment (Collier, 1988, 1998, 2003; Collier and Hooker, 1999). Metabolism allows organisms to exploit the environment long enough to reproduce, but inheritance is what allows the exploration of variable and changing conditions (Agosta and Brooks, 2020). Inheritance is the essence of being evolvable.

To be a good exploiter requires functioning well enough in the current conditions. To be a good explorer requires being able to cope when those conditions change (Kováč, 2007; Popadiuk, 2012). To be evolvable, organisms need to be able to do both (Brooks et al., 2019; Agosta and Brooks, 2020; and see Page, 2011; Popadiuk, 2012). At first this might seem a paradox: there is a long history of assuming an evolutionary tradeoff between being especially good at doing one thing (specializing) and being able to do multiple things (generalizing)—the "jack-of-all-trades is master of none" principle. The assumption is based on the premise that in the evolution of specialization, natural selection whittles variation down to such a degree that species (inheritance systems) lose the capacity to do anything else, like colonize a new host. This perspective fails to recognize that the capacity for exploration emerges and grows naturally in living systems because of the conservative but evolvable nature of the inheritance system and despite persistent natural selection for better-performing variants (Agosta and Brooks, 2020).

A conservative system of indefinitely growing capacities

Compared to metabolism, the portion of the lifetime energy budget of an organism allocated to inheritance is very small. Reproduction requires only a small fraction of the metabolic budget because replication is a recycling process of "copy from a template, rinse and repeat" and because producing gametes is inexpensive compared to maintaining an organism throughout its lifetime. Staying alive amounts to staying organized, and this is very expensive and ephemeral in an entropic universe governed by the second law of thermodynamics (Lotka, 1913, 1925; Schrödinger, 1945). But producing offspring is relatively cheap and persistent (Brooks and Wiley, 1988; Agosta and Brooks, 2020). This means that once the combined metabolic-inheritance system emerged, evolution was both highly "affordable" and probable given the routine overproduction of similar but varied propagules.

The relatively low cost of reproduction is critical for evolution, but the inherent nature of the information in the inheritance system is what produces the indefinite variation and capacity to respond when conditions change. First, mutation coupled with mistakes during replication (imperfect copying) provide a background source of de novo variation. Second, the information encoded by DNA is both digital and combinatorial. Organisms are cohesive analog wholes, but they are *digital replicators* from which the information encoding for a new organism can be recombined and read at multiple levels of the genome and at multiple times. This generates the potential for an enormous amount of information, both expressed and unexpressed, to be stored in inheritance systems (Brooks and Wiley, 1988; Smith, 1988, 1998, 2000; Maynard Smith and Szathmáry, 1995, 1999; Szathmáry, 2000, 2015; de Vladar et al., 2017). There are, of course, constraints that arise from correlations between parts of the integrated system, including molecular affinities, cell-to-cell adhesion, genetic correlations, mate recognition systems, and symbiosis (Brooks and Wiley, 1988). If, for example, one gene requires another gene to function, then the potential information (variation) that could be expressed by inheritance is constrained. If one species requires another species for survival, then entire genomes are linked, again constraining the information that can be expressed. But despite these constraints, digital replication generates enormous amounts of variation that is the source of the preexisting capacities that evolution relies on when conditions change (Brooks and Wiley, 1988; Smith, 1988, 1998, 2000; Maynard Smith and Szathmáry, 1995, 1999; Szathmáry, 2000, 2015; de Vladar et al., 2017).

Third, as the inheritance system evolves and diversifies (Figure 1.1), the information contained within it expands and so does the *difference* between the information that is expressed—"what's realized"—and the information that could be expressed—"what's possible"—at any given time (Brooks and Wiley, 1988; Brooks and Agosta, 2012; Agosta and Brooks, 2020). As Darwin recognized, "diversity begets diversity"—in an expanding system of accumulating information/variation/capacities, the realization of one

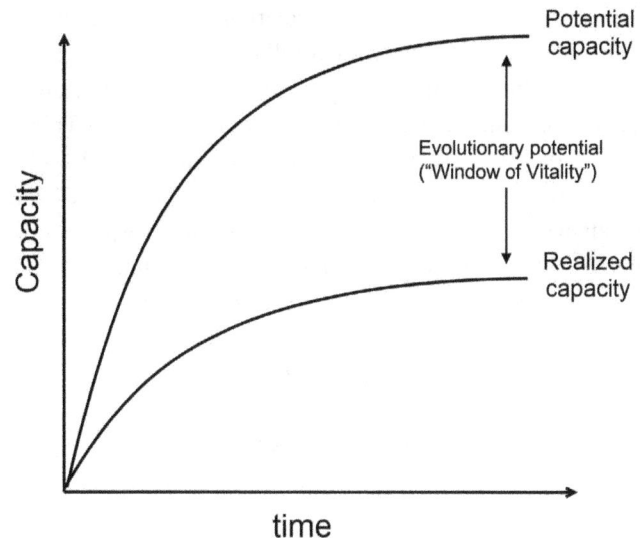

Figure 1.1. The Brooks-Wiley dynamic. As evolution unfolds and life diversifies, the realized capacities for organisms to engage functionally with their surroundings grow. The realization of one possibility always gives rise to new possibilities, so the potential for new capacities grows in tandem. The difference between realized and potential capacity equates to evolutionary potential, the "Window of Vitality" where reproduction and survival could occur, and it too grows as the system evolves. While the growth of all three components is indefinite, the rate of growth slows over time because of the buildup of historical correlations in the inheritance system (e.g., genetic correlations that constrain the expression of information). Modified from Agosta and Brooks (2020).

possibility always produces new possibilities. In this way, life creates and grows its own *capacity space* to explore, or what Ulanowicz (1997) called the "Window of Vitality." Capacity space represents a realm of possibilities for organisms to engage functionally with the environment, where the difference between *realized capacities* and *potential capacities* (Figure 1.1) is tantamount to *evolutionary potential*. The greater the difference between "what's realized" and "what's possible," the greater the potential for the inheritance system to do something new, like switch to a new host, when conditions change (Brooks and Agosta, 2012; Agosta and Brooks, 2020).

Upon life's inception, it had some minimal but sufficient capacity to achieve some minimal but sufficient level of functional engagement with the surroundings (Moreno and Ruiz-Mirazo, 2009). In this moment, life emerged as a combined metabolic-inheritance system capable of exploiting the surroundings long enough for reproduction and evolution to occur, thereby catalyzing the growth of the capacity space that it continues to explore. While the buildup of historical constraints (genetic correlations, mate recognition systems, symbiotic relationships, etc.) in the inheritance system slows the overall rate of growth of diversity over time (Figure 1.1), the difference between potential capacity and realized capacity continues to grow. This is the Brooks-Wiley dynamic (Brooks and Wiley, 1988), and it guarantees that as evolution unfolds, "what's realized" will be an ever smaller subset of "what's possible" *regardless of persistent natural selection for better-performing variants.* From this perspective, the risk space for host-switching and EID is truly large and indefinite (Brooks and Ferrao, 2005; Brooks et al., 2014, 2019; Boeger et al., 2022).

A conservative system of retained evolutionary potential

The growth of capacity space as evolution unfolds is indefinite but not unlimited. The conservative nature of inheritance produces extremely high levels of historical cohesion in biological systems. As previously mentioned, as a system expands, cohesive forces form constraints that slow the expansion. For example, gravity acts as a cohesive force that slows the expansion of the universe, allowing for the emergence of structures like stars, planets, and galaxies. In biology, the demand that all parts of the system be functionally integrated plays an analogous role (Brooks and Wiley, 1988). The essential point is that, while the inheritance system is open to change, it is severely constrained by the requirement for functional integration with the rest of the system. This is the reason that the concept of "selfish" genes driving evolution (Dawkins, 1976) is largely irrelevant— once integrated into an inheritance system, genes are part of a larger functional whole.

Across the biological hierarchy, historical correlations among various parts of the combined metabolic-inheritance system build up as evolution unfolds (Brooks and Wiley, 1988). This is the "cost of integration," and it places severe constraints on both the rate of evolution and the realm of possibilities that evolution can explore. At the same time, it also facilitates evolution by lowering the "cost of innovation" in two key ways. First, conservatism significantly reduces the threshold for generating novel information because inheritance mainly recycles and recombines old information (Jacob, 1977; Gould and Lewontin, 1979; Gould and Vrba, 1982; Janzen and Martin, 1982; Brooks and McLennan, 2002; McLennan, 2008). For parasites, this can significantly lower the threshold for acquiring new hosts because genetic changes are not required for colonization (Agosta, 2006; Agosta and Klemens, 2009; Agosta et al., 2010; Araujo et al., 2015; Brooks et al., 2019; Lajoie and Parfrey, 2022). The ability to co-opt and combine

preexisting traits for new functions alleviates the inability to simply produce the "right adaptation at the right time" in response to change.

Second, conservatism slows down the entire evolutionary process. While severely limiting on one hand, this "buys time" for preexisting parts of the system to meet and become integrated into a new functional whole (Maynard Smith and Szathmáry, 1995; Brooks and Agosta, 2012; Agosta and Brooks, 2020). The evolution of herbivory in animals is a symbiosis with microbes that required the concatenation of numerous evolutionary phenomena, all of which arose at different times and were retained by conservative inheritance long enough to be combined into a true herbivore (Agosta and Brooks, 2020). This includes the evolution of microbes that could digest plant cellulose, the ability for animals to ingest but not digest the microbes, the ability for the microbes to live inside the animal's gut, the emergence of gut-living microbes separated from their free-living ancestors, and a way for animals to pass the microbes to their offspring.

Conservative inheritance "stores history" long enough to produce a constant and growing lag between "what's possible" and "what's realized" for living systems (Figure 1.1). Because inheritance is conservative, no living system can be perfectly fit to its current conditions. And this is key to continued survival. Persisting indefinitely relies far more on having the potential to cope when conditions change and far less on how fit an inheritance system is to the current conditions. For the SP, the essential point is that all inheritance systems retain this potential to some degree (Agosta and Brooks, 2020), even parasites with a very narrow host-range (Agosta et al., 2010; Brooks et al., 2019, 2022).

Nature of the conditions: opportunities in sloppy fitness space

The capacity to exploit and explore emerges from the "nature of the organism," but the opportunities to do so are a function of the conditions in which organisms find themselves. The "nature of the conditions" gives rise to *opportunity space* where organisms find chances for survival dependent on their inherited capacities to exploit and explore (Brooks et al., 2019; Agosta and Brooks, 2020). For a given species, only a subset of the global opportunity space for the biosphere is available for survival and reproduction. This subset of opportunity space available to any given species is its *realized opportunity space* or what is more commonly referred to as *fitness space.*

Fitness space emerges when capacity space is imposed on opportunity space (Figure 1.2). Fitness describes how well organisms cope with their surroundings; fitness space, therefore, is analogous to "niche space" and represents the set of conditions in which survival and reproduction can occur (Hutchinson, 1957). Organisms are "fit" for any conditions in which they have the capacity to survive and reproduce. The fundamental demographic of evolution is not "survival of the *fittest*" but "survival of the *fit*" (Brooks and Agosta, 2012; Brooks et al., 2019; Agosta and Brooks, 2020). If only a single or few fittest variants survived, the stock of standing variation for evolution to act on when conditions change would be very small. There would be little potential to respond because today's fittest variant may not be fit at all in the new conditions, leaving little room for inheritance to explore new options in fitness space. Fortunately, survival does not require being the absolute best, it requires only being *good enough* to cope with the conditions at hand. All organisms that reproduce are fit, and while some are fitter than others, they all compose a *fittest collective*—a distribution of variants with adequate capacity to survive the current conditions. This fittest collective includes a range of variants—a genotypic-phenotypic distribution—that represent the evolutionary potential to cope when conditions change.

The need to achieve positive fitness dictates that organismal capacity must always complement ecological opportunity, but it does not need to perfectly match (Agosta and Brooks, 2020). Moreover, capacity cannot perfectly match opportunity because conservative inheritance ensures that evolution always lags behind the conditions (Maynard Smith, 1976), while also storing a history of past success. Conservative inheritance therefore all but guarantees that fitness space will be "sloppy" (Agosta and Klemens, 2008) not tightly optimized to any particular set of conditions. The sloppiness is proportional to the difference between *realized fitness space*—what organisms are currently doing to survive and reproduce—and *fundamental fitness space*—what organisms could be doing if given the opportunity (Figure 1.2). And this difference is proportional to the capacity to cope with change.

To be evolvable, all species including parasites must maintain an inheritable difference between the actual and the possible in fitness space. This appears to be a universal feature of life emerging from the conservative but mutable nature of the organism (Brooks and Wiley, 1988; Agosta and Klemens, 2008; Daniels et al., 2008; Brooks and Agosta, 2012; Soberón and Arroyo-Peña, 2017; Agosta and Brooks, 2020). This is how the collective biosphere has coped with constant change over the past 4 billion years.

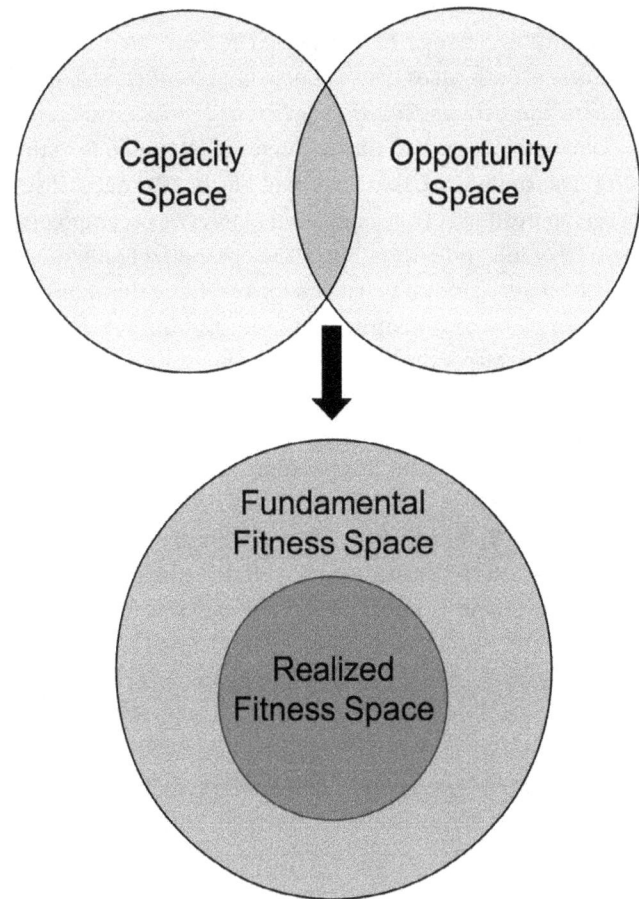

Figure 1.2. Fitness space emerges from the interaction of organismal capacities to engage functionally with the environment—capacity space—and the ecological chances to use those capacities—opportunity space. The difference between what organisms are doing to survive and reproduce—realized fitness space—and what they could be doing—fundamental fitness space—makes fitness space inherently "sloppy." Modified from Agosta and Brooks (2020).

Darwinian conflict resolution in sloppy fitness space

Understanding how species mount an initial response when conditions change is key for managing EID (Agosta et al., 2010; Brooks et al., 2019). It boils down to a more general understanding of the dynamics of inheritance systems in sloppy fitness space (Agosta and Brooks, 2020). Since organisms will produce as many highly similar offspring as possible regardless of the conditions, reproductive overrun is inevitable. This produces Darwin's constant "struggle for survival," routinely putting organisms in *conflict* with their surroundings, including other organisms. Superfluous reproduction means that all inheritance systems may grow too much and become *victims of their own success*. If the conditions change, they may also become *victims of circumstance*. Both cases trigger *Darwinian conflict resolution* (Agosta and Brooks, 2020).

Darwin recognized that a constant "struggle for survival" was an inescapable feature of life, but for natural selection to produce the accumulated biodiversity composing the tree of life and coexisting in the entangled bank, there must be a persistent mechanism for resolving this conflict. This begins by organisms using inherited information to explore new opportunities in fitness space. Known as *ecological fitting* (Janzen, 1985), this is the general mechanism behind host-switching and the default response for all living systems when conditions change (Brooks and McLennan, 2002; Agosta, 2006; Brooks et al., 2006; Agosta and Klemens, 2008; Agosta et al., 2010; Araujo et al., 2015; Malcicka et al., 2015; Braga et al., 2018; Brooks et al., 2019, 2022; Agosta and Brooks, 2020). The capacity for ecological fitting emerges from phylogenetic conservatism (Brooks and McLennan, 2002) and other related universal aspects of inheritance, including phenotypic plasticity (West-Eberhard, 2003) and evolutionary trait co-option (the ability of existing traits to be co-opted and combined in novel ways to perform novel functions) (McLennan, 2008).

The capacity for ecological fitting affords inheritance systems critical degrees of freedom for coping with changing environments by exploring new options in under-used, less preferred, or previously inaccessible portions of fitness space. The capacity to *move away* from portions of fitness space that are densely populated, deteriorating, or disappearing into new portions of fitness space is key to indefinite persistence (see, e.g., paleontological studies by Stigall et al., 2017, 2019; Stigall, 2019), even if this leads to reduced fitness. In evolution, being fit is what matters; a marginal existence is better than not existing. *Ecological fitting in sloppy fitness space* is how life continues to apply what Agosta and Brooks (2020) called "biological assumption zero": organisms will do what they can, where they can, when they can, within the constraints of evolutionary history (inheritance) and ecological opportunity.

Reproductive overrun of highly similar offspring means there will always be a tendency for inheritance systems to be specialized in regions of highly preferred fitness space, implying the under-use of more marginal but still survivable regions (Figure 1.3a). When conditions are stable (e.g., when a parasite is isolated with a single host species), exploiting the surroundings as much as possible takes precedence over exploring them. During these times, conditions are largely predictable, allowing many variants to survive

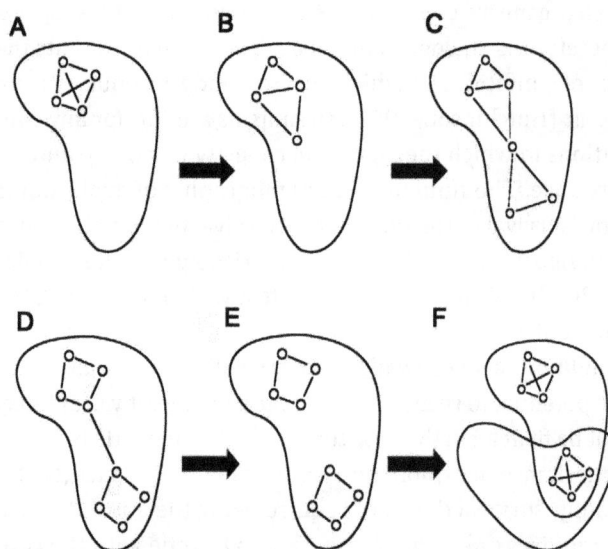

Figure 1.3. Darwinian conflict resolution in sloppy fitness space. The outer shape represents the fitness space of a single inheritance system. Circles represent members of the inheritance system, and lines represent connections between them; the number of connections indicates how cohesive and isolated members of the system are in fitness space. (A) When conditions are stable, preferred portions of fitness space are exploited as much as possible, causing inheritance systems to become more connected and *specialized* in isolation. (B, C) When conditions change, conflicts in fitness space manifest as a "struggle for survival," leading some members of the inheritance system to *move away* and explore new opportunities for survival, causing the system to become more disconnected and *generalized* in fitness space. (D, E) Exploration leads to the exploitation of local conditions, producing diverging subsystems, each specializing in a different part of fitness space. (F) Selection for increased cohesion with local conditions leads to speciation, producing a new inheritance system with its own fitness space, including an overlap with the original inheritance system resulting from common ancestry. Modified from Agosta and Brooks (2020).

and reproduce by doing much of the same thing again and again, exploiting as much energy and materials as possible to make as many offspring as possible, with natural selection favoring fitter variants but with all variants that are fit enough surviving. What happens when conditions change? Exploration of new opportunities for survival becomes paramount. Inheritance systems begin to spread out in fitness space, becoming more generalized as preexisting variation "wanders" into more marginal or previously inaccessible parts of it (Figure 1.3b, c). Extinction occurs when an inheritance system runs out of options in fitness space, when its capacity no longer complements opportunity.

Within the SP, "specialist" and "generalist" are not nouns describing the static traits of static species but verbs describing dynamic states of inheritance systems within fitness space (Agosta and Brooks, 2020). All inheritance systems have the capacity to become more specialized or generalized in their fitness space, and, moreover, all have the capacity to oscillate between these states depending on the conditions. Species that are specializing in their fitness space, in fact, have more potential to respond to a change in conditions than those that are generalizing because they have more sloppy fitness space to explore. Specializing in fitness space is an indicator of *evolutionary potential being stored* when conditions are stable. Generalizing in fitness space is an indicator of *evolutionary potential being spent* when conditions change (Agosta and Brooks, 2020).

Ecological fitting provides the means for inheritance systems to explore "what's possible" when the conditions change. In the extreme, this may completely exhaust the capacities of an inheritance system so that realized fitness space equals fundamental fitness space, leaving no more possibilities to explore. The propensity for organisms to exploit their surroundings as much as possible predominates, producing new conflict with no possibility of resolution. While critical for the initial response to change, ecological fitting alone does not produce the selective accumulation of diversity represented by the tree of life and coexisting in the entangled bank. Conflict resolution requires an additional mechanism that results in previously linked subgroups permanently splitting into two or more novel inheritance systems (Figure 1.3d–f). When a recently generalized inheritance system becomes isolated in a new part of fitness space, selection for cohesion or *co-accommodation* (Brooks, 1979) with the new environment, including other organisms, leads to increased functional integration and increased connections within the subsystem as it becomes more specialized in its new portion of fitness space (Figure 1.3d). Through speciation, this may produce a newly emergent inheritance system with its own fitness space that may overlap with the ancestral inheritance system because of shared evolutionary history (Figure 1.3f). At this point, the original conflict has been resolved by diversifying and co-accommodating with the new conditions, but because the nature of the organism always predominates, conflict resolution always leads to new conflict.

Darwinian evolution is an iterative process of conflict and conflict resolution (Agosta and Brooks, 2020). Diversity accumulates rather than replaces itself because the criterion for indefinite persistence is simply to be good enough to reproduce (it is survival of the *fit*) and because each bout of conflict resolution sets the stage for new conflict. Evolution is *conflict resolution by ecological fitting in sloppy fitness space, followed by co-accommodation with the new conditions, all reinforced by natural selection.* The fuel that drives this forward is the potential emerging from preexisting capacities to first explore and then exploit new opportunities in fitness space. When conditions are stable, living systems build this potential. When conditions change, this potential is spent. Generating, maintaining, and deploying this potential is the key to individual inheritance systems coping with change, and for our purposes, is the fundamental source of EID (Brooks et al., 2019).

Part 2. How Ecosystems Cope with Change

Within the SP, basic Darwinian evolutionary dynamics played out in the context of sloppy fitness space explain how individual inheritance systems, including parasites with very narrow host ranges, explore new ecological opportunities for survival when conditions change. The recent work of Araujo et al. (2015), Braga et al. (2018), and Feronato et al. (2021) provides a quantitative demonstration of this exploration by modeling opportunities for parasites to infect new hosts. But this is only part of the story. The context of the EID crises is the interconnected web of ecosystems—the entangled bank—that comprises the biosphere and that we depend on for survival and have the capacity to alter in ways that favor EID (Brooks et al., 2019). Understanding EID as a phenomenon related to global change requires a broad eco-evolutionary framework that describes how the collective biosphere responds when it is perturbed (Brooks and Ferrao, 2005; Hoberg and Brooks, 2008; Agosta et al., 2010; Brooks et al., 2014, 2019, 2022; Hoberg et al., 2015, 2017; Boeger et al., 2022).

In the early 20th century, some believed that the entangled bank was a "superorganism"—a single entity composed of individual species working together for their collective survival, like a colony of bees or the cells that compose organisms. This view was advanced most prominently by Clements (1905, 1916). Unlike a colony of bees or a multicellular organism, however, the species that make up ecosystems do not share a singular inheritance system. Therefore, the assemblage of species that compose an ecosystem cannot be an "organism" (Tansley, 1935). Each member of an ecosystem works toward its own survival using the capacities it has inherited, including the capacity for ecological fitting. As each member degrades the surroundings to meet its own requirements, it alters the surroundings in such a way—from producers to consumers to decomposers—that converts them into new biomass that may meet the functional needs of other members. Within ecosystems, otherwise independent inheritance systems are therefore linked by a complex set of trophic interactions, with each species ensuring its own survival by indirectly providing the means for other species to survive. Each organism's fitness space represents potential fitness space for another organism; the sheer diversity of parasites exemplifies this.

Ecosystem function is an emergent property of each member's activities, each with its own inherited capacities to exploit the current conditions as best it can. Ecosystems are therefore "centers of exploitation" for their constituent members—places where organisms use metabolism to stay alive long enough to reproduce while also retaining the capacity to explore new opportunities when conditions change. This gives rise to a *collective evolutionary potential stored within ecosystems*, or what Agosta and Brooks (2020) called an *evolutionary commons*. When perturbed, expression of the evolutionary commons may result in ecological rewiring of trophic connections within ecosystems (e.g., host-switching), but on a large enough scale perturbations may cause ecosystems to break apart. When this happens, the capacity for ecological fitting stored within the evolutionary commons means there is the potential for new ecosystems to form out of the remnants of the old.

The evolutionary commons makes the biosphere extremely robust and resilient to perturbations, not fragile as some have assumed. This has dual implications for humans. No matter how much we perturb it, the biosphere is unlikely to collapse. After all, the aftermath of each great mass extinction event has been *mass evolutionary renewal*, the production of new biodiversity emerging from the species that survived. At the same time, since the biosphere is indifferent to the fate of any given species, the expression of the potential stored in the evolutionary commons may not always work in our favor. The ecosystems we depend on may disappear or change too much to support our survival, or they might give rise to the next EID. Understanding how the potential stored in the evolutionary commons is expressed when the biosphere is perturbed is the key to assessing the risk space for EID in a period of global change (Brooks et al., 2019, 2022). To do so requires going beyond explaining the dynamics of individual inheritance systems in a relatively stable, unchanging fitness space.

What happens when external perturbations change the nature of the conditions so much that the dimensions of fitness space itself change for multiple inheritance systems, including entire biotas, at once? When the dimensions of fitness space itself are altered, old opportunities to survive and reproduce may disappear and new opportunities may arise. The nature of the organism is to take advantage of new opportunities by exploring new parts of fitness space that were previously inaccessible or nonexistent. As mentioned previously, bouts of ecological fitting in static fitness space alone would be self-limiting, leading to episodes of expansion that simply fill fitness space, and this would not lead to the indefinite diversification that characterizes the tree of life or the complexity that characterizes the entangled bank. To build a diverse, complex, persistent biosphere requires an essential ingredient: repeated systemic perturbations that routinely alter the fitness space of multiple species. These perturbations catalyze the dynamics that allow living systems to fluctuate between exploiting and exploring both the *geographical* and *functional* dimensions of fitness space (Brooks et al., 2019; Agosta and Brooks, 2020).

Fluctuating in the geographical dimensions of fitness space

For Darwin and Wallace, cofounders of the theory of natural selection, where a species lived was the primary component of its fitness space. Darlington (1943) extended this to include not only the places where species lived, but the movement of species to and from those places catalyzed by external changes in the conditions. He concluded that species arose in "centers of diversification," where external perturbations caused geographic ranges to fluctuate around a continuously occupied core. These fluctuations might be driven by the formation of barriers to dispersal, producing episodes of isolation leading to speciation, and the breakdown of those barriers, producing episodes of biotic expansion from the core and, as Darlington saw it, setting the stage for new species to replace older species.

The notion of new species arising in geographic centers of origin and then expanding into new areas to *replace* older species is based on the idea that the species-area relationship (Cain, 1938) is a result of the environment comprising a limited number of niches for species to fill. In this way of thinking, once an area becomes saturated, new species can be added only if another species leaves or goes extinct (e.g., MacArthur, 1969; Roughgarden and Feldman, 1975; Hairston, 1980; Case, 1981). Niches, however, are an emergent property of the nature of the organism, not the conditions, synonymous with fitness space (Hutchinson, 1957; Colwell and Rangel, 2009). No fixed number of niches are in the environment for species to fill; each organism's fitness space represents potential fitness space for another organism.

Following Darlington, Wilson (1959, 1961) proposed the *taxon cycle*. Multiple species from a given area may colonize new areas when a change in conditions expands the amount of suitable habitat and then may contract their ranges when another change in conditions reduces the amount of suitable habitat. MacArthur and Wilson (1963, 1967) extended this to produce the equilibrium theory of island biogeography. They proposed a one-way dynamic in which "islands" were colonized by "source areas" that contained a preexisting pool of species. Like Darlington's, their theory relied heavily on the species-area relationship and the idea of an equilibrium number of species that could fill a limited number of niches. When an island had fewer than the equilibrium number of species, it was open to colonization. When the equilibrium number of species was reached, it was closed to colonization, unless a new species displaced an old one or an old species went extinct. For MacArthur and Wilson, the available fitness space on an island was a fixed, static quantity and only pre-existing species could fill it.

Erwin (1979, 1981, 1985) proposed a biogeographical theory that included both a mechanism for speciation and allowance for fitness space to be a dynamic property of the organism. The taxon pulse hypothesis posits that groups of species (biotas) experience fluctuations in their geographic ranges catalyzed by repeated changes in environmental conditions that drive diversification. When changing conditions cause dispersal barriers to break down, species expand geographically but also in fitness space as the inheritance systems generalize, spending potential in their exploration of new areas. When conditions change again, new barriers to dispersal may arise, isolating populations both geographically and in fitness space, promoting diversification and speciation. When changing conditions break down dispersal barriers again, a new phase of geographic expansion and generalization in fitness space is initiated—and so the cycle is repeated—as changing conditions catalyze new bouts of isolation and expansion, variously isolating and mixing together species geographically, producing the historically contingent complex patterns of species distributions and coexistence that we observe.

The empirical evidence for taxon pulse–driven biotas is extensive (e.g., Spironello and Brooks, 2003; Bouchard and Brooks, 2004; Brooks and Ferrao, 2005; Halas et al., 2005; Folinsbee and Brooks, 2007; Hoberg and Brooks, 2008; Lim, 2008; Eckstut et al., 2011). A history of repeated taxon pulses, especially over large areas, produces biotas that are idiosyncratic, composed of many species that have associated with each other for varying lengths of times and have arrived under different conditions. These dynamics in the geographical dimensions of fitness space, catalyzed by external perturbations, cause complex mosaics of species co-existing and interacting, forming dynamic ecosystems that can break apart and reform based on the inherited capacities of the constituent members. And because ecological fitting is a sorting process in which only the "fits" that work persist, the results of these dynamics for interacting species can look highly similar to the results expected from a long history of coevolution and cospeciation. In a Costa Rican forest, for example, there is remarkably close morphological matching between flower corolla tube depth and proboscis length in a community of plants and their moth pollinators even though none of the species are endemic to the area, and they evolved elsewhere at different times and places, and immigrated to the site at different times and under different conditions (Agosta and Janzen, 2005).

Fluctuating in the functional dimensions of fitness space

Fitness space is composed of not only the places where species can live but also the things they can do in those places. Taxon pulses lead to the mixing and matching of species in different locations, which in turn catalyzes phases of exploration of new opportunities for survival during periods of geographic expansion followed by phases of exploitation of those opportunities during periods of isolation. Thus, along with fluctuations in geographical fitness space, the oscillation hypothesis posits that inheritance systems fluctuate between generalizing and specializing in functional fitness space (Janz et al., 2006; Janz and Nylin, 2008; Nylin and Janz, 2009; Nylin et al., 2014).

The oscillation hypothesis was originally proposed to explain the evolution of host range in plant-feeding insects, but its applicability is far more general (Agosta et al., 2010; Brooks et al., 2019, 2022; Agosta and Brooks, 2020). Nonetheless, the dynamics of changes in functional fitness space are relatively easy to visualize for groups of organisms that form highly specific interactions with each other, like insects and plants or parasites and hosts. For these organisms, it is relatively easy to draw a connection between "host" and "function," and therefore between "host space"—the range of hosts that can be used—and the functional dimensions of fitness space. And since host space is a part of fitness space, there will always be a difference between *realized host range* (the set of hosts that are being used) and *fundamental host range* (the set of all hosts that could be used). As with fitness space in general, host space is inherently sloppy (Agosta, 2006; Agosta and Klemens, 2008, 2009; Agosta et al., 2010; Brooks et al., 2019, 2022).

The capacity for ecological fitting means that despite forming highly specific associations, parasites always have opportunities to explore new hosts. Parasitism is not an inherent evolutionary dead end. The capacity for host-switching manifests as changes in host range that do not track phylogenetic patterns of host-relatedness directly but are constrained by the phylogenetic distribution of necessary resources found in related hosts (Brooks and McLennan, 2002; Brooks et al., 2006). The literature contains a mountain of evidence that shows this pattern, so phylogenetically conservative host-switching should be viewed as a routine phenomenon (Brooks and McLennan, 2002; Agosta, 2006; de Vienne et al., 2013; Nylin et al., 2018; Trivellone and Panassiti, 2022; see Trivellone, Araujo, et al., 2022 for recent advances in R-based statistical methods for its detection in co-phylogenetic studies.) Moreover, published studies reveal phylogenetically conservative patterns of associations over long periods of time that alternate between increases and decreases in host range, in accordance with the oscillation hypothesis (e.g., Janz et al., 2006; Nylin et al., 2014; Jorge et al., 2018; Brooks et al., 2015; Boeger et al., 2022).

The evolution of pocket gophers (Geomyidae) and their parasitic lice (*Geomydoecus*) in North America provides a particularly instructive example (Brooks et al., 2015). Transmission of lice occurs when gophers are in their nests, so the opportunities for the parasite to switch to new hosts are limited even when external perturbations catalyze geographic expansion. Nevertheless, host-switching driven by external perturbations is a common theme in the history of these associations (Brooks et al., 2015). A burst of diversification in pocket gophers and lice 4.2 million to 1.8 million years ago coincided with a period of substantial climate and habitat change (Spradling et al., 2004). Around half of the associations that emerged during this time were the result of host-switching, followed by episodes of cospeciation. The pattern shows clear evidence of the oscillation dynamic, alternating between episodes of host range expansion by lice—generalizing in fitness space—followed by episodes of isolation and diversification—specializing in fitness space—in association with their new hosts (Brooks et al., 2015). Computer simulations have reproduced these patterns, showing how easily oscillations in sloppy fitness space can emerge from basic Darwinian dynamics in the context of opportunity space that includes the chance to encounter new hosts (Araujo et al., 2015; Braga et al., 2018; Feronato et al., 2021).

Avoiding extinction: fluctuating in combined geographical-functional fitness space

External perturbations like climate change alter both the geographical and functional dimensions of fitness space for multiple species, affecting parts of or even the entire biosphere (Brooks and Agosta, 2012; Brooks et al., 2019; Agosta and Brooks, 2020). Even without external perturbations, as dynamic complex systems, ecosystems experience continuous internal change because of "autonomous turnover" of species distributions, abundances, and interactions (O'Sullivan et al., 2021). If ecosystems were fragile, these perturbations would simply break them apart, with limited potential to form new connections within the system or to form new systems. But because of the collective capacity for coping with change retained within ecosystems—the evolutionary commons—they are relatively immune to the "butterfly effect" (Agosta and Brooks, 2020).

The butterfly effect—in which a small change in one part of a system has a large effect on the whole system (Lorenz, 1972)—has been used as a metaphor to describe the supposed fragility of the biosphere. It is, however, a truly poor way to describe a complex evolutionary system. Robustness, resiliency, and responsiveness in the face of internal and external change are hallmarks of complex systems (Kitano, 2004, 2007; Page, 2011). Complex systems are multilevel, hyperdiverse, hyperconnected, functionally redundant, and modular. All these features contribute to the *anti–butterfly effect* being a fundamental property of the biosphere (Agosta and Brooks, 2020). While ecosystems all have vulnerability thresholds beyond which perturbations can no longer be absorbed and "tipping points" at which they may abruptly shift to some new state (Dakos et al., 2019), they do not simply collapse in response to the loss or addition of new species. Moreover, the biosphere does not collapse when ecosystems are lost or change. If living systems were this brittle, the complex persistent biosphere would not exist. Again, recall that so far the aftermath of each mass extinction has been mass evolutionary renewal.

Ecosystems are resilient and changeable in proportion to their diversity and capacities of their constituent members for ecological fitting in sloppy fitness space (Agosta and Brooks, 2020; and see Dakos et al. 2019), which provides species degrees of freedom for exploring new opportunities by moving to new geographic areas and by co-opting existing functions for new functions, including forming new trophic connections with other species. When perturbations lead to geographic expansion, they also catalyze intense periods of exploration of new opportunities for survival, causing species to spend evolutionary potential by generalizing in fitness space. When perturbations lead to geographic isolation, exploitation of the new opportunities is reinforced by natural selection, leading species to specialize functionally in fitness space. And since new diversity emerges in isolation when conditions are stable, new evolutionary potential to exploit new opportunities can build before the next perturbation catalyzes a new episode of geographic expansion. In this way, alternations in geographical and functional fitness space are coupled but out of phase. Each new perturbation leads to a pulse of geographic expansion, the mixing of different species, including hosts and parasites, and the spending of potential as each species generalizes, followed by episodes of isolation and then specialization when potential is restored.

The entangled bank is constructed from and persists indefinitely because of repeated and overlapping cycles of taxon pulses and correlated oscillations in functional fitness space across multiple temporal and spatial scales. The built-in capacity of individual inheritance systems to alternate between periods of specialization and generalization in combined geographical-functional fitness space is how they, and by extension the collective biosphere, avoid extinction. The biosphere is robust, resilient, and responsive. But there are limits to its evolvability. The prodigious amounts of informational capacity produced by living systems affords them the potential to cope with change, but they cannot predict the future and are therefore vulnerable to extinction. No matter how fit a species is in today's conditions, it can still be unfit tomorrow when conditions change. Individual inheritance systems and the connections among them therefore routinely go extinct, but the biosphere persists because of the capacity to absorb even massive perturbations arising from the dual exploiter-explorer nature of its constituent members and collective potential stored in the evolutionary commons (Agosta and Brooks, 2020).

Concluding Remarks

The EID crisis fundamentally involves three interrelated variables: humans, the ecosystems on which we depend and which contain the organisms that are the sources of EID, and external factors such as climate change, land conversion, urbanization, and global trade and travel that cause changes in the conditions that catalyze EID (Brooks et al., 2019). The overarching message of this chapter is that we cannot disentangle EID from the overall response of ecosystems to climate change and that EID is fundamentally an expression of the same evolutionary potential that allows the rest of the biosphere to cope with change by

changing. Or more simply, we cannot understand EID without a more general understanding of how nature works. This understanding is rooted in Darwinism and synthesized by the SP.

The SP tells us that in a period of global change, we should expect more host-switching and therefore more EID. It tells us the risk space for EID is very large and that the realization of these risks increases the more ecosystems are perturbed. The anti–butterfly effect, mediated by the capacity for ecological fitting in sloppy fitness space, means that cospeciation and coextinction are unlikely to be major factors structuring parasite-host interactions, a prediction that has been corroborated by two large meta-analyses of cophylogenetic studies (de Vienne et al., 2013; Trivellone and Panassiti, 2022). This implies that species of pathogens (or species that can can be pathogenic) routinely persist longer through evolutionary time than their original species of host. For SARS-CoV-2 not only is the ACE2 receptor phylogenetically conserved among mammals, specialization on binding to the ACE2 receptor is conserved among coronaviruses (Low-Gan et al., 2021; Ruiz-Aravena et al., 2022), suggesting substantial sloppy fitness space in these interactions and, thus, great potential for host-switching.

The risk space for EID is substantial, but the SP also tells us it is not random. Conservative inheritance means the traits involved in enabling parasites to be transmitted among potential hosts, like the ACE2 binding site, are highly specific. Therefore, the community structure of parasite-host interactions should be highly conservative, even without pervasive cospeciation/coextinction dynamics. Brooks et al. (2006) examined six assemblages of lung flukes (platyhelminths) that parasitize frog species in temperate forests and grasslands in the USA and tropical dry and wet forests in Mexico and Costa Rica. They predicted that if ecological fitting was the dominant factor structuring the communities, as opposed to cospeciation/coextinction, then (1) associations should be largely determined by conservative traits related to parasite transmission rather than host phylogenetic relatedness and (2) communities should exhibit similar patterns of associations at the generic and family levels, even though they are from widely separated areas and very different habitats. Their data corroborated both predictions: conservatism in parasite and host traits related to transmission (i.e., habitat and feeding preferences) was the primary determinant of the associations, not host phylogeny, with each community converging on a similar phylogenetic structure. The shared requirement for aquatic habitats of tadpoles for all frog species allows potential colonization by essentially any lung fluke species if given the opportunity; additional lung fluke species are associated with frog hosts largely as a function of how much time the adults spend in the aquatic habitat.

The SP tells us that while the potential for EID is large, where and when EID will emerge is predictable based on knowledge of the conservative traits of parasites and hosts, suggesting proactive measures rooted in evolutionary principles (Brooks et al., 2014, 2015, 2019, 2022; Boeger et al., 2022; Hoberg, Boeger, et al., 2022; Hoberg, Trivellone, et al., 2022). Knowing, for example, that malaria (1) is caused by organisms in the genus *Plasmodium* and (2) is spread to humans exclusively through bites of adult mosquitoes in the genus *Anopheles*, all of which (3) have aquatic larvae and feed on vertebrates is a large step toward narrowing the "malaria risk space." Basic natural history knowledge set in a proper theoretical framework (Hoberg et al., 2015; Hoberg, Boeger, et al., 2022; Hoberg, Trivellone, et al., 2022) provides humans with significant capacity to predict and preempt where the next case of emergence may occur. This is the fundamental objective of the DAMA protocol (Brooks et al., 2014, 2015, 2019; Boeger et al., 2022; Hoberg, Boeger, et al., 2022; Hoberg, Trivellone, et al., 2022), which stands for Document, Assess, Monitor, and Act. DAMA is a direct application of the SP to the problem of emerging disease. It recognizes that the EID risk space is large but nonrandom because of phylogenetic conservatism. It is designed to anticipate EID before it happens, or when it does happen, to be better prepared to mitigate the spread not just among humans but among nonhuman species that may be potential suitable hosts. The combination of the SP—providing a theoretical foundation for understanding emerging disease in the context of global change—and DAMA—providing a framework for applying the SP to emerging disease—can be a powerful approach for transforming our response to the EID crisis to a more proactive, evolutionarily informed effort.

Literature Cited

Agosta, S.J. 2006. On ecological fitting, plant-insect associations, herbivore host shifts, and host plant selection. Oikos 114: 556–565.

Agosta, S.J.; Brooks, D.R. 2020. The Major Metaphors of Evolution: Darwinism Then and Now. Springer International Publishing, Cham, Switzerland.

Agosta, S.J.; Janzen, D.H. 2005. Body size distributions of large Costa Rican dry forest moths and the underlying relationship between plant and pollinator morphology. Oikos 108: 183–193.

Agosta, S.J.; Klemens, J.A. 2008. Ecological fitting by phenotypically flexible genotypes: implications for species associations, community assembly and evolution. Ecology Letters 11: 1123–1134.

Agosta, S.J.; Klemens, J.A. 2009. Resource specialization in a phytophagous insect: no evidence for genetically based performance trade-offs across hosts in the field or laboratory. Journal of Evolutionary Biology 22: 907–912.

Agosta, S.J.; Janz, N.; Brooks, D.R. 2010. How specialists can be generalists: resolving the "parasite paradox" and implications for emerging infectious disease. Zoologia 27: 151–162.

Araujo, S.B.L.; Braga, M.P.; Brooks, D.R.; Agosta, S.J.; Hoberg, E.P.; von Hartenthal, F.W.; et al. 2015. Understanding host-switching by ecological fitting. PLOS ONE 10: e0139225.

Boeger, W.A.; Brooks, D.R.; Trivellone, V.; Agosta, S.J.; Hoberg, E.P. 2022. Ecological super-spreaders drive host-range oscillations: Omicron and risk space for emerging infectious disease. Transboundary and Emerging Diseases 69: e1280–e1288.

Bouchard, P.; Brooks, D.R. 2004. Effect of vagility potential on dispersal and speciation in rainforest insects. Journal of Evolutionary Biology 17: 994–1006.

Braga, M.P.; Araujo, S.B.L.; Agosta, S.J.; Brooks, D.; Hoberg, E.P.; Janz, N.; et al. 2018. Host use dynamics in a heterogeneous fitness landscape generates oscillations in host range and diversification. Evolution 72: 1773–1783.

Brooks, D.R. 1979. Testing the context and extent of host-parasite coevolution. Systematic Zoology 28: 299–307.

Brooks, D.R.; Agosta, S.J. 2012. Children of time: the extended synthesis and major metaphors of evolution. Zoologia 29: 497–514.

Brooks, D.R.; Boeger, W.A.; Hoberg, E.P. 2022. The Stockholm paradigm: lessons for the emerging infectious disease crisis. MANTER: Journal of Parasite Diversity 22. https://doi.org/10.32873/unl.dc.manter22

Brooks, D.R.; Ferrao, A.L. 2005. The historical biogeography of coevolution: emerging infectious diseases are evolutionary accidents waiting to happen. Journal of Biogeography 32: 1291–1299.

Brooks, D.R.; Hoberg, E.P. 2008. Darwin's necessary misfit and the sloshing bucket: the evolutionary biology of emerging infectious diseases. Evolution: Education and Outreach 1: 2–9.

Brooks, D.R.; Hoberg, E.P.; Boeger, WA. 2015. In the eye of the Cyclops: the classic case of cospeciation and why paradigms are important. Comparative Parasitology 83: 1–8.

Brooks, D.R.; Hoberg, E.P.; Boeger, W.A. 2019. The Stockholm Paradigm: Climate Change and Emerging Disease. University of Chicago Press, Chicago.

Brooks, D.R.; Hoberg, E.P.; Boeger, W.A.; Gardner, S.L.; Araujo, S.B.L.; Botero-Cañola, S.; et al. 2020. Before the pandemic ends: making sure it never happens again. World Complexity Science Academy Journal 1: 1–10.

Brooks, D.R.; Hoberg, E.P.; Boeger, W.A.; Gardner, S.L.; Galbreath, K.E.; Herczeg, D.; et al. 2014. Finding them before they find us: informatics, parasites and environments in accelerating climate change. Comparative Parasitology 81: 155–164.

Brooks, D.R.; León-Règagnon, V.; McLennan, D.A.; Zelmer, D. 2006. Ecological fitting as a determinant of the community structure of platyhelminth parasites of anurans. Ecology 87(Supplement): S76–S85.

Brooks, D.R.; McLennan, D.A. 1993. Parascript: Parasites and the Language of Evolution. Smithsonian Institution Press, Washington, DC.

Brooks, D.R.; McLennan, D.A. 2002. The Nature of Diversity: An Evolutionary Voyage of Discovery. University of Chicago Press, Chicago.

Brooks, D.R.; Wiley, E.O. 1988. Evolution as Entropy: Toward a Unified Theory of Biology. 2nd ed. University of Chicago Press, Chicago.

Cain, S.A. 1938. The species-area curve. American Midland Naturalist 19: 573–581.

Carlson, C.J.; Albery, G.F.; Merow, C.; Trisos, C.H.; Zipfel, C.M.; Eskew, E.A.; et al. 2022. Climate change increases cross-species viral transmission risk. Nature 607: 555–562.

Case, T.J. 1981. Niche packing and coevolution in competition communities. Proceedings of the National Academy of Sciences, USA 78: 5021–5025.

Cepelewicz, J. 2020. Nature versus nurture? Add "noise" to the debate. Quanta. https://quantamagazine.org/nature-versus-nurture-add-noise-to-the-debate-20200323

Clements, F.E. 1905. Research Methods in Ecology. University Publishing Company [now University of Nebraska Press].

Clements, F.E. 1916. Plant succession: an analysis of the development of vegetation. Publication no. 242. Carnegie Institution of Washington, Washington, DC.

Collier, J. 1988. The dynamics of biological order. In: Information, Entropy and Evolution: New Perspectives on Physical and Biological Evolution. B.H. Weber, D.H. Depew, and J.D. Smith (eds.). MIT Press. pp. 227–242.

Collier, J. 1998. Information increase in biological systems: how does adaptation fit? In: Evolutionary Systems: Biological and Epistemological Perspectives on Selection and Self-organization. G. van de Vijver, S.N. Salthe, M. Delpos (eds.). Kluwer Academic Publishers. pp. 129–140.

Collier, J. 2003. Hierarchical dynamical information systems with a focus on biology. Entropy 5: 100–124.

Collier, J.; Hooker, C. 1999. Complexly organised dynamical systems. Open Systems and Information Dynamics 6: 241–302.

Colwell, R.K.; Rangel, T.F. 2009. Hutchinson's duality: the once and future niche. Proceedings of the National Academy of Sciences, USA 106: 19651–19658.

Dakos, V.; Matthews, B.; Hendry, A.P.; Levine, J.; Loeuille, N.; Norberg, J.; et al. 2019. Ecosystem tipping points in an evolving world. Nature Ecology & Evolution 3: 355–362.

Daniels, B.C.; Chen Y.-J.; Sethna, J.P.; Gutenkunst, R.N.; Myers, C.R. 2008. Sloppiness, robustness, and evolvability in systems biology. Current Opinion in Biotechnology 19: 389–395.

Darlington, P.J., Jr. 1943. Carabidae of mountains and islands: data on the evolution of isolated faunas, and on atrophy of wings. Ecological Monographs 13: 37–61.

Darwin, C. 1872. The Origin of Species. 6th ed. John Murray, London.

Dawkins, R. 1976. The Selfish Gene. Oxford University Press, New York.

de Vienne, D.M.; Refregier, G.; Lopez-Villavicencio, M.; Tellier, A.; Hood, M.E.; Giraud, T. 2013. Cospeciation vs host-shift speciation: methods for testing, evidence from natural associations and relation to coevolution. New Phytologist 198: 347–385.

de Vladar, H.P.; Santos, M.; Szathmáry, E. 2017. Grand views of evolution. Trends in Ecology and Evolution 32: 324–334.

Eckstut, M.E.; McMahan, C.D.; Crother, B.I.; Ancheta, J.M.; McLennan, D.A.; Brooks, D.R. 2011. PACT in practice: comparative historical biogeographic patterns and species-area relationships of the Greater Antillian and windward Hawaiian Island terrestrial biotas. Global Ecology and Biogeography 20: 545–557.

Erwin, T.L. 1979. Thoughts on the evolutionary history of ground beetles: hypotheses generated from comparative faunal analyses of lowland forest sites in temperate and tropical regions. In: Carabid Beetles. T.L. Erwin, G.E. Ball, D.R. Whitehead, A.L. Halpern (eds.). Springer. pp. 539–592.

Erwin, T.L. 1981. Taxon pulses, vicariance, and dispersal: an evolutionary synthesis illustrated by carabid beetles. In: Vicariance Biogeography: A Critique. G. Nelson, D.E. Rosen (eds.). Columbia University Press, New York. pp. 159–196.

Erwin, T.L. 1985. The taxon pulse: a general pattern of lineage radiation and extinction among carabid beetles. In: Taxonomy, Phylogeny, and Zoogeography of Beetles and Ants. G.E. Ball (ed.). W. Junk. pp. 437–472.

Feronato, S.G.; Araujo, S.; Boeger, W.A. 2021. 'Accidents waiting to happen'—insights from a simple model on the emergence of infectious agents in new hosts. Transboundary and Emerging Diseases 69: 1727–1738.

Folinsbee, K.; Brooks, D.R. 2007. Miocene hominoid biogeography: pulses of dispersal and differentiation. Journal of Biogeography 34: 383–397.

Gánti, T. 1979. A Theory of Biochemical Supersystems and Its Application to Problems of Natural and Artificial Biogenesis. University Park Press, Baltimore, MD.

Gánti, T. 2003. The Principles of Life. Oxford University Press, New York.

Gould, S.J.; Lewontin, R.C. 1979. The spandrels of San Marco and the Panglossian paradigm: a critique of the adaptationist programme. Proceedings of the Royal Society of London, Series B 205: 581–598.

Gould, S.J.; Vrba, E.S. 1982. Exaptation—a missing term in the science of form. Paleobiology 8: 4–15.

Hairston, N.G. 1980. Species packing in the salamander genus *Desmognathus*: what are the interspecific interactions involved? American Naturalist 115: 354–366.

Halas, D.; Zamparo, D.; Brooks D.R. 2005. A protocol for studying biotic diversification by taxon pulses. Journal of Biogeography 32: 249–260.

Hoberg, E.P.; Agosta, S.J.; Boeger, W.A.; Brooks, D.R. 2015. An integrated parasitology: revealing the elephant through tradition and invention. Trends in Parasitology 3: 128–133.

Hoberg, E.P.; Boeger, W.A.; Molnár, O.; Földvári, G.; Gardner, S.L.; Juarrero, A.; Kharchenko, V.; Ortiz, E.; Preiser, W.; Trivellone, V.; Brooks, D.R. 2022. The DAMA protocol, an introduction: finding pathogens before they find us. MANTER: Journal of Parasite Diversity 21. https://doi.org/10.32873/unl.dc.manter21

Hoberg, E.P.; Brooks, D.R. 2008. A macroevolutionary mosaic: episodic host-switching, geographic colonization, and diversification in complex host-parasite systems. Journal of Biogeography 35: 1533–1550.

Hoberg, E.P.; Brooks, D.R. 2015. Evolution in action: climate change, biodiversity dynamics and emerging infectious disease. Philosophical Transactions of the Royal Society of London, Series B 370: 20130553.

Hoberg, E.P.; Cook, J.A.; Agosta, S.J.; Boeger, W.A.; Galbreath, K.E.; Laaksonen, S.; Kutz, S.J.; Brooks, D.R. 2017. Arctic systems in the quaternary: ecological collision, faunal mosaics and the consequences of a wobbling climate. Journal of Helminthology 91: 409–421.

Hoberg, E.P.; Trivellone, V.; Cook, J.A.; Dunnum, J.L.; Boeger, W.A.; Brooks, D.R.; et al. 2022. Knowing the biosphere: documentation, specimens, archives, and names reveal environmental change and emerging pathogens. MANTER: Journal of Parasite Biodiversity 26. https://doi.org/10.32873/unl.dc.manter26

Hutchinson, G.E. 1957. Concluding remarks. Cold Spring Harbor Symposium in Quantitative Biology 22: 415–427.

Jacob, F. 1977. Evolution and tinkering. Science 196: 1161–1166.

Janz, N. 2011. Ehrlich and Raven revisited: mechanisms underlying codiversification of plants and enemies. Annual Review of Ecology, Evolution, and Systematics 42: 71–89.

Janz, N.; Nylin, S. 2008. The oscillation hypothesis of host-plant range and speciation. In: Specialization, Speciation, and Radiation: The Evolutionary Biology of Herbivorous Insects. K.J. Tilmon (ed.). University of California Press, Oakland, CA. pp. 203–215.

Janz, N.; Nylin, S.; Wahlberg, N. 2006. Diversity begets diversity: host expansions and the diversification of plant-feeding insects. BMC Evolutionary Biology 6: 4.

Janzen, D.H. 1979. New horizons in the biology of plant defenses. In: Herbivores: Their Interaction with Secondary Plant Metabolites. G.A. Rosenthal, M.R. Berenbaum (eds.). Academic Press, Cambridge, MA. pp. 331–350.

Janzen, D.H. 1980. When is it coevolution? Evolution 34: 611–612.

Janzen, D.H. 1985. On ecological fitting. Oikos 45: 308–310.

Janzen, D.H.; Martin, P.S. 1982. Neotropical anachronisms: the fruits the gomphotheres ate. Science 215: 19–27.

Jorge, F.; Perera, A.; Poulin, R.; Roca, V.; Carretero, M.A. 2018. Getting there and around: host range oscillations during colonization of the Canary Islands by the parasitic nematode *Spauligodon*. Molecular Ecology 27: 533.

Kelley, S.T.; Farrell, B.D. 1998. Is specialization a dead end? The phylogeny of host use in *Dendroctonus* bark beetles (Scolytidae). Evolution 52: 1731–1743.

Kitano, H. 2004. Biological robustness. Nature Review Genetics 5: 826–837.

Kitano, H. 2007. Towards a theory of biological robustness. Molecular Systems Biology 3: 137.

Kováč, L. 2007. Information and knowledge in biology: time for reappraisal. Plant Signaling and Behavior 2: 65–73.

Lajoie, G.; Parfrey, L.W. 2022. Beyond specialization: re-examining routes of host influence on symbiont evolution. Trends in Ecology & Evolution 37: 590–598.

Lamarck, J.B. 1809. Philosophie Zoologique ou Exposition des Considérations Relatives à l'Histoire Naturelle des Animaux. Musée d'Histoire Naturelle, Paris.

Lim, B.K. 2008. Historical biogeography of New World emballonurid bats (tribe Diclidurini): taxon pulse diversification. Journal of Biogeography 35: 1385–1401.

Lorenz, E.N. 1972. Predictability: does the flap of a butterfly's wings in Brazil set off a tornado in Texas? Presentation given at the 139th meeting for the American Association for the Advancement of Science, Washington, DC.

Lotka, A.J. 1913. Evolution from the standpoint of physics, the principle of the persistence of stable forms. Scientific American Supplement 75: 345–346, 354, 379.

Lotka, A.J. 1925. Elements of Physical Biology. Williams & Wilkins, Baltimore, MD.

Low-Gan, J.; Huang, R.; Kelley, A.; Warner Jenkins, G.; McGregor, D.; Smider, V.V. 2021. Diversity of ACE2 and its interaction with SARS-CoV-2 receptor binding domain. Biochemical Journal 478: 3671–3684.

MacArthur, R. 1969. Species packing, and what competition minimizes. Proceedings of the National Academy of Sciences, USA 64: 1369–1371.

MacArthur, R.H.; Wilson, E.O. 1963. An equilibrium theory of insular zoogeography. Evolution 17: 373–387.

MacArthur, R.H.; Wilson, E.O. 1967. The Theory of Island Biogeography. Princeton University Press, Princeton, NJ.

Malcicka, M.; Agosta, S.J.; Harvey, J.A. 2015. Multi-level ecological fitting: indirect life cycles are not a barrier to host switching and invasion. Global Change Biology 21: 3210–3218.

Maynard Smith, J. 1976. What determines the rate of evolution? American Naturalist 110: 331–338.

Maynard Smith, J.; Szathmáry, E. 1995. The Major Transitions in Evolution. Oxford University Press, New York.

Maynard Smith, J.; Szathmáry, E. 1999. The Origins of Life. Oxford: Oxford University Press, New York.

McLennan, D.A. 2008. The concept of co-option: why evolution often looks miraculous. Evolution: Education and Outreach 1: 247–258.

Moran, N.A. 1988. The evolution of host-plant alternation in aphids: evidence for specialization as a dead end. American Naturalist 132: 681–706.

Moreno, A.; Ruiz-Mirazo, K. 2009. The problem of the emergence of functional diversity in prebiotic evolution. Biology and Philosophy 24: 585–605.

Nylin, S.; Agosta, S.J.; Bensch, S.; Boeger, W.A.; Braga, M.P.; Brooks, D.R.; et al. 2018. Embracing colonizations: a new paradigm for species association dynamics. Trends in Ecology and Evolution 33: 4–14.

Nylin, S.; Janz, N. 2009. Butterfly host plant range: an example of plasticity as a promoter of speciation? Evolutionary Ecology 23: 137–146.

Nylin, S.; Slove, J.; Janz, N. 2014. Host plant utilization, host range oscillations, and diversification in nymphalid butterflies: a phylogenetic investigation. Evolution 68: 105–124.

O'Sullivan, J.D.; Terry, J.C.D.; Rossberg, A.G. 2021. Intrinsic ecological dynamics drive biodiversity turnover in model metacommunities. Nature Communications 12: 1–11.

Page, S.E. 2011. Diversity and Complexity. Princeton University Press, Princeton, NJ.

Popadiuk, S. 2012. Scale for classifying organizations as explorers, exploiters or ambidextrous. International Journal of Information Management 2032: 75–87.

Price, P.W. 1980. Evolutionary Biology of Parasites. Princeton University Press, Princeton, NJ.

Roughgarden, J.; Feldman, M. 1975. Species packing and predation pressure. Ecology 56: 489–492.

Ruiz-Aravena, M.; McKee, C.; Gamble, A.; Lunn, T.; Morris, A.; Snedden, C.E; et al. 2022. Ecology, evolution and spillover of coronaviruses from bats. Nature Review Microbiology 20: 299–314.

Schrödinger, E. 1945. What is Life? Cambridge University Press, New York.

Skulan, J. 2000. Has the importance of the amniotic egg been overstated? Zoological Journal of the Linnean Society 130: 235–261.

Smith, J.D.H. 1988. A class of mathematical models for evolution and hierarchical information theory. Institute for Mathematics and Its Applications Preprint Series 396: 1–13.

Smith, J.D.H. 1998. Canonical ensembles, competing species, and the arrow of time. In: Evolutionary Systems: Biological and Epistemological Perspectives on Selection and Self-organization. G. Van de Vijver, S.N. Salthe, M. Delpos (eds.). Kluwer Academic, Amsterdam. pp. 141–154.

Smith, J.D.H. 2000. On the evolution of semiotic capacity. In: Semiotics, Evolution, Energy. E. Taborsky (ed.). Shaker Verlag, Aachen, Germany. pp. 283–309.

Soberón, J.; Arroyo-Peña, B. 2017. Are fundamental niches larger than the realized? Testing a 50-year-old prediction by Hutchinson. PLOS ONE 12: e0175138.

Spironello, M.; Brooks, D.R. 2003. Dispersal and diversification in the evolution of Inseliellium, an archipelagic dipteran group. Journal of Biogeography 30: 1563–1573.

Spradling, T.A.; Brant, S.V.; Hafner, M.S.; Dickerson, C.J. 2004. DNA data support a rapid radiation of pocket gopher genera (Rodentia: Geomyidae). Journal of Mammalian Evolution 11: 105–125.

Stigall, A.L. 2019. The invasion hierarchy: ecological and evolutionary consequences of invasions in the fossil record. Annual Review of Ecology, Evolution and Systematics 50: 355–380.

Stigall, A.L.; Bauer, J.E.; Lam, A.R.; Wright, D.F. 2017. Biotic immigration events, speciation, and the accumulation of biodiversity in the fossil record. Global and Planetary Change 148: 242–257.

Stigall, A.L.; Edwards, C.T.; Freeman, R.L.; Rasmussen, C.M.Ø. 2019. Coordinated biotic and abiotic change during the Great Ordovician Biodiversification Event: Darriwilian assembly of early Paleozoic building blocks. Palaeogeography, Palaeoclimatology, Palaeoecology 530: 249–270.

Szathmáry, E. 2000. The evolution of replicators. Philosophical Transactions of the Royal Society of London, Series B 355: 1669–1676.

Szathmáry, E. 2015. Toward major evolutionary transitions theory 2.0. Proceedings of the National Academy of Sciences, USA 112: 10104–10111.

Tansley, A.G. 1935. The use and abuse of vegetational concepts and terms. Ecology 16: 284–307.

Trivellone, V.; Araujo, S.B.L.; Panassiti, B. 2022. HostSwitch: an R package to simulate the extent of host-switching by a consumer. The R Journal 14: 179–194.

Trivellone, V.; Hoberg, E.P.; Boeger, W.A.; Brooks, D.R. 2022. Food security and emerging infectious disease: risk assessment and risk management. Royal Society Open Science 9: 211687.

Trivellone, V.; Panassiti, B. 2022. A field synopsis, systematic review, and meta-analyses of cophylogenetic studies: what is affecting congruence between phylogenies? MANTER: Journal of Parasite Biodiversity 24. https://doi.org/10.32873/unl.dc.manter24

Ulanowicz, R.E. 1997. Ecology, the Ascendent Perspective. Columbia University Press, New York.

West-Eberhard, M.J. 2003. Developmental Plasticity and Evolution. Oxford University Press, New York.

Wiegmann, B.M.; Mitter, C.; Farrell, B. 1993. Diversification of carnivorous parasitic insects—extraordinary radiation or specialized dead-end? American Naturalist 142: 737–754.

Wilson, E.O. 1959. Adaptive shift and dispersal in a tropical ant fauna. Evolution 13: 122–144.

Wilson, E.O. 1961. The nature of the taxon cycle in the Melanesian ant fauna. American Midland Naturalist 95: 169–193.

2

The Stockholm Paradigm: The Conceptual Platform for Coping with the Emerging Infectious Disease Crisis

Daniel R. Brooks, Walter A. Boeger, and Eric P. Hoberg

Abstract

The emerging infectious disease (EID) crisis represents an immediate existential threat to modern humanity. Current policies aimed at coping with the EID crisis are ineffective and unsustainably expensive. They have failed because they are based on a scientific paradigm that produced the *parasite paradox*. The Stockholm Paradigm (SP) resolves the paradox by integrating four elements of evolutionary biology: ecological fitting, sloppy fitness space, coevolution, and responses to environmental perturbations. It explains why and how the EID crisis occurs and is expanding and what happens after an EID emerges that sets the stage for future EIDs. The SP provides a number of critical insights for changing scientific and public policy in a manner that allows us to begin coping with the EID crisis in an effective manner. It provides hope that we can anticipate EIDs and prevent them or at least mitigate their impacts.

Keywords: Stockholm Paradigm, emerging infectious disease, ecological fitting, sloppy fitness space, oscillation hypothesis, coevolution, climate change, environmental perturbation

Introduction

Across the entire biosphere, encompassing wildlands and managed landscapes and the ecological interfaces created by them, emerging infectious diseases (EIDs) caused by viruses, bacteria, fungi, protists, and metazoans that infect humans, livestock, crops, and wildlife are increasing in number and socioeconomic impact. The response by health agencies—public, agricultural, veterinary, and wildlife—has been not as effective as we might wish, but it's not because of lack of effort. The shortcomings are due in large part to the fact that health specialists are guided by three expectations: (1) EIDs will be rare; (2) because they will be rare,

EIDs can be handled with traditional palliative measures (medication, vaccination), or in very rare cases with crisis response; and (3) EIDs cannot be predicted, so regardless of cost, crisis response is the best we can do. The first two expectations are contradicted by experience and the third has not been taken seriously because we have not recognized the contradiction between experience and expectation in the first two cases. And all three expectations stem from a failed paradigm of pathogen-host evolution and ecology.

The belief that EIDs ought to be rare and, in any event, are unpredictable stems from a core principle of the accepted framework for pathogen evolution—pathogens are so strongly co-adapted to particular host species that they cannot change hosts unless specific genetic mutations arise—hence the rare and unpredictable—that allow a new host to be colonized. Empirical evidence supports the notion that pathogens are highly specialized with respect to their hosts, and yet comparisons of pathogen and host phylogenies indicate that pathogens have often changed hosts in evolutionary history, consistent with contemporary experience with EIDs (Brooks and Hoberg, 2000; Hoberg and Brooks, 2008, 2015; Brooks et al., 2015, 2019). These inconsistencies in the standard paradigm produced the *parasite paradox* (Agosta et al., 2010; Brooks et al., 2019).

When preparation and palliation are not effectively coping with any kind of disease, prevention becomes necessary. EIDs have achieved that status. By many estimates, more than half the species on this planet are pathogens of some form or another. EIDs are much more than just a few viruses affecting human beings that make occasional headlines. EIDs include all pathogens that affect humans and every other species—wild and domestic—upon which humans depend for survival and well-being, and includes diseases we have never seen before but are seeing now as well as diseases that we thought we had contained or eradicated that are reemerging. EIDs are also costly; conservative estimates prior to the SARS-CoV-2 pandemic assessed the combined treatment costs and production losses due to EIDs at US$1.3 trillion per year (Agosta et al., 2010; Boeger et al., 2022; Brooks et al., 2022; Hoberg, Boeger, Brooks, et al., 2022; Trivellone et al., 2022). Most of those costs are hidden, rolled into the cost of doing business, leading to increased costs for health care and food. But there is now clear evidence that EIDs are unsustainably expensive; at the current pace, the costs of EIDs associated with food availability, sustainability, and safety in the US will exceed the projected US gross domestic product (GDP) within 80 years (Trivellone et al., 2022).

The Stockholm Paradigm (SP) (Agosta et al., 2010; Brooks et al., 2014; Hoberg and Brooks, 2015; Brooks et al., 2019; Agosta and Brooks, 2020; Agosta, 2022) allows us to see the EID crisis as an expected outcome of climate change and anthropogenic impacts on the biosphere. The SP explains why and how the EID crisis occurs and is expanding, and what happens after an EID emerges that sets the stage for future EIDs. The SP combines the effective elements of various preexisting perspectives on evolutionary diversification into a novel and broadly integrative framework (Agosta and Brooks, 2020; Agosta, 2022). As an explanatory platform for EID, and beyond its Darwinian foundation, early insights that led to the SP can be found in the heyday of the orthogenetic movement in parasitology (e.g., Wenrich, 1935) and the coevolutionary arms race movement of the late twentieth century (e.g., Brooks, 1979; Janzen, 1985). (See Brooks et al., 2019, for a more detailed historical account.)

As a unified conceptual framework, the SP provides a number of critical insights for changing scientific and public policy in a manner that allows us to begin effectively coping with the EID crisis. Above all, the SP enables us to anticipate EIDs and prevent them, or at least mitigate their impacts. Adopting this new perspective will require some rethinking on the part of those trained in the traditional paradigm.

Ecological Fitting

Ecological fitting (EF) (Janzen, 1985; Brooks and McLennan, 2002; Brooks et al., 2019; Agosta and Brooks, 2020; Agosta, 2023, this volume) is an umbrella term for the way in which inheritance systems use built-in evolutionary capacities to cope with changing conditions. The major elements of EF are phylogenetic conservatism, co-option, and phenotypic plasticity. These are not "hidden traits" but rather well-known elements of the foundation of Darwinian evolution—the inherited capacity to cope with changing

conditions by changing. As Darwin suggested, the most powerful of these is phylogenetic conservatism. Phylogenetic conservatism in resource specialization, transmission dynamics, and microhabitat preferences allow pathogens to be highly specialized and yet flexible. Pathogens with specific host resource requirements may be capable of infecting a broad range of hosts if the specific resource is phylogenetically conservative and widespread (Janzen, 1985; Brooks and McLennan, 2002; Brooks and Boeger, 2019; Agosta and Brooks, 2020; Agosta, 2023, this volume). It is as simple and fundamental as recognizing that SARS-CoV-2 requires some level of compatibility between viral S1 and the host's ACE2 receptors for infection (Conceicao et al., 2020; Damas et al., 2020); therefore, almost all mammals must be considered at risk for infection if exposed (Boeger et al., 2022; Hoberg, Boeger, Brooks, et al., 2022).

A robust modeling platform has shown that EIDs can occur easily in this way (Araujo et al., 2015; Braga et al., 2018; Feronato et al., 2022). A second implication of EF is that pathogens exhibit pronounced specificity for certain characteristics of host species but do not exhibit specificity for particular host *species*. This specificity for characteristics has long been acknowledged by researchers working with insect-plant associations (e.g., Nylin et al., 2018), but within parasitology and disease research disciplines, the belief that pathogens are somehow specialized on particular host species persists. This belief is an archaic holdover from orthogenetic views of evolution that impedes efforts to understand how many hosts might be susceptible to a given pathogen and why (Brooks and McLennan, 1991, 1993, 2002; Brooks et al., 2019). We strongly encourage that the term *host specificity* be replaced with the term *host range* to avoid continuing confusion and misunderstanding.

Sloppy Fitness Space

The SP asserts that increasing host range is a matter of pathogens having the opportunity to encounter susceptible but previously unexposed hosts. For explaining EID, the SP thus places at least as much emphasis on the opportunities for pathogens to encounter susceptible but previously unexposed hosts as on the inherited capacities of the pathogen or host. In evolutionary biology, opportunity space is fitness space. *Fundamental pathogen fitness space* (FFS) encompasses all conditions worldwide in which a pathogen could survive, while *realized pathogen fitness space* (RFS) encompasses those conditions in which a pathogen is currently surviving. The greater the difference is between FFS and RFS, the sloppier the fitness space and the greater the potential for persistence during changing conditions (Agosta and Klemens, 2008; Agosta, 2016; Agosta and Brooks, 2020; Agosta, 2023, this volume).

The host range of a pathogen species has been traditionally called *host specificity* but is more appropriately termed *host fitness space. Fundamental host fitness space* encompasses the phylogenetic distribution of the essential resources needed for the pathogen to survive, at least minimally. *Realized host fitness space* encompasses how much of the fundamental host fitness space is accessible to the pathogen at any given place and time. The SP therefore predicts that both realized and fundamental host fitness space will be variable (see Matthews et al., 2023 and Sanchez et al., 2023 for excellent recent examples). Traditional views of pathogen-host associations assume there is a tight fit between fundamental and realized host fitness space. As noted above, the SP assumes that fundamental host fitness space is more extensive than realized host fitness space, the difference being a manifestation of how "sloppy" the host fitness space is, and therefore what potential there might be for host range expansion, given the opportunity. The SP thus predicts that there will be a positive correlation between the known host range and the number of published reports for most pathogens (e.g., Brooks and McLennan, 1993; Sanchez et al., 2023).

All pathogens are specialists with respect to the inherited elements of EF: required environmental (mostly host) resources, microhabitat preferences (called *site specificity* by parasitologists and *tissue tropisms* by disease specialists), and transmission dynamics. Equally, all pathogens are relatively generalized or specialized in proportion to how much of their FFS is RFS. A broad host range does not make a pathogen a generalist. As the number of documented hosts for SARS-CoV-2 has increased, researchers have begun to refer to the virus as an ecological generalist, despite the fact that it remains highly specialized on ACE2 and is restricted to hosts having that resource. In reality, SARS-CoV-2 is an example of a pathogen that is a specialist with respect to a host resource that is phylogenetically conservative and widespread (across all mammals—an entire class of vertebrates). The COVID-19 pandemic is a powerful exemplar of a pathogen that has generalized in host fitness space as a result of being given many opportunities to come into contact with many susceptible but previously unexposed hosts (Boeger et al., 2022; Hoberg, Boeger, Brooks, et al., 2022). The increase in host range by SARS-CoV-2 is a classic case of ecological fitting in sloppy fitness space (SFS).

Increasing host range by EF increases the distribution of the pathogen in fitness space based on preexisting capacities. By generalizing in host fitness space, however, the chances that new variants will be able to emerge and survive increases. Although initial colonization of a new host is achieved through preexisting variation, each new host species represents a new selective regime for the colonizing pathogen population. Novel variants thus emerge *after the colonization of new hosts*, not before. The evolution of post-infection variation creates significant potential for new clades of pathogens with distinct epidemiological characteristics to emerge simply by chance. The emergence of multiple novel variants of SARS-CoV-2 following its initial emergence and subsequent geographic spread is a prime example. This aspect of the SP dynamic gives rise to the unpredictable aspects of new EIDs, including greater or lesser virulence, or even greater transmissibility, as appears to be the case for Omicron (Boeger et al., 2022; Hoberg, Boeger, Brooks et al., 2022).

Pathogens are not "enemies"

The metaphor of pathogens as enemies and agents of disease and death is powerful. But it is misleading. Pathogens are normal components of the biosphere. Their effects on humans and species that humans care about are not a personal insult to humanity; they are simply part of how pathogens make a living. It is convenient but shortsighted to try to understand pathogens by focusing on how they discomfort us and our lives. If we are to understand how to cope with the EID crisis, we must focus on the distribution of pathogen diversity and how human activities intersect with pathogens' biological "business as usual." There may be places where particular hosts have reduced the number of pathogens compared to other places, but this does not mean there is any evolutionary process of enemy escape or enemy release, much less places that are enemy-free zones. It is simply an indication that pathogen and host fitness space are dynamic and relatively independent of each other.

Infectious disease is not an inherited trait

Disease and death preoccupy health professionals, while survival and persistence are the focus of evolutionary biology. There is no question that diseased organisms may be less successful evolutionarily than those that are not diseased. However, there is no natural capacity called "causing disease"—"disease" is a human interpretation of a variety of natural conditions. If an infected host is diseased, it is an indication that the host represents marginal fitness space for the pathogen. The more marginal the new host, the greater the misfit, often manifested as disease. At the same time, marginal fitness space provides the strong selection for variants that are better accommodated to the new host fitness space. Such new variants may not evolve rapidly, and indeed in a Darwinian world there is no guarantee it will ever happen. Even if there is disease associated with the pathogen-host association, if both pathogen and host continue to survive—even though they do not thrive—their association will persist (Anderson and May, 1982, 1985; Araujo et al., 2015). Pathogenicity and virulence are equally not inherited traits but outcomes of living in marginal fitness space. When we speak of the evolution of pathogenicity, of virulence, or disease, we are at best speaking metaphorically; at worst, we are wasting our time.

Pathogens do not magically appear and disappear

Pathogens are present even when they are not producing disease or headlines, a fundamental insight explored by J.R. Audy more than 60 years ago (Audy, 1958) but largely ignored today. If a host is heavily affected by disease, that host represents marginal fitness space for the pathogen, and the SP suggests there must be at least one other host that is not diseased. Thus, when there is disease emergence we must not simply focus on the newly diseased hosts but must immediately search for other hosts—sometimes called reservoirs but more appropriately called original hosts—which were the source of the emergence in the first place (Brooks et al., 2014; Brooks and Hoberg, 2015; Brooks et al., 2019; Hoberg, Boeger, Molnár, et al., 2022; Hoberg, Boeger et al., 2023, this volume). Likewise, when the disease emergence subsides or seems to disappear from the newly infected host, we must not waste time celebrating the demise of the pathogen. We can be certain that it still exists with the potential to reemerge if we do not take steps to limit exposure between the original host and the host of interest to us. The periodic reemergence of pathogens, such as Ebola, is an example. When there are no active Ebola cases in human beings for a month, Ebola has not disappeared from the planet only to reappear magically in the same place or elsewhere at a later time.

The Oscillation Hypothesis: Where Does Coevolution "Fit"

Exploiting environmental resources is the essence of evolutionary specialization and explains why pathogens are so specialized with respect to required host resources,

microhabitat references, and transmission dynamics. Coevolutionary interactions describe the dynamics by which pathogens exploit hosts. They may be passive (*Resource Tracking*) (Jermy, 1976, 1984) or active; if active, they may be symmetrical (*Coevolutionary Arms Race*) (Mode, 1958, 1961, 1962, 1964) or asymmetrical (*Red Queen Dynamic*) (Hamilton, 1980). These different possibilities produce different selection filters that allow novel variants to emerge or rare variants to be amplified through progressive evolutionary specialization (Brooks et al., 2019; Brooks and Boeger, 2019). They do not, however, provide an explanation for how specialized pathogen-host associations can become generalized; in other words, they lead to the parasite paradox rather than resolving it (Agosta et al., 2010).

The more intense the exploitation, the more localized the specific interactions. And the more localized the association, the smaller the amount of fundamental fitness space occupied. And if the specialized traits are phylogenetically conservative and widespread, the greater the specialization through exploitation, the sloppier the fitness space becomes and the greater the potential for increases in host range if new opportunities present themselves. Coevolutionary interactions may help create the conditions under which novel variants emerge and rare variants are then amplified, but they do not restrict the pathogen's fundamental fitness space. This dynamic interaction among pathogens and actual and potential hosts is how pathogens can be highly co-adapted to a particular host in a particular place and still retain the capacity to expand their host range, given the opportunity.

Pathogens, like all species, are both exploiters and explorers. They constantly exploit their immediate surroundings but retain the capacity to explore new potential fitness space. Conditions favoring exploitation should be associated with restricted host ranges and emergence of diverse novel variants; conditions favoring exploration should be associated with increasing host range and the spread of preexisting capacities within pathogen fitness space. Alternating between exploitation-biased and exploration-biased behavior produces the evolutionary patterns of increasing and decreasing host range that provided insight leading to the *oscillation hypothesis* (Janz and Nylin, 2008). Some pathogens can be deemed "virulent but not dangerous" to humans because humans represent marginal host fitness space to which the pathogen cannot be easily transmitted (Guth et al., 2022), while others—such as Omicron—can be the opposite (Boeger et al., 2022; Hoberg, Boeger, Brooks, et al., 2022).

Evolution does not affect what is not happening, so no matter how intense the local co-adapted responses by pathogens to given hosts may be, they will have no direct effect on susceptible but unexposed hosts in other places. Thus, when new opportunities occur and pathogens encounter novel hosts, the locally co-adapted associations cannot function as evolutionary firewalls against expanding their host range and producing EIDs.

One implication of the SP is that coextinction of pathogens and hosts is not likely to have been a major phenomenon in the history of life (Hoberg and Brooks, 2008). Altered environmental conditions that might lead to the extinction of an original host may also create new connections in pathogen fitness space, increasing the opportunity for pathogens to be exposed to susceptible but previously unexposed hosts. We need not fear that host extinction will lead to catastrophic pathogen extinction, nor can we hope that coextinction will serve as a form of evolutionary biological control.

Climate Change and EID: Creating Opportunities for Exploration

The SP highlights the fact that the presence of a pathogen in a host is a result of both capacity and opportunity. No matter how specialized an association between a specific pathogen and its localized host may be, there are other possibly susceptible hosts elsewhere that have not yet been exposed to the pathogen, perhaps simply because it is geographically too distant. The capacity for expanding into new hosts, therefore, is always present, waiting for something to create new opportunities, unleashing the exploratory capacity of pathogens to drive EF in SFS.

Oscillations in host range do not occur at random but rather in bursts correlated with alternating environmental perturbations and environmental stability, and every study we have been able to do on this shows that such bursts are always associated with episodes of regional or global climate change (Hoberg and Brooks, 2008; Hoberg et al., 2017; Brooks et al., 2019). Climate change may trigger opportunities for pathogens to expand their host range simply by allowing species to move (for a recent example, see Kafle et al., 2020); if a pathogen lives somewhere wet and it becomes dry, it will move away and vice versa. Such movements bring pathogens into contact with hosts that are susceptible and have never before been exposed to them.

The SP demonstrates a direct relationship between climate change and EID. The emerging infectious disease phenomenon is an aspect of global climate change in a simple yet fundamental way, and that is that climate change initiates movement among species. Environmental perturbations catalyze biotic expansions as species leave areas where conditions have changed to an extent that they are no longer inhabitable. In effect, they create massive numbers of invasive species. That gives opportunistic species like pathogens more opportunities to be opportunistic, increasing the odds that they will encounter susceptible hosts that have never before been exposed.

Environmental perturbations create new episodes of expansion based on preexisting capacities (*ecological fitting in sloppy fitness space*). Those episodes are manifested by biotic expansions that create geographic and ecological mosaics and increased host range (Hoberg and Brooks, 2008). Environmental stability creates new episodes of isolation and specialization during which novel variants can emerge, aided by founder effect. Biotic isolation creates geographic and ecological mosaics of exploiters and decreased host range (Hoberg and Brooks, 2008). As geographic and host range expands, the pathogen generalizes in fitness space; once established in a new host or new place under stable conditions, the pathogen specializes in fitness space, and new variants will emerge.

There is no inherent directionality in host range changes; that depends on multiple factors that influence ecological opportunities for exposure between pathogens and susceptible hosts. Host range changes can be achieved directly or by a stepping-stone dynamic, for example, from wildlife to domestic animals to humans and vice versa (Araujo et al., 2015; Boeger et al., 2020; Morens and Fauci, 2020; Hoberg, Boeger, Brooks, et al., 2022), and fast-evolving pathogens, such as viruses, retain the capacity to retrocolonize a previous host species (Feronato et al., 2022). Terms such as pathogen spillover, spillback, host switches, host jumps, and host changes all refer to singular outcomes of host colonization and represent particular portions of the more inclusive SP dynamic.

As pathogens take advantage of new opportunities by becoming more generalized in host fitness space, the proportion of diseased hosts will be reduced (Audy, 1958). This phenomenon has recently been termed the *dilution effect* and has been used as evidence that increased biodiversity in some way could reduce the risk of disease outbreaks (Andreazzi et al., 2023). Generalizing in host fitness space, however, sets the stage for pathogen diversification by allocating pathogen variation among multiple new hosts, many of which will not be diseased and each of which may be a source of novel variants. A short-term reduction in disease overall is offset by an increased potential for later disease outbreaks.

Human Civilization Created New Opportunities for EID

In the past 15,000 years, humans have modified the face of the earth in ways that have created isolated habitats within a matrix of wildlands. These modifications created isolated spaces—agricultural landscapes, urban landscapes, green spaces within urban landscapes—within which humans and their domestic species lived, bordered by interfaces with the wildlands. For almost all of that time, no one recognized that such anthropogenic changes were altering opportunity space for pathogens. Pathogens maintained by humans within those new spaces existed in conditions that allowed novel variants to emerge and rare variants to be amplified. Those variants could be transmitted into the wildlands across the habitat interfaces created by humans, and they could be transmitted from one part of the human landscape to another by cooperative efforts involving trade and travel as well as noncooperative efforts involving warfare and colonization (Brooks et al., 2019; Brooks et al., 2022; Trivellone et al., 2022). We have now emerged as the planet's primary ecological super-spreaders of disease (Boeger et al., 2022; Hoberg, Boeger, Brooks, et al., 2022) (Figure 2.1). Poultry and livestock, crops, anthropophilic rodents, and wild boar are additional examples of ecological super-spreaders. Bats, interestingly, are not in themselves super-spreaders. Their nocturnal habits and highly isolated roosting behavior keeps them isolated from other hosts. But their high diversity of pathogens makes them excellent sources of pathogen emergence via the stepping-stone dynamic and illustrates the downside of the dilution effect.

The risk space for EID is thus far larger than imagined, even recently (Carlson et al., 2022), and is continuously replenished and expanded in time by evolution in what

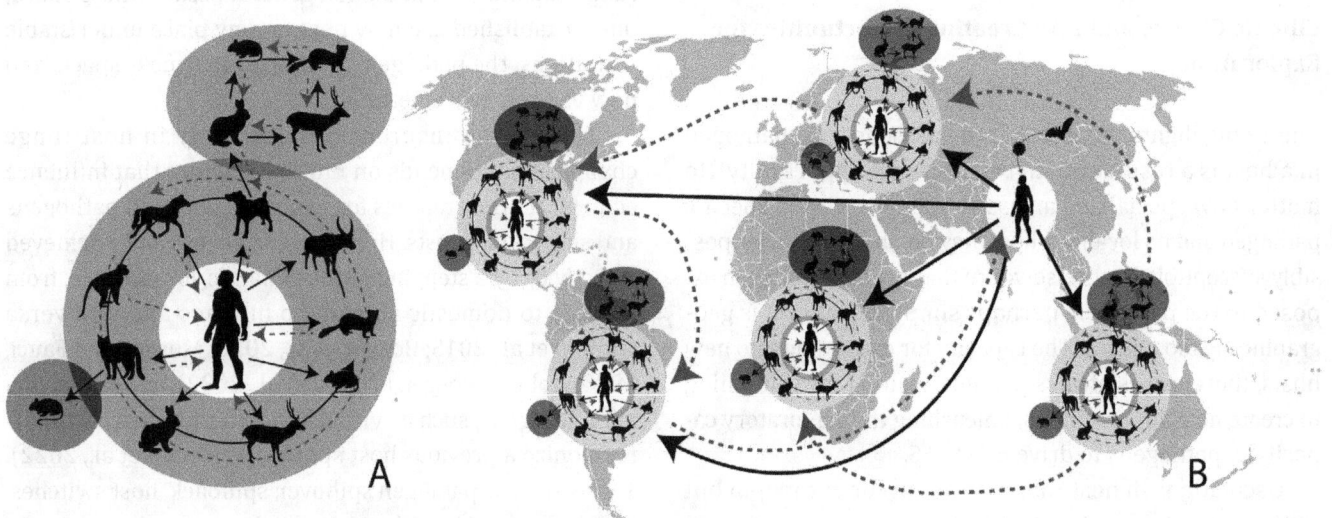

Figure 2.1. Dynamics of a pandemic from the perspective of the Stockholm Paradigm, exemplified by COVID-19. **A.** Diagrammatic representation of circulation of a pathogen with humans as ecological super-spreaders, involving transmission among realms of urban, peri-urban, and wildlife species (*circles and ovate spaces*). Overlaps between realms (*represented by darker gray*) represent interface zones, where pathogen exchanges may occur between realms—a process which may vary spatially, temporally, and at local scales because of inherent characteristics of the mammalian assemblages (e.g., diversity, behavior of peri-urban species, environmental characteristics) and humans (e.g., culture, traditions, economics). **B.** SARS-CoV-2 emergence in Asia was likely associated with stepping-stone dynamics apparently involving a species of mammal (yet to be definitively identified) that bridged the ecological distance, providing the opportunity between the donor (bats) and the recipient species (humans). Initial stages of the pandemic were driven by human movements around the planet, spreading the virus across regions and continents (*solid arrows*). Connectivity mediated by humans disseminated or inserted SARS-CoV-2 into new systems of exploration, initially into urban and peri-urban realms and subsequently forming a complex network of transmission and emergence also involving the wildlife realm. Emergence across new realms, with distinct geographic and environmental contexts, resulted from multiple trajectories (events) of expansion and exploration over time, with subsequent potential for isolation and exploitation spatially and temporally, processes that have been demonstrated empirically. These dynamics are postulated in origins of novel variants (under different regimes of selection and isolation) of the pathogen. Given opportunity, those variants (including Delta and Omicron) became disseminated among susceptible mammals, driving secondary retrocolonization in humans. Continued expansions linked to globalized travel by humans (*dashed arrows*) during the course of the pandemic resulted in subsequent spread of each successive new variant and continued cycles of oscillation.

has been characterized as *Audy space* (Araujo et al., 2015; Hoberg and Brooks, 2015; Braga et al., 2018; Brooks and Boeger, 2019; Brooks et al., 2019; D'Bastiani et al., 2020; Feronato et al., 2022; Hoberg, Boeger et al., 2023, this volume). The biosphere is replete with a growing number of evolutionary "accidents waiting to happen" (Brooks and Ferrao, 2005), which are pathogens circulating globally in ecosystems and managed landscapes (Carlson et al., 2022). The existence of these pathogenic species that are ready and able to expand their host ranges explains why traditional approaches for coping with EID have failed. Responding only after the fact for any emergence, no matter how rapidly, is ultimately ineffective and unsustainably costly (Brooks et al., 2019; Brooks et al., 2022; Trivellone et al., 2022). Even adequately managed EIDs may recycle in the risk space and reemerge as distinct lineages with unique epidemiological features. Quarantine lockdowns create restricted host space, amplifying the intensity of coevolutionary interactions and allowing novel variants to emerge and rare variants to be amplified. Should the lockdowns be relaxed before transmission rates drop to below sustainable levels within the lockdown area, the effect of the relaxation will be a minibiotic expansion event, with the rapid spread of novel variants. These events have been a recurring global theme during the SARS-CoV-2 pandemic.

Establishing the Paradigm

If the SP is to be accepted as the basis for understanding and coping with EID, three conditions must be met. First, we need a sound conceptual framework, which is provided by the SP, which itself, as suggested by Agosta and Brooks (2020) and Agosta (2023, this volume), is the dynamic explanation for Darwin's entangled bank. Second, we need a sound modeling platform for testing the feasibility of the various aspects of the paradigm and for finding unexpected possibilities. Such a platform exists and has been critical in providing support for the feasibility of the SP in producing the dynamic evolutionary patterns associated with EID (Araujo et al., 2015; Braga et al., 2018; Feronato et al., 2021; Boeger et al., 2022; D'Bastiani et al., 2022; for an overview, see Souza et al., 2023, this volume and references therein). Finally, we need a robust empirical database covering a variety of interconnected studies.

This work begins with documenting not only that host range changes have been common throughout the evolutionary history of pathogen lineages but also that those host range changes follow a pattern. That host range expansions are generally the result of ecological fitting is demonstrated by findings that newly colonized hosts share specific resources critical for the establishment and persistence of the pathogen. Host fitness space is sloppy to the extent that phylogenetic conservatism in required host resources includes both plesiomorphic and apomorphic traits. This means that colonized hosts may be closely related to the original hosts, though not sister groups. Host range expansion as a result of persistent plesiomorphic host resources thus results in a significant degree of phylogenetic signal that could be mistaken for cryptic cospeciation. This led Brooks, Hoberg, and Boeger (2019) to suggest that cophylogenetic studies could be adapted for use in assessing the extent of host fundamental fitness space and the sloppiness of host realized fitness space. We would thus expect to find that studies reporting high levels of cospeciation would be derived only from studies using methods of analysis that treat host range expansion as cryptic cospeciation (see Trivellone and Panassiti, 2023, this volume and references therein). In addition, we would expect to find evidence that host ranges oscillate through evolutionary history, as suggested by the oscillation hypothesis (Janz and Nylin, 2008). What has been called the classic case of cospeciation actually shows evidence of widespread and oscillating host range expansion (Brooks et al., 2015).

Second, we need evidence that the biogeographic histories of pathogens and hosts are dominated by the taxon pulse dynamic. A protocol exists for documenting the taxon pulse explicitly (Halas et al., 2005), and studies of multiple taxa inhabiting a common set of areas has provided consistent support for the hypothesis that the taxon pulse is responsible for the complex geographic associations of species on this planet (Brooks and Folinsbee, 2005, 2012; Folinsbee and Brooks, 2007; Eckstutt et al., 2011; Dominguez-Dominguez et al., 2019). We would thus expect to find that studies reporting patterns dominated only by geographic isolation, or geographic isolation and random dispersal events, would be derived from studies using methods of analysis that treat geographic expansion as cryptic isolation (Brooks and Van Veller, 2008). Reanalysis of published studies in light of the taxon pulse is clearly indicated.

Third, we need evidence that host ranges are correlated in a particular manner with taxon pulses. That is to say, both geographic and host range expansions are highly correlated in space and time. Host range expansions occur during periods of biotic expansion affecting hosts and pathogens. In a complementary manner, host range contractions—with the possibility of some cospeciation—occur during periods of biotic isolation. One study that explicitly makes those connections has been published (Brooks and Ferrao, 2005; Folinsbee and Brooks, 2007; Brooks and Folinsbee, 2012), but we believe many other published studies also show the same predicted patterns without making an explicit connection with either the SP or the taxon pulse (reviewed in Brooks et al., 2019).

Finally, we need evidence that geographic and host range expansions are manifestations of ecological fitting; that is, they are associated with preexisting capacities that take advantage of new opportunities. This general Darwinian expectation has been documented extensively for both host and pathogen groups using contemporary phylogenetic methods beginning nearly 40 years ago (Brooks, 1984; see examples and references in Brooks and McLennan, 1991, 1993, 2002) and augmented by recent studies (see Brooks, Hoberg, and Boeger, 2019; Agosta and Brooks, 2020 and references therein). The evidence of preexisting capacity for geographic and host range expansions is clearly seen in all contemporary emerging disease outbreaks—notably COVID—from which new variants arise *only after* the pathogen has colonized a new host. This includes examples of the stepping-stone dynamic (Braga et al., 2015; Araujo et al., 2015; Patella et al., 2017; Boeger et al., 2022) creating ecological super-spreaders of emerging disease (Boeger et al., 2022; Hoberg et al. 2022).

More robust examples would involve evidence of pathogen-host associations structured by phylogenetic conservatism in structural and functional traits—in particular, microhabitat preference (site specificity or tissue tropism) and transmission dynamics (life cycle patterns, a subset of which is sometimes referred to as ecological niches). One of the earliest examples that ecological fitting could be responsible for the structuring of entire pathogen communities occurring in various habitat types was provided by Brooks et al. (2000). They showed that ecological fitting in the form of phylogenetic conservatism in transmission dynamics and microhabitat preference for the pathogens and phylogenetic conservatism in feeding habits and relative time spent near water on the part of the hosts explained the community structure of platyhelminth parasites infecting frogs in temperate deciduous forests in central Europe and the eastern United States, the temperate grasslands known as the Great Plains in the United States, and the tropical and tropical wet forests in Mexico and Costa Rica. Furthermore, two studies showed explicitly how ecological fitting could allow even pathogens with complex life cycles to become established in new geographic areas where none of their original intermediate or definitive hosts lived (Brooks et al., 2006; Malcicka et al., 2015). Contemporary examples of the influence of ecological fitting in conjunction with episodic expansion and isolation of taxon pulse dynamics include many pathogen species whose geographic ranges have changed according to changes in precipitation or temperature or other manifestations of global climate change (e.g., Hoberg et al., 2012; Galbreath and Hoberg, 2015; Hoberg et al., 2017; Haas et al., 2020).

At the moment, we lack formal meta-analyses for all but the case of host range changes (Trivellone and Panassiti, 2023, this volume). We believe, however, that there are abundant published studies supporting each of these described elements of the SP. Studies beginning with Hoberg and Brooks (2008, 2010) give strong indications that corroborate the full expectations of the SP.

Conclusions

The EID crisis represents an immediate existential threat to modern humanity. Current policies aimed at coping with the EID crisis are ineffective and unsustainably expensive. They have failed because they are based on a scientific paradigm that produced the parasite paradox. The SP resolves the parasite paradox. Pathogens are ecologically specialized, but those specializations are phylogenetically conservative. A wide range of pathogens may have similar transmission dynamics and microhabitat preferences within hosts. There may be many susceptible but unexposed hosts that need only a change in geographic distribution or trophic structure to acquire a new pathogen (Brooks et al., 2019). Throughout evolutionary history, climate perturbations have allowed pathogens to oscillate between exploring fitness space that is inherently sloppy (Agosta and Klemens, 2008; Agosta, 2016; Agosta and Brooks, 2020), encountering a diverse assemblage of susceptible hosts, and exploiting hosts during periods of unstable environmental conditions (Araujo et al., 2015; Hoberg and Brooks, 2015; Braga et al., 2018; Brooks et al., 2019; D'Bastiani et al., 2020; Feronato et al., 2022; Trivellone et al., 2022). New pathogen-host associations resulting from changing opportunities set the stage for *subsequent* emergence of genetic innovations. Empirical evidence from deep- and shallow-time phylogenetic studies show clearly that environmental change and associated geographic expansion are correlated with host range expansion, which leads to emerging diseases (Brooks and Ferrao, 2005; Hoberg and Brooks, 2008; Brooks et al., 2015; Brooks et al., 2019; Carlson et al., 2022; Guth et al., 2022).

The evidence is manifest, but failure to take notice of the long-term evolutionary dynamics of pathogens has led to short-sighted assumptions about the nature of EIDs, which in turn has limited our efforts to cope with them. That, in turn, has limited our ability to fully understand the emerging disease crisis as an evolutionary phenomenon and to produce effective measures for coping with it. Evolution leads to indefinite survival in an ever-changing world because it generates and stores vast amounts of *evolutionary potential* in living organisms and the ecosystems they form. They express only a fraction of that potential in any given place at any given time; the rest is what gives the biosphere a built-in capacity to persist in the face of a changing environment by changing itself.

The SP produces a clear message for a world experiencing accelerating global climate change: the risk space is immense, and novel hosts are easily accessible when environmental perturbations—anthropogenic or not—occur. By underestimating the risk space, we have fooled ourselves into thinking that crisis response can be sustainable, even in the face of massive evidence to the contrary. But this dire depiction contains real hope. If expanding host range, the essence of EID, is based on preexisting capacities, we can use that information to anticipate and prevent, or at least mitigate the socioeconomic impacts of EIDs. Reducing the cost of EIDs to sustainable levels is the goal of the DAMA protocol (Brooks et al., 2014, 2019; Hoberg et al., 2022; Molnár, Hoberg, et al., 2022; Molnár, Knickel, et al., 2022; Hoberg, Boeger, et al., 2023, this volume; Hoberg, Trivellone, et al., 2023, this volume), the public policy extension of the SP. If humanity is to survive indefinitely into the future, people must adopt public policies that better mimic the biological systems that have produced both climate change and the EID crisis (Agosta and Brooks, 2020; Brooks and Agosta, forthcoming 2024).

Literature Cited

Agosta, S.J. 2016. On ecological fitting, plant-insect associations, herbivore host shifts, and host plant selection. Oikos 114: 556–565.

Agosta, S.J. 2022. The Stockholm paradigm explains the dynamics of Darwin's entangled bank, including emerging infectious disease. MANTER: Journal of Parasite Biodiversity 27. https://doi.org/10.32873/unl.dc.manter27

Agosta, S.J. 2023. The Stockholm Paradigm explains the eco-evolutionary dynamics of the biosphere in a changing world, including emerging infectious disease. In: An Evolutionary Pathway for Coping with Emerging Infectious Disease. S.L. Gardner, D.R. Brooks, W.A. Boeger, E.P. Hoberg (eds.). Zea Books, Lincoln, NE.

Agosta, S.J.; Brooks, D.R. 2020. The Major Metaphors of Evolution: Darwinism Then and Now. Springer International, New York. https://doi.org/10.1007/978-3-030-52086-1

Agosta, S.J.; Janz, N.; Brooks, D.R. 2010. How specialists can be generalists: resolving the "parasite paradox" and implications for emerging infectious disease. Zoologia (Curitiba) 27: 151–162. https://doi.org/10.1590/S1984-46702010000200001

Agosta, S.J.; Klemens, J.A. 2008. Ecological fitting by phenotypically flexible genotypes: implications for species associations, community assembly and evolution. Ecology Letters 11: 1123–1134.

Anderson, R.M.; May, R.M. 1982. Coevolution of hosts and parasites. Parasitology 85: 411–426.

Anderson, R.M.; May, R.M. 1985. Epidemiology and genetics in the coevolution of parasites and hosts. Proceedings of the Royal Society of London B 219: 281–283.

Andreazzi, C.S.; Martinez-Vacquero, L.A.; Winck, G.R.; Cardoso, T.S.; Texeira, B.R.; Xavier, S.C.C.; et al. 2023. Vegetation cover and biodiversity reduce parasite infection in wild hosts across ecological levels and scales. Ecography 2023: e06579. https://doi.org/10.1111/ecog.06579

Araujo, S.B.; Braga, M.P.; Brooks, D.R.; Agosta, S.J.; Hoberg, E.P. 2015. Understanding host-switching by ecological fitting. PLOS ONE 10: e0139225.

Audy, J.R. 1958. The localization of disease with special reference to the zoonoses. Transactions of the Royal Society of Tropical Medicine and Hygiene 52: 309–328. https://doi.org/10.1016/0035-9203(58)90045-2

Boeger, W.A.; Brooks, D.R.; Trivellone, V.; Agosta, S.J.; Hoberg, E.P. 2022. Ecological super-spreaders drive host-range oscillations: Omicron and risk space for emerging infectious disease. Transboundary and Emerging Diseases 69: e1280–e1288. https://doi.org/10.1111/tbed.14557

Braga, M.P.; Araujo, S.B.L.; Agosta, S.; Brooks, D.R.; Hoberg, E.P. 2018. Host use dynamics in a heterogeneous fitness landscape generates oscillations in host range and diversification. Evolution 72: 1773–1783. https://doi.org/10.1111/evo.13557

Braga, M.P.; Razzolini, E.; Boeger, W.A. 2015. Drivers of parasite sharing among Neotropical freshwater parasites. Journal of Animal Ecology 84: 487–497. https://doi.org/10.1111/1365-2656.12298

Brooks, D.R. 1979. Testing the context and extent of host-parasite coevolution. Systematic Zoology 28: 299–307.

Brooks, D.R.; Agosta, S.J. Forthcoming 2024. A Darwinian Survival Guide: Hope for the Twenty-first Century. MIT Press, Cambridge MA.

Brooks, D.R.; Boeger, W.A. 2019. Climate change and emerging infectious diseases: evolutionary complexity in action. Current Opinons in Systems Biology 13: 75–81. https://doi.org/10.1016/j.coisb.2018.11.001

Brooks, D.R.; Ferrao, A. 2005. The historical biogeography of coevolution: emerging infectious diseases are evolutionary accidents waiting to happen. Journal of Biogeography 32: 1291–1299.

Brooks, D.R.; Folinsbee, K.E. 2005. Paleobiogeography: documenting the ebb and flow of evolutionary diversification. B. S. Lieberman and A. Stigall Rode, eds. Paleontological Society Papers 11: 15-43.

Brooks, D.R.; Folinsbee, K.E. 2012. Phylogenetic methods in palaeobiogeography: changing from simplicity to complexity without losing parsimony. In: Palaeogeography and Palaeobiogeography: Biodiversity in Space and Time. P. Upchurch, A.J. McGowan, C.S.C. Slater (eds.). CRC Press, Boca Raton, FL. 13–38 p.

Brooks, D.R.; Hoberg, E.P. 2000. Triage for the biosphere: the need and rationale for taxonomic inventories and phylogenetic studies of parasites. Comparative Parasitology 67: 1–25.

Brooks, D.R.; Hoberg, E.P.; Boeger, W.A. 2015. In the eye of the cyclops: the classic case of cospeciation and why paradigms are important. Comparative Parasitology 82: 1–8. https://doi.org/10.1654/4724C.1

Brooks, D.R.; Hoberg, E.P.; Boeger, W.A. 2019. The Stockholm Paradigm: Climate Change and Emerging Disease. University of Chicago Press, Chicago.

Brooks, D.R.; Hoberg, E.P.; Boeger, W.A.; Gardner, S.L.; Galbreath, K.E.; Herczeg, D.; et al. 2014. Finding them before they find us: informatics, parasites, and environments in accelerating climate change. Comparative Parasitology 81: 155–164. https://doi.org/10.1654/4724b.1

Brooks, D.R.; Hoberg, E.P.; Boeger, W.A.; Trivellone, V. 2022. Emerging infectious disease: an underappreciated area of strategic concern for food security. Transboundary and Emerging Diseases 69: 254–267. https://doi.org/10.1111/tbed.14009

Brooks, D.R.; León-Règagnon, V.; McLennan, D.A.; Zelmer, D. 2006. Ecological fitting as a determinant of the community structure of platyhelminth parasites of anurans. Ecology 87 (Supplement): S76–S85. https://doi.org/10.1890/0012-9658(2006)87[76:EFAADO]2.0CO;2

Brooks, D.R.; McLennan, D.A. 1991. Phylogeny, Ecology and Behavior: A Research Program in Comparative Biology. University of Chicago Press, Chicago.

Brooks, D.R.; McLennan, D.A. 1993. Parascript: Parasites and the Language of Evolution. Smithsonian Institution Press, Washington, DC.

Brooks, D.R.; McLennan, D.A. 2002. The Nature of Diversity: An Evolutionary Voyage of Discovery. University of Chicago Press, Chicago.

Brooks, D.R.; McLennan, D.A.; León-Règagnon, V.; Hoberg, E.P. 2006. Phylogeny, ecological fitting and lung flukes: helping solve the problem of emerging infectious diseases. Revista Mexicana de Biodiversidad 77: 225–234.

Brooks, D.R.; Van Veller, M.G.P. 2008. Assumption 0 analysis: comparative phylogenetic studies in the age of complexity. Annals of the Missouri Botanical Garden 95: 201–223. https://doi.org/10.3417/2006017

Carlson, C.J.; Albery, G.F.; Merow, C.; Trisos, C.H.; Zipfel, C.M.; Eskew, E.A.; et al. 2022. Climate change increases cross-species viral transmission risk. Nature 607: 555–562. https://doi.org/10.1038/s41586-022-04788-w

Conceicao, C.; Thakur, N.; Human, S.; Kelly, J.T.; Logan, L.; Bialy, D.; et al. 2020. The SARS-CoV-2 Spike protein has a broad tropism for mammalian ACE2 proteins. PLOS Biology 18: e3001016. https://doi.org/10.1371/journal.pbio.3001016

Damas, J.; Hughes, G.M.; Keough, K.C.; Painter, C.A.; Persky, N.S.; Corbo, M.; et al. 2020. Broad host range of SARS-CoV-2 predicted by comparative and structural analysis of ACE2 in vertebrates. Proceedings of the National Academy of Sciences USA 117: 22311–22322. https://doi.org/10.1073/pnas.2010146117

D'Bastiani, E.; Campião, K.M.; Boeger, W.A.; Araújo, S.B.L. 2020. The role of ecological opportunity in shaping host-parasite networks. Parasitology 147: 1452–1460. https://doi.org/10.1017/S003118202000133X

Domínguez-Domínguez, O.; Pedraza-Lara, C.; Gurrola-Sánchez, N.; Perea, S.; Pérez-Rodríguez, R.; Israde-Alcántara, I.; et al. 2010. Historical biogeography of the Goodeinae (Cyprinodontiforms). In Viviparous Fishes II. M.C. Uribe, H.J. Grier (eds.). New Life Publications, Homestead, FL. 19–61 p.

Eckstut, M.E.; McMahan, C.D.; Crother, B.I.; Ancheta, J.M.; McLennan, D.A.; Brooks, D.R. 2011. PACT in practice: comparative historical biogeographic patterns and species-area relationships of the Greater Antillean and Hawaiian Island terrestrial biotas. Global Ecology and Biogeography 20: 545–557. https://doi.org/10.1111/j.1466-8238.2010.00626x

Feronato, S.G.; Araujo, S.; Boeger ,W.A. 2022. "Accidents waiting to happen"—Insights from a simple model on the emergence of infectious agents in new hosts. Transboundary and Emerging Diseases 69: 1727–1738. https://doi.org/10.1111/tbed.14146

Folinsbee, K.E.; Brooks D.R. 2007. Miocene hominoid biogeography: pulses of dispersal and differentiation. Journal of Biogeography 34: 383–397.

Galbreath, K.E.; Hoberg, E.P. 2015. Host responses to cycles of climate change shape parasite diversity across North America's Intermountain West. Folia Zoologica 64: 218–232. https://doi.org/10.25225/fozo.v64.i3.a4.2015

Guth, S.; Mollentze, N.; Renault, K.; Streicker, D.G.; Visher, E.; Boots, M.; Brook, C.E. 2022. Bats host the most virulent—but not the most dangerous—zoonotic viruses. Proceedings of the National Academy of Sciences USA 119: e2113628119. https://doi.org/10.1073/pnas.2113628119

Haas, G.M.S.; Hoberg, E.P.; Cook, J.A.; Haukisalmi, V.; Makarikov, A.A.; Gallagher, S.R.; et al. 2020. Taxon pulse dynamics, episodic dispersal and host colonization across Beringia drive diversification of a Holarctic tapeworm assemblage. Journal of Biogeography 47: 2457-2471. https://doi.org/10.1111/jbi.13949

Halas, D.; Zamparo, D.; Brooks, D.R. 2005. A historical biogeographical protocol for studying biotic diversification by taxon pulses. Journal of Biogeography 32: 249–260. https://doi.org/10.1111/j.1365-2699.2004.01147.x

Hamilton, W.D. 1980. Sex versus non-sex versus parasite. Oikos 35: 282–290. https://doi.org/10.2307/3544435

Hoberg, E.P.; Boeger, W.A.; Brooks, D.R.; Trivellone, V.; Agosta, S.J. 2022. Stepping-stones and mediators of pandemic expansion: a context for humans as ecological super-spreaders. MANTER: Journal of Parasite Biodiversity 18. https://doi.org/10.32873/unl.dc.manter18

Hoberg, E.P.; Boeger, W.A.; Molnár, O.; Földvári, G.; Gardner, S.L.; Juarrero, A.; et al. 2022. The DAMA protocol, an introduction: finding pathogens before they find us. MANTER: Journal of Parasite Biodiversity 21. https://doi.org/10.32873/unl.dc.manter21

Hoberg, E.P.; Boeger, W.A.; Molnár, O.; Földvári, G.; Gardner, S.L.; Juarrero, A.; et al. 2023. The DAMA protocol: anticipating to prevent and mitigate emerging infectious diseases. In: An Evolutionary Pathway for Coping with Emerging Infectious Disease. S.L. Gardner, D.R. Brooks, W.A. Boeger, E.P. Hoberg (eds.). Zea Books, Lincoln, NE.

Hoberg, E.P.; Brooks, D.R. 2008. A macroevolutionary mosaic: episodic host-switching, geographical colonization and diversification in complex host-parasite systems. Journal of Biogeography 35: 1533–1550. https://doi.org/10.1111/j.1365-2699.2008.01951.x

Hoberg, E.P.; Brooks D.R. 2010. Beyond vicariance: integrating taxon pulses, ecological fitting and oscillation in evolution and historical biogeography. In: The Geography of Host-Parasite Interactions. S. Morand, B. Krasnov (eds.). Oxford University Press, Oxford, UK. 7–20 p.

Hoberg, E.P.; Brooks, D.R. 2015. Evolution in action: climate change, biodiversity dynamics and emerging infectious disease. Philosophical Transactions of the Royal Society B 370: 20130553. https://doi.org/10.1098/rstb.2013.0553

Hoberg, E.P.; Cook, J.A.; Agosta, S.J.; Boeger, W.; Galbreath, K.E.; Laaksonen, S.; et al. 2017. Arctic systems in the Quaternary: ecological collision, faunal mosaics and the consequences of a wobbling climate. Journal of Helminthology 91: 409–421. https://doi.org/10.1017/S0022149X17000347

Hoberg, E.P.; Galbreath, K.E.; Cook, J.A.; Kutz, S.J.; Polley, L. 2012. Northern host-parasite assemblages: history and biogeography on the borderlands of episodic climate and environmental transition. D. Rollinson, S.I. Hays (eds.). Elsevier. Advances in Parasitology 79: 1–97.

Hoberg, E.P.; Trivellone, V.; Cook, J.A.; Dunnum, J.L.; Boeger, W.B.; et al. 2023. Document: pathogen diversity—finding them before they find us. In: An Evolutionary Pathway for Coping with Emerging Infectious Disease, S.L. Gardner, D.R. Brooks, W.A. Boeger, E.P. Hoberg (eds.). Zea Books, Lincoln, Nebraska.

Janz, N.; Nylin, S. 2008. The oscillation hypothesis of host plant-range and speciation. In: Specialization, Speciation, and Radiation: The Evolutionary Biology of Herbivorous Insects. J. Tilmon (ed.). University of California Press, Berkeley. 203–215 p.

Janzen, D.H. 1985. On ecological fitting. Oikos 45: 308–310.

Jermy, T. 1976. Insect–host-plant relationship—co-evolution or sequential evolution? Symposium Biologica Hungarica 16: 109–113.

Jermy, T. 1984. Evolution of insect/host plant relationships. American Naturalist 124: 609–630.

Kafle, P.; Peller, P.; Massolo, A.; Hoberg, E.P.; Leclerc, L.-M.; Tomaselli, M.; Kutz, S. 2020. Range expansion of muskox lungworms track rapid Arctic warming: implications for geographic colonization under climate forcing. Scientific Reports 10: 17323. https://doi.org/10.1038/s41598-020-74358-5

Malcicka, M.; Agosta, S.J.; Harvey, J.A. 2015. Multi level ecological fitting: indirect life cycles are not a barrier to host switching and invasion. Global Change Biology 21: 3210–3218. https://doi.org/10.1111/gcb/12928

Matthews, A.E; Wijeratne, A.J.; Sweet, A.D.; Hernandes, F.A; Toews, D.P.L.; Boves, T.J. 2023. Dispersal-limited symbionts exhibit unexpectedly wide variation in host specificity. Systematic Biology: syad014. https://doi.org/10.1093/sysbio/syad014

Mode, C.J. 1958. A mathematical model for the co-evolution of obligate parasites and their hosts. Evolution 12: 158–165.

Mode, C.J. 1961. Generalized model of a host-pathogen system. Biometrics 17: 386–404.

Mode, C.J. 1962. Some multi-dimensional birth and death processes and their applications in population genetics. Biometrics 18: 543–567.

Mode, C.J. 1964. A stochastic model of the dynamics of host-pathogen systems with mutation. Bulletin of Mathematical Biophysics 26: 205–233.

Molnár, O.; Hoberg, E.P.; Földvári, G.; Trivellone, V.; Boeger, W.B.; Brooks, D.R. 2022. The 3P framework: a comprehensive approach to coping with the emerging infectious disease crisis. MANTER: Journal of Parasite Biodiversity 23. https://doi.org/10.32873/unl.dc.manter23

Molnár, O.; Hoberg, E.P.; Trivellone, V.; Földvári, G.; Brooks, D.R. 2023. Prevent-Prepare-Palliate: the 3P framework—integrating the DAMA protocol into global public health systems. In: An Evolutionary Pathway for Coping with

Emerging Infectious Disease. S.L. Gardner, D.R. Brooks, W.A. Boeger, E.P. Hoberg (eds.). Zea Books, Lincoln, NE.

Molnár, O.; Knickel, M.; Marizzi, C. 2022. Taking action: turning evolutionary theory into preventive policies. MANTER: Journal of Parasite Biodiversity 28. https://doi.org/10.32873/unl.dc.manter28

Molnár, O.; Knickel, M.; Marizzi, C. 2023. All hands on deck: turning evolutionary theory into preventive policies. In: An Evolutionary Pathway for Coping with Emerging Infectious Disease. S.L. Gardner, D.R. Brooks, W.A. Boeger, E.P. Hoberg (eds.). Zea Books, Lincoln, NE.

Morens, D.M.; Fauci, A.S. 2020. Emerging pandemic diseases: how we got to COVID-19. Cell 182: 1077–1092. https://doi.org/10.1016/j.cell.2020.08.021

Nylin, S.; Agosta, S.; Bensch, S.; Boeger, W.A.; Braga, M.P.; Brooks, D.R.; et al. 2018. Embracing colonizations: a new paradigm for species association dynamics. Trends in Ecology and Evolution 33: 4–14. https://doi.org/10.1016/j.tree.2017.10.005

Patella, L.; Brooks, D.R.; Boeger, D.A. 2017. Phylogeny and ecology illuminate the evolution of associations under the Stockholm Paradigm. Vie et Milieu 67: 91–102.

Sanchez, J.P.; López Berrizbeitia, M.F.; Ezquiaga, M.C. 2023. Host specificity of flea parasites from mammals of the Andean Biogeographic Region. Medical and Veterinary Entomology 2023: 1–12. https://doi.org/10.1111/mve.12649

Souza, A.T.C, Araujo, S.B.L., and Boeger, W.A. 2022. The evolutionary dynamics of infectious diseases on an unstable planet: insights from modeling the Stockholm aradigm. MANTER: Journal of Parasite Biodiversity 25. https://doi.org/10.32873/unl.dc.manter25

Souza, A.T.C, Araujo, S.B.L., and Boeger, W.A. 2023. Modeling the Stockholm Paradigm: insights for the nature and dynamics of emerging infectious diseases. In: An Evolutionary Pathway for Coping with Emerging Infectious Disease. S.L. Gardner, D.R. Brooks, W.A. Boeger, E.P. Hoberg (eds.). Zea Books, Lincoln, NE.

Trivellone, V.; Hoberg, E.P.; Boeger,W.A.; Brooks, D.R. 2022. Food security and emerging infectious disease: risk assessment and risk management. Royal Society Open Science 9: 211687. https://doi.org/10.1098/rsos.211687

Trivellone, V.; Panassiti, B. 2022. A field synopsis, systematic review, and meta-analyses of cophylogenetic studies: What is affecting congruence between phylogenies? MANTER: Journal of Parasite Biodiversity 24. https://doi.org/10.32873/unl.dc.manter24

Trivellone, V.; Panassiti, B. 2023. Pathogen-host phylogenetic congruence varies with paradigmatic assumptions, analytical method, and type of association. In: An Evolutionary Pathway for Coping with Emerging Infectious Disease. S.L. Gardner, D.R. Brooks, W.A. Boeger, E.P. Hoberg (eds.). Zea Books, Lincoln, NE.

Wenrich, D.H. 1935. Host-parasite relations between parasitic protozoa and their hosts. Proceedings of the American Philosophical Society 75: 605–650.

3

Pathogen-Host Phylogenetic Congruence Varies with Paradigmatic Assumptions, Analytical Method, and Type of Association

Valeria Trivellone and Bernd Panassiti

Abstract

We conducted a global meta-analysis to explore patterns of cophylogeny among pathogens and their hosts. We evaluated the influence of three factors—namely, cophylogenetic method, association, and ecosystem type—on the outcome of the analyses, that is, the degree of congruence between phylogenies of interacting species. The published papers were identified using 4 different databases and 13 keywords; we included all studies for which statistical approaches to compare phylogenies (cophylogenetic analyses) of interacting lineages were used. After the initial screening, 296 studies were selected to extract response variables (outcome of the cophylogenetic analyses, i.e., congruent, incongruent, or both) and code information from the three selected factors (method of analyses, association, and ecosystem type). The final dataset included 485 scored cophylogenetic results. The data were analyzed using the chi-square test and regression techniques. We provided evidence for the outcome to be strongly dependent on the method and association type. In particular, cophylogenies in mutualistic associations are more frequently congruent when using global-fit methods, and in parasitic associations they are more incongruent when using event-based methods than expected under independence. The generalized mixed model, including two non-nested factors (method and association), yielded a higher probability of congruent results for parasites, mutualistic, and commensal when using global-fit (83%, 97%, and 99%, respectively) and event-based methods (58%, 89%, and 96%, respectively). These results are concordant with the prevalence null hypothesis derived from the literature, which suggests that parasitic and mutualistic associations show patterns of congruence when compared with the hosts' phylogenies. However,

an underlying misleading assumption driving theoretical frameworks and inductive analytical methods (namely global-fit distance-based and event-cost-based methods) have been discussed here to largely affect the results of cophylogenetic analyses. We therefore suggest the use of an alternative theoretical framework, the Stockholm Paradigm (SP), to reanalyze published raw data, and the integration of the cophylogenetic analyses into a workbench (DAMA protocol, the policy extension of SP) aimed to anticipate emerging infectious diseases.

Keywords: cophylogeny, codiversification, DAMA protocol, herbivory, host, pathogen, pollinator, Stockholm Paradigm

Coevolution: Concurrent Speciation With or Without Mutual Modifications

Coevolution, cospeciation, and codivergence, concepts often incorrectly used interchangeabley, embrace mechanisms that are thought to be driving much of the diversity in the tree of life (Hembry et al., 2014; Laine, 2009; Raguso, 2021). Since the founding idea by Darwin about the factors that generate diversity—"namely, the nature of the organism, and the nature of the conditions. The former seems be much more the important; . . ." (Darwin, 1872)—a relentlessly increasing number of papers have tried to build theories and operational framework for the assessment of the processes that shape the associations among interacting species. The idea of cospeciation seems to have originated in the early twentieth century (Fahrenholz, 1913; Kellogg, 1913) with a seminal intuition about parasite phylogenies often mirroring host phylogenies. Using parasitic associations as study models, more than half a century later the term cospeciation was defined by Brooks as "cladogenesis of an ancestral parasite species as a result of, or concomitant with, host cladogenesis" (1979). Interestingly, in his original work, Brooks provided an interpretation of the concept of coevolution by concatenating two main processes, co-accommodation and cospeciation, the former being "the mutual adaptation of a given parasite species and its host(s) through time [...] co-accommodation refers to the relationship between a parasite species and its host during the period in which the parasite exhibits no cladogenesis" (Brooks, 1979). Concurrently, the idea of coevolution stemmed from the studies by Ehrlich and Raven (1964) on plant-insect herbivore interactions that used a primordial method for coevolutionary studies using phylogenetic information. In doing so, they provided evidence for insect-plant associations being shaped by similarities in plant chemical cues that "do not necessarily indicate the plants'1 overall phenetic or phylogenetic relationships." A more articulated, formal definition of coevolution arrived later with Janzen (1980) as "an evolutionary change in a trait of the individuals of a population, followed by an evolutionary response by the second population to the change in the first" and further developed by Thompson (1982, 1994). Subsequently, the term cospeciation has been revised repeatedly to expand its application to various types of associations, changing its interpretation to support specific testing models, which have included several other processes (e.g., host switching, independent speciation or duplication, extinction, failure to diverge, or missing the boat). Among them, some examples include:

> Cospeciation is the joint speciation of two or more lineages that are ecologically associated, the paradigm example being a host and its parasite. (Page, 2003)

> Process whereby a symbiont speciates at the same time as another species (this may result from vicarious events or from narrow host specificity). This is a pattern and does not assume causal relationships. (de Vienne et al., 2013)

> The process in which a lineage speciates as a result of another speciation event: more specific than codivergence, it is concerned only with species. (Charleston, 2016)

Untangling Correlative and Causative Processes Driving Cophylogenetic Patterns

Along with the increasing controversy about how to define and concatenate all these concepts and processes in a single unified theory, various methods emerged to test which of the processes play the major role in shaping interacting communities. The most popular approach is to use cophylogenetic analyses—that is, the comparison of phylogenies of interacting lineages to uncover patterns of mutual descent with or without mutual modification or mutual speciation (D.R. Brooks, pers. comm.). In this area of comparative phylogenetics, the main aim is to test the congruence among phylogenies and the significance of the cophylogenetic structure. Brooks provided the first formal method to quantify the degree of cospeciation and co-accommodation (Brooks, 1979, 1981, 1985, 1988, 1990). Nevertheless, simultaneous cospeciation does not necessarily imply dependency and mutuality of the modifications and speciation. Unwarranted assumptions claiming that the congruence between phylogenies and the time estimates may be conclusive for the actual cospeciation reconstruction (i.e., cladogenesis of an ancestral species because of the cladogenesis of another interacting species) among taxa were often inherited without reflection (de Vienne et al., 2013). Even in some cases for which cospeciation may seem likely (such as vertically transmitted symbionts and their hosts), prior assumptions may unnecessarily cloud the conclusions of cophylogenetic studies. These assumptions support a causal inference, and few examples of methods based on deductive reasoning are available (e.g., Phylogenetic Analysis for Comparing Trees—PACT algorithm, Wojcicki and Brooks, 2004, 2005). This biased assumption builds upon the reasoning that the pathogen phylogeny mirrors host phylogeny. As a result, pathogens will always follow the evolutionary history of their hosts—that is, they will speciate as a consequence of host cladogenesis (or causative cladogenesis), and they will go extinct when they are not able to adapt to their host, or they will duplicate sympatrically into the same host. In this scenario, host switches are rare, and the pathogen tends to be specialized on a single host species. A major consequence was the emergence of an unrealistic optimism about the very low likelihood that a pathogen would suddenly acquire a new host, as cospeciation, revealed by cophylogeny, is the dominant process. This process would represent an evolutionary firewall that would make emerging infectious diseases (EIDs) rare events; however, an increasing body of literature is providing evidence for host switching being as probable as other processes with no extra costs (Brooks et al., 2019; Boeger et al., 2022; Trivellone et al., 2022).

Previous cophylogenetic methods are grouped in two main categories: (1) global-fit and (2) event-based. Global-fit methods quantify the degree of congruence between phylogenies and significance of the overall associations or of each single link. These methods are based on statistical tests and do not infer about the importance of different evolutionary processes possibly involved and revealed by congruent or incongruent phylogenies. Event-based methods measure the fit between phylogenies and define the likelihood for numbers of single evolutionary events that may have caused the observed associations. These methods in general deliver the most probable reconstruction of the cophylogenetic history of the interacting lineages. All methods in both categories have computational or theoretical limits, and researchers often apply several of them to the same data set to take advantage of desirable characteristics of each.

Recently, a plethora of revisionary studies provided comprehensive discussion on terminology and theoretical approaches underlying the cophylogenetic analyses (Hoberg and Brooks, 2008, 2015; Suchan and Alvarez, 2015; Charleston, 2016; Hembry and Althoff, 2016; Marquis et al., 2016; Kariñho Betancourt, 2018; Doña and Johnson, 2019; Harmon et al., 2019; Maron et al., 2019; Morris and Moury, 2019; Sagoff, 2019; Zohdy et al., 2019; Blasco-Costa et al., 2021; Medina et al., 2022). Other papers provided overviews of statistical frameworks to test for coevolutionary diversification or available cophylogenetic methods (Brooks, 2003; Charleston, 2003; de Vienne et al., 2013; Althoff et al., 2014; Charleston and Libeskind-Hadas, 2014; Poisot, 2015; Filipiak et al., 2016; Martínez-Aquino, 2016; Groussin et al., 2020; Dismukes et al., 2022; Hernández-Hernández et al., 2021). Historically rooted and consistent with specialization on single taxa, several reviews evaluated overall patterns of codiversification, cospeciation, and coevolution of various groups of organisms representing specific association types, grouped as parasitic, mutualistic, and commensal (Clayton et al., 2004; Jackson, 2004; Aliouat-Denis et al., 2008; Mattiucci and Nascetti, 2008; Araújo and Hughes, 2016; Arbuckle et al., 2017; Anderson and de Jager, 2020; Anholt, 2020).

Research Questions and Prevalent Hypotheses

In the present review, we evaluated all previous papers that compared phylogenies, concurrent diversification, and mutual adjustment of interacting lineages. We performed an updated field synopsis for the evolution of cophylogenetic studies applied to symbiotic (*sensu lato*) associations. We investigated the influence of three factors, including **cophylogenetic method**, **association**, and **ecosystem type,** on the outcome of the statistical cophylogenetic analyses. We addressed the following questions:

(Q1) How has the usage of words such as "cophylogeny" and/or "codiversification" and quantitative cophylogenetic analyses of interacting lineages changed over time?

(Q2) What is the proportion of studies that yield congruent versus incongruent outcomes in cophylogenetic analyses with respect to the three factors of the present meta-analysis?

(Q3) Do cophylogenetic method, association, and ecosystem type significantly affect the outcome of cophylogenetic analyses?

Our hypotheses are mainly based on the field synopsis and are used in our meta-analysis as a baseline to compare alternative results (in particular for research questions Q2 and Q3). As the association type is concerned, parasitic association (Hartmann et al., 2019) and mutualistic or commensal associations (especially those that involve symbionts that are thought to be exclusively vertically transmitted or dispersal-limited, Bronstein et al., 2006; Groussin et al., 2020; Hayward et al. 2021; Matthews et al., 2023) show more congruent cophylogeny than expected by chance because the cospeciation events are thought to drive micro-evolutionary trajectories for these types of associations. A few alternative hypotheses were supported in the literature for parasitic associations, and evidence of incongruence was revised in Poulin (2021) (literature therewith). In addition, incongruence rather than congruence between phylogenies is expected to happen more often under a changing environment (Runghen et al., 2021), as also predicted by ecological fitting theory (Agosta, 2006; Agosta et al., 2010). In our meta-analysis, we considered herbivory as a special case for parasitic associations, a relationship that is hypothesized to show higher episodes of incongruences between phylogenies (as revised in Hoberg and Brooks, 2008). For pollination as a special example of mutualism, Hembry and Althoff (2016) previously reported: "We find that most species-rich brood pollination mutualisms show significant phylogenetic congruence at high taxonomic scales, but there is limited evidence for the processes of both cospeciation and duplication, and there are no unambiguous examples known of strict-sense contemporaneous cospeciation." This finding is also in agreement with Lieuter et al. (2017).

We also hypothesized that global-fit methods may yield more congruent results than expected by chance because of the overuse during the last decades of distance-based cophylogenetic methods, which are prone to type I error (i.e., rejection of the H_0, independence between phylogenies, when it is true) (Balbuena et al., 2013). However, some event-based reconciliation methods may yield more congruent results because the assumption is that cospeciation is expected to be more likely than any other event, and the congruence is interpreted as evidence for cospeciation (Ronquist, 1995), resulting in a tendency to maximize the number of cospeciation events (Page, 1994). We believe that each main category has an idiosyncratic risk to provide either a congruent or an incongruent outcome. We further hypothesized that the ecosystem type has an influence on the outcome of the phylogenetic analyses regardless of the category of the method used because aquatic habitats are considered more stable compared to terrestrial habitats, and the interaction between lineages would be the major constraint with which to cope. In Table 3.1, we report the expected cophylogenetic patterns for each of the three factors.

Based on the results, we suggest a reanalysis of published raw data sets using an alternative theoretical framework (i.e., the Stockholm Paradigm [SP]) that will aid in shedding light on the fundamental biological mechanisms involved in coevolutionary processes. We also discuss how to integrate cophylogenetic analyses into the policy extension for SP—that is, DAMA (Document, Assess, Monitor, Act), which is a workbench for the implementation of strategies to anticipate EIDs.

Methods: Field Synopsis, Systematic Review, and Meta-analysis

A systematic and quantitative global-level overview of the current state of knowledge from studies that used different statistical approaches to compare phylogenies of two interacting groups of organisms was carried out. Prior studies were collected, screened and evaluated using the established guidelines in Moher et al. (2009).

Search strategy

We carried out a literature search using four different databases: PubMed, ScienceDirect, Scopus, and Web of Science. The databases were searched on 12 February 2022. In order to eliminate the high ambiguity generated by some keywords used singularly and to include the maximum number of relevant studies, we used a defined set of single keywords and combinations of them. We selected 2 main keywords, "cophylogeny" and "codiversification," and 11 companion keywords were linked to them using the logical operator "AND," as follows: "cophylogeny AND coevolution AND symbiosis," "cophylogeny AND generalist," "cophylogeny AND herbivore," "cophylogeny AND host AND cladogram," "cophylogeny AND host AND switching," "cophylogeny AND pathogen," "cophylogeny AND phytophagous," "cophylogeny AND pollinator," "cophylogeny AND specificity," "cophylogeny AND symbiosis," "codiversification AND

Table 3.1. Cophylogenetic patterns (congruence vs incongruence) tested in this study. Three factors (cophylogenetic method, association, and ecosystem type) are evaluated as drivers of the outcome of the cophylogenetic analysis.

Predicted cophylogenetic patterns					
Congruence			**Incongruence**		
Association	Method	Ecosystem	Association	Method	Ecosystem
• Parasitic • Mutualistic • Commensal	• Global-fit (distance-based methods) • Event-based (reconciliation methods)	Aquatic	• Herbivore (special case parasite) • Pollinator (special case mutualism)	Both	Terrestrial

coevolution AND symbiosis," "codiversification AND generalist," "codiversification AND herbivore," "codiversification AND host AND cladogram," "codiversification AND host AND switching," "codiversification AND pathogen," "codiversification AND phytophagous," "codiversification AND pollinator," "codiversification AND specificity," and "codiversification AND symbiosis."

Additional studies from the gray literature recommended by experts were also considered.

Collection, screening, and eligibility
We collected a total of 5,970 papers (Scopus: 4,879; ScienceDirect: 365; Web of Science: 251; PubMed: 475). The screening was handled with a script written in R (revtools v. 0.4.1 [Westgate, 2019] and rbibutils v. 2.2.8 [Boshnakov and Putman, 2022] R packages). A total of 1,595 unique papers were selected. The performance of each database was summarized using the ggVennDiagram R-package v. 1.2.0 (Gao, 2021).

Initial evaluation was based on title and abstract, when available; however, for most of the published papers, examining the full text was necessary to retrieve relevant data. The criteria of inclusion were based on: (1) papers that used at least a pair of phylogenies (either molecular or morphological) to investigate the degree of congruence between groups of interacting lineages and (2) papers that either evaluated congruence or incongruence and/or attempted to reconcile phylogenies by using one or more of the cophylogenetic methods grouped in two main classes or categories (event-based and global-fit) based on statistical inference and formalized algorithms for which software or webtools are available. The criteria of exclusion are summarized as follows: (1) monographs, syntheses, and literature reviews, not including original cophylogenetic studies; (2) studies on methodological approaches that used either toy data, data from other papers (unless analyzed with different analytical approaches or software), or any other kind of simulation. The total number of eligible papers was 296.

Data extraction and database creation
The selected papers were scored according to three main explanatory variables (factors) related to the research questions: type of association (hereafter Association), type of ecosystem (Ecosystem), and type of method for cophylogenetic analysis (Method). Association is a categorical factor that includes five main levels: mutualistic (*mut*), commensal (*com*), parasitic (*par*), herbivory (*herb*), and pollination (*pol*). Mutualistic associations are those in which two different interacting species benefit from the relationship,

commensal refers to one species benefiting while the other neither benefits nor is harmed, and parasitic occurs when one benefits and the other is harmed. While herbivory and pollination may be included, respectively, in the broader categories of parasitic and mutualistic/commensal, we kept them as separate levels to further explore the specific hypotheses of this study. To clarify these associations further, another level was created—*mixed*—to refer to association types that were defined by the authors as including more than one main level of the association type (e.g., organisms of one species that may be either parasitic or commensal with another species) or when the authors used a phylogeny for a broad group encompassing species from more than one type of association.

Ecosystem is a categorical factor with two levels: *terrestrial* and *aquatic*. Method is a categorical factor including the cophylogenetic methods used in the revised papers. These methods were grouped into two main levels: *event-based* and *global-fit* (Figure 3.1). In each main level, we evaluated the proportion of methods that are considered to yield more congruent patterns (cost-based reconciliation methods, Recon*, and distance-based methods, Dist-bas*, Figure 3.1) and others considered not affected by this bias (other reconciliation methods, Recon, and other global-fit methods, Figure 3.1).

For each factor, levels were assigned based on what the authors of the paper stated or on information retrieved from associated literature (i.e., from the reference list).

The response variable was scored as a categorical value based on the main **outcome** provided in the evaluated paper which resulted either from an analysis of overall fit (or fit of each single species-species association or link) between the two phylogenies and/or from either a reconciliation or cost-based method. Three outcomes were retrieved from the literature: the phylogenies were mainly congruent (*c*), mainly incongruent (*i*), or partially congruent and incongruent (*ic*). According to the literature evaluated, the last outcome is mainly driven by the specific methods of analysis used; for example, if a global-fit method suggests overall congruence between phylogenies, and the whole contribution is driven by few links, then some authors prefer to interpret the outcome as both congruent and incongruent.

When authors used more than one cophylogenetic method to analyze the phylogenies, we recorded the corresponding outcome for each analysis. The final dataset includes DOI, publication year, Method, Association, Ecosystem, and Outcome.

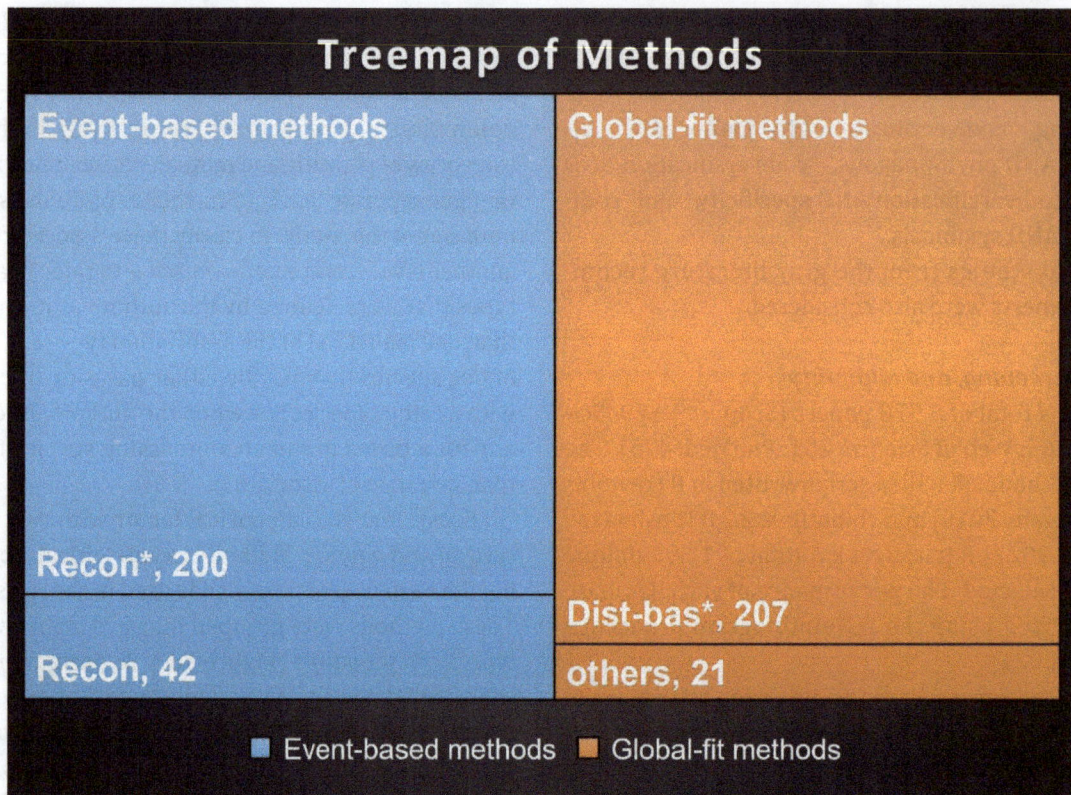

Figure 3.1. Treemap chart of the two categories of the cophylogenetic methods, event-based (in blue) and global-fit (in orange). The subcategories (Recon: reconciliation, Dist-bas: distance-based, and others) and the number of papers using that method are reported in the bottom left of each box. The full table of the methods used in the papers analyzed here is available in Trivellone and Panassiti, 2022.

References for the methods reviewed are: Jeffreys, 1961; Mantel, 1967; Robinson and Foulds, 1981; Penny and Hendy, 1985; Kishino and Hasegawa, 1989; Ronquist and Nylin, 1990; Page, 1994; Farris et al., 1995; Critchlow et al., 1996; Shimodaira and Hasegawa, 1999; Goldman et al., 2000; Ho et al., 2015; Lieberman, 2001; Swofford, 2001; Charleston and Page, 2002; Goudet, 2002; Legendre et al., 2002; Ronquist, 2002; Shimodaira, 2002; Lieberman, 2003a, 2003b; Wojcicki and Brooks, 2004, 2005; Merkle and Middendorf, 2005; Bollback, 2006; Guimarães, Jr. and Guimarães, 2006; de Vienne, Giraud, and Martin, 2007; Huson et al., 2007; Meier-Kolthoff et al., 2007; Stamatakis et al., 2007; Hagberg et al., 2008; Hommola et al., 2009; Ulrich et al., 2009; Conow et al., 2010; Hadfield, 2010; Merkle et al., 2010; Borcard et al., 2011; Poulin, 2011; Schliep, 2011; Scornavacca et al., 2011; Charleston, 2012; Huson and Scornavacca, 2012; Ronquist et al., 2012; Balbuena et al., 2013; Szöllősi, Rosikiewicz, et al., 2013; Szöllősi, Rosikiewicz, Boussau, et al., 2013; Yu et al., 2013; Hadfield et al., 2014; Jacomy et al., 2014; Marquitti et al., 2014; Baudet et al., 2015; Kuhner and Yamato, 2015; Oksanen et al., 2015; Yu et al., 2015; Beaulieu, 2017; Hutchinson et al., 2017; Libeskind-Hadas, 2019; Paradis and Schliep, 2019; Balbuena et al., 2020; Minh et al., 2020; Baudet, 2021; Maddison and Maddison, 2021; Santichaivekin et al., 2021; Llaberia-Robledillo et al., 2022; Santichaivekin et al., 2022; SICSG, 2022a, 2022b; Sinaimeri et al., 2022; and Szöllosi, 2022.

Statistical analyses

Field synopsis

To obtain an overview of the state of knowledge reflected by studies that addressed the topic of cophylogeny and co-diversification (Q1), we considered the studies selected in the initial screening (N = 1,595). Using a paired t-test, we compared the sample means of two groups of studies: those that did not satisfy the eligibility criteria (i.e., discarded studies, N = 1,299) and studies retained for the meta-analyses which applied a quantitative cophylogenetic analysis (i.e., selected studies, N = 296). To evaluate the usage of co-phylogenetic analyses over time, a linear regression was applied to publications that used cophylogenetic analyses expressed as a function of years. This was written as: Number of publications = $b_0 + b_1 \times$ Publication year, where b_0 is the intercept and b_1 is the slope.

Systematic review and meta-analyses

To study the relationship between the outcome of the analyses and each factor (Q2), we used a goodness-of-fit chi-square test and the Bayes Factor (Jeffreys, 1961) using the function "ggbarstats" from the ggstatsplot R-package v. 0.9.1 (Patil, 2021). For both tests, the null hypothesis is that two compared categorical variables are independent (H_0). The three categorical variables—outcome, cophyloge-netic method, and association type—were first arranged in a structured contingency table using the function "structa-ble" from the vcd R-package v. 1.4-9 (Meyer et al., 2006). Dependencies among variables were explored using contingency table frequencies and log-linear models as explained by Zeileis et al. (2007). To further analyze the independence between the outcome and our factors, we used a mosaic plot (vcd R-package) and inferred the departure from independence using Pearson standardized residuals according to Friendly (1994).

To answer Q3, we fitted a generalized linear mixed model (GLMM) and specified a binomial error distribution and a logit-link function. We estimated the probability of receiving an incongruent (0) or congruent (1) outcome as a function of three predictors: **cophylogenetic method, association**, and **ecosystem type**. Our predictors were factors with two levels for the Method (*event-based* and *global-fit*), two levels for Ecosystem (*terrestrial* and *aquatic*), and six levels for Association (*mut, com, par, herb, pol, mixed*). Moreover, we included DOI (i.e., the study ID) as a nested random effect (a.k.a. mixed model, which allows the intercept to vary with DOI) to consider the nonindependence between observations within the same study that applied more than one method on the same dataset. In this way we consider the possible bias introduced by the tendency of the same dataset analyzed with different methods to provide the same result (pseudoreplication) (Hurlbert, 1984).

We fitted a total of six GLMMs: the full model, including all three predictors and their interactions, and five parsimonious models. We ranked our models using the second-order Akaike Information Criterion (AIC) scores, and the final model with the lower value of AIC was selected. The AIC value indicates a more parsimonious model (Burnham and Anderson, 2002). The GLMM was fitted using "glmer"-function from the lme4 R-package v. 1.1-27.1 (Bates et al., 2015). Finally, we inspected the distribution of simulated model residuals using the DHARMa R-package v. 0.4.5 (Hartig, 2022). All statistical analyses were conducted using R software v. 4.1.2 (R Core Team, 2019).

Results

Field Synopsis

The four databases yielded 1,327 (Scopus), 113 (ScienceDirect), 133 (Web of Science), and 342 (PubMed) papers. Scopus detected the highest number of unique citations (1,112, 70% of the total), PubMed found 203 (13%), ScienceDirect 62 (4%), and Web of Science 3 (0.2%). The highest overlap was among PubMed, Scopus, and Web of Science (4% of shared published papers), between Pubmed and Scopus (4%), and Scopus and Web of Science (3%) (Figure 3.2).

After eligibility screening, we included and extracted data from 296 papers published from 1997 to 2022 (for the last year only the first two months), reporting cophylogenetic analyses that test significance of the congruence between phylogenies of interacting lineages and/or estimates of coevolutionary events.

(Q1) How has the usage of words such as "cophylogeny" and/or "codiversification" and quantitative cophylogenetic analyses of interacting lineages changed over time?

The usage of the words *cophylogeny* and *codiversification*, used to query the databases, ranged from 1997 to 2022. After the selection of published papers that used statistical analyses to study the cophylogeny of interacting lineages, the temporal range was narrowed by four years (2001–2022). In each year, the proportion of published papers merely mentioning the two keywords rather than statistically analyzing cophylogeny or codiversification was significantly higher (t = 5.3907, df = 20, *p-value* < 0.001) (Figure 3.3A). For the selected published papers, usage of the

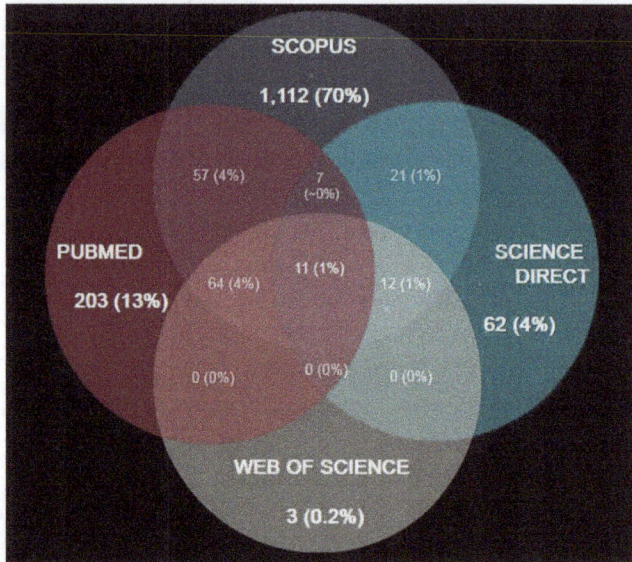

Figure 3.2. Venn diagram reporting the results of the literature search using four different databases: PubMed, ScienceDirect, Scopus, and Web of Science (redrawn and modified from Trivellone and Panassiti, 2022).

keywords steadily increased over the years, showing a significant positive linear trend (*p-value* < 0.001, Figure 3.3B). The number of published papers released from 2001 to 2021 ranged from 1 to 35 papers per year.

Systematic review and meta-analyses

(Q2) What is the proportion of studies that yield congruent versus incongruent outcomes in cophylogenetic analyses with respect to the three factors of the present meta-analysis?

The final dataset includes five columns: three factors (**cophylogenetic method**, **association**, and **ecosystem type**), the dependent variable (Outcome), and the random variable (DOI). For each of the three factors, Table 3.2 shows the proportion of studies yielding congruent, incongruent, or both outcomes. Overall, a higher number of studies reported congruent phylogenies (56%) compared to incongruent (34%), and only 10% of the studies reported both outcomes for the same analyses. The higher number of the reviewed studies investigated **parasitic associations** (58%); among them 52% yielded congruent results and 36% incongruent. The 31% of the studies that focused on mutualistic associations yielded a higher proportion of congruent results (66%) compared to incongruent (26%). Commensal associations showed a similar trend with all studies but one yielding congruent result. The two special cases of parasitic and mutualistic associations, herbivory and pollination, yielded predominantly incongruent results: 81% and 42%, respectively. The large majority of the

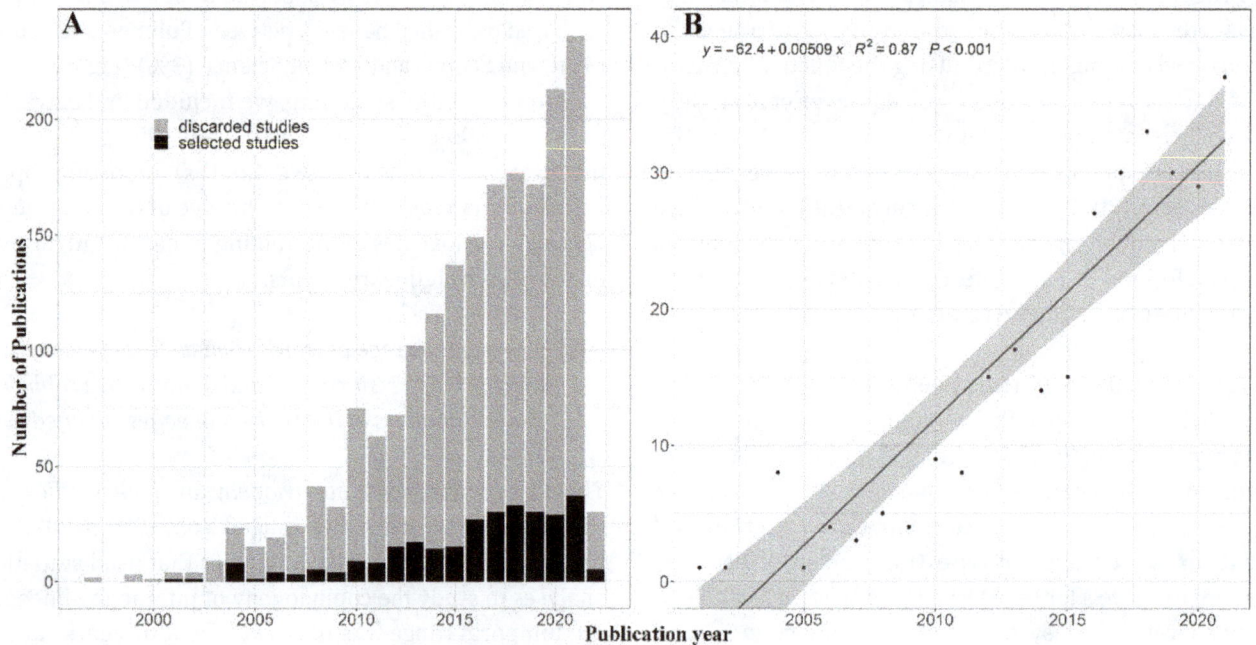

Figure 3.3. Number of published papers per year that include cophylogenetic studies of interacting lineages, selected using four scientific databases (Pubmed, ScienceDirect, Scopus, and Web of Science). **(A)** The proportion of studies that addressed the topics of cophylogeny, codiversification, and coevolution (discarded studies, gray bars) was compared with studies that carried out formal cophylogenetic analyses on real data (selected studies, black bars) over a time period of 25 years (from 1997 to 2022). **(B)** Linear increase (R^2 = 0.87) of the number of published papers that used one or more cophylogenetic methods to evaluate phylogenies of interacting lineages (from 2001 to 2022) (reprinted from Trivellone and Panassiti, 2022).

Table 3.2. Overview of the number of studies that carried out cophylogenetic analyses of interacting species. The proportion of the three outcomes (*c* = congruence, *i* = incongruence, *ic* = both) is reported in relation to Method, Association, and Ecosystem. Percentages for each outcome are based on the total row marginals (modified from Trivellone and Panassiti, 2022).

		Outcome			
		c	*i*	*ic*	Total
Method	*Event-based*	117 (48%)	94 (39%)	33 (14%)	244
	Global-fit	153 (63%)	71 (29%)	17 (7%)	241
Association*	*par*	148 (52%)	101 (36%)	33 (12%)	281
	mut	99 (67%)	39 (26%)	10 (7%)	148
	com	6 (86%)	1 (14%)	0 (0%)	7
	pol	7 (33%)	9 (43%)	5 (24%)	21
	herb	2 (18%)	9 (82%)	0 (0%)	11
	mixed	8 (50%)	6 (38%)	2 (12%)	16
Ecosystem	*aquatic*	40 (56%)	20 (28%)	11 (16%)	71
	terrestrial	230 (56%)	145 (35%)	39 (9%)	413
				Total	**485**

* Abbreviations for the type of associations: *com*, commensal; *herb*, herbivory; *mixed*, a combination of more than two of the other levels; *mut*, mutualistic; *par*, parasitic; *pol*, pollination

reviewed studies focused on terrestrial ecosystems (85%). Nonetheless, congruent results were obtained in about half of the analyses carried out for each Ecosystem type.

The probability of independence between the Outcome of the analysis and both Method and Association is lower than expected (*p-value* < 0.001)—that is, there is a high probability that the outcome significantly depends on the method of analyses used and the type of association. On the other hand, the probability of independence is higher than expected for Ecosystem type (*p-value* = 0.22) but not significant. The Cramér's V value measures the degree of association between categorical variables and varies from 0 to 1. Our results indicated a weak relationship of the outcome with both Method and Association type, with Cramér's V values of 0.15 and 0.14, respectively. The relationship between Outcome and Ecosystem was negligible (0.05). The Bayes Factor (log(BF)) tests were both null (H_0 = the variables are independent) and alternative hypotheses (H_1) and values greater than 2.30 indicate strong evidence for H_0, whereas values lower than –2.30 strongly support H_1. The outcome of cophylogenetic analyses is strongly dependent on the Method used. Similarly, the Bayesian Cramér's V effect sizes (Cramér's V posterior) yielded the same result of the Cramér's V values.

When including the Method and Association as nested factors, mutualistic associations analyzed with global-fit methods yielded more congruent results than expected by chance, whereas herbivory yielded more incongruent results (Pearson residuals > 2). Using event-based methods, parasitic associations yielded more incongruent or mixed results than expected. The area of each box also gives an indication of its proportion to the whole, relative to the same row. The same analysis was carried out using pollination, commensal, and mixed types of Associations, and all Pearson residuals fell between 2 and –2, indicating independence between variables.

(Q3) Do method, association, and ecosystem type significantly affect the outcome of cophylogenetic analyses?

All models were built using a binomial GLMM by eliminating 50 out of 485 entries of the collected metadata, which included the *ic* outcome. Among the discarded entries, only 7 studies that used more than one method yielded the same mixed outcome.

Among the six candidate models tested, we selected the most parsimonious model (AIC = 512.1) with additive main effects of Method and Association and without interaction (simpler model). We found that the type of Method and Association significantly influence the outcome of cophylogenetic analyses (Table 3.3). The continuous values of the linear predictor are transformed to the range between 0 and 1 using the inverse logit, where 1 is the probability of obtaining a congruent outcome.

Table 3.3. Summary statistics of coefficients of fixed effects from a binomial generalized linear mixed model (GLMM) with outcome of cophylogenetic analyses (incongruent, congruent) as a function of method and association. Coefficient estimates are on logit (log-odds) scale.

Model 6	Estimate	Std. Error	z value	Pr(>|z|)
(Intercept)	–3.79	1.65	–2.29	0.022*
Method—*Event-based*	–1.28	0.42	–3.03	0.003**
Association—*com*	8.23	3.25	2.53	0.011*
Association—*mixed*	5.72	2.59	2.20	0.027*
Association—*mut*	7.15	2.29	3.12	0.002**
Association—*par*	5.38	1.91	2.81	0.005**
Association—*pol*	2.78	2.05	1.35	0.176

Overall, we found a slightly higher probability of global-fit methods to yield congruent results compared to event-based independently of type of association under study. The associations with the highest probability (ranging from 0.88 to 0.99, for event-based and global-fit methods) of a congruent outcome were commensal (*com*) and mutualistic (*mut*). On the other hand, the probability of obtaining incongruence between phylogenies is higher (ranging from 0.07 and 0.28) for plant-pollinator associations (*pol*) than plant-herbivore associations (*herb*). For parasitic (*par*) associations, event-based methods yield a lower probability (~ 0.57) of a congruent result compared to global-fit (~0.83) (Figure 3.4).

In Figure 5 an overview of the overall results is presented.

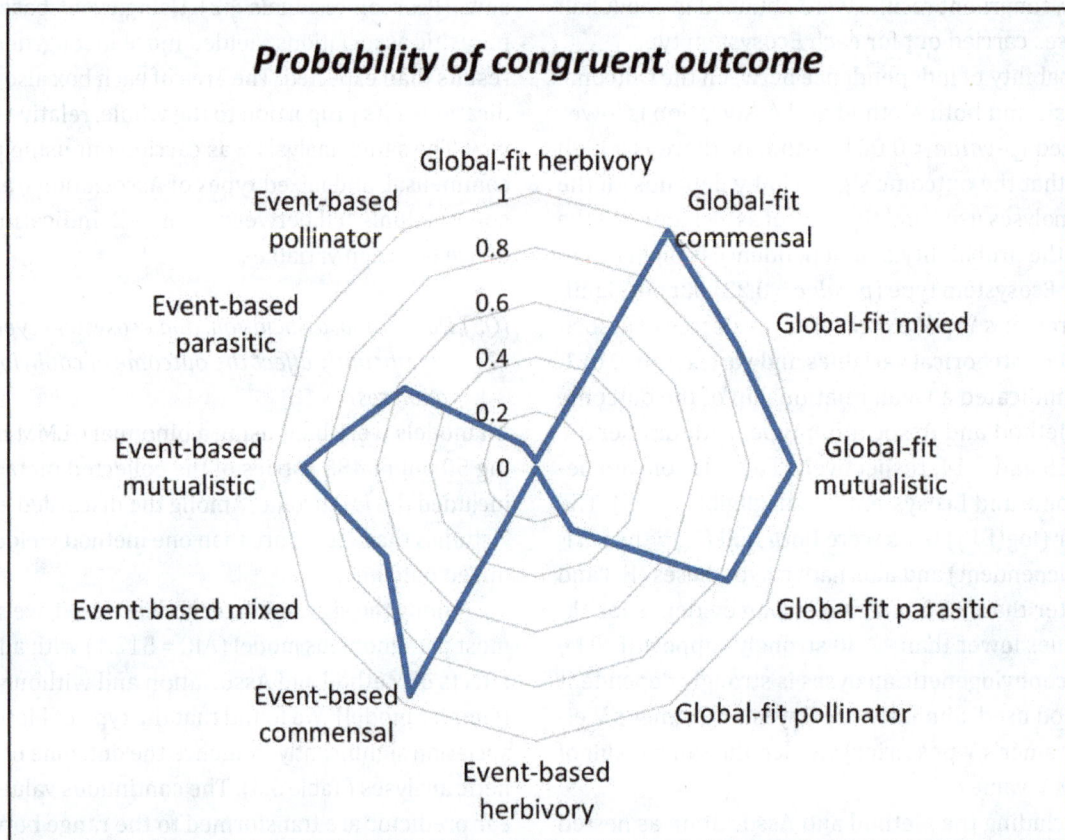

Figure 3.4. Predicted probabilities of congruent outcomes as a function of association and method (redrawn and modified from Trivellone and Panassiti, 2022).

	Take-home message 1	Take-home message 2	Take-home message 3	Overall Trend
Q1 Usage of cophylogenetic methods	Significantly more papers use the words cophylogeny and codiversification rather than statistically analyze the patterns formally	In the last 21 years, the number of papers using formal statistical cophylogenetic analyses increased by 5 times		**Increasing** number of **qualitative methods** rather than cophylogenetic analyses to infer coevolution of interacting lineages
Q2 Proportion of congruent vs. incongruent results and test of independence	Outcome of cophylogenetic analyses depends on the **Method** and **Association type**, with the first being the most important factor	**Parasitic** associations are expected to be significantly more **incongruent** using **Event-based** methods **Mutualistic** are expected to be significantly more **congruent** using **Global-fit** methods	Some proportion of congruent and incongruent were yielded in each habitat type with no significant difference	**Cophylogenetic pattern** detected significantly **depends on** analytical **Method** The **Method affects** the outcome **differently based on the type of Association** under study
Q3 Probability of congruent vs. incongruent	**Analytical Method** and **Association type** significantly influence the outcome of cophylogenetic analyses	**Global-fit** methods have a higher probability of yielding **congruent** when compared with Event-based	**Parasitic**, **commensal**, and **mutualistic** datasets have a higher probability of yielding **congruent** with **Global-fit** methods	Overall, the **highest probabilities of congruence** are yielded by **mutualistic**, **commensal**, and **parasitic** associations compared to herbivory and pollinators

Conclusion

1–Cophylogenetic patterns in parasitic associations tend to adhere with the hypothesis from the literature for parasites mirroring hosts' phylogenies.

2–The overuse of Global-fit distance-based and Event-cost-based methods may result in bias toward congruent outcomes.

Figure 3.5. Overview of the results for the global meta-analysis of cophylogenetic studies carried out in 2022.

Discussion

A previous attempt to review studies reporting cophylogenetic analyses was provided by de Vienne et al. (2013). In that study, the authors reviewed 103 published papers retrieved from the ISI Web of Knowledge with the main aim to evaluate convincing cases of cospeciation by attributing a qualitative score (1–5) that summarized their evaluation varying from convincing case of cospeciation (1) to unclear results (5). Their conclusion is that cases of "true" cospeciation are rare (7%) and that cophylogenetic methods overestimate the occurrence of such events. Although we strongly agree with these authors regarding the different biases introduced by available statistical approaches and by overused assumptions, in our review we wanted to provide a systematic meta-analysis of the main results in order to quantify the outcomes and provide a more objective evaluation.

To our knowledge, this is the first systematic review of cophylogenetic studies using four different search databases and the first quantitative meta-analysis to test the most popular assumption in the literature (usually used as H_0, congruence between phylogenies) against alternatives (H_1, incongruences).

Our systematic search confirmed that Scopus provides about 84% more coverage than PubMed, ScienceDirect, and Web of Science, which is a percentage four times higher than that reported in another revisionary study on biomedical sciences (Falagas et al., 2008). This discrepancy is possibly due to the multidisciplinary nature of our research topic and the keywords used, as pointed out by AlRyalat et al. (2019).

An interesting result emerging from our systematic review of cophylogenetic analyses of interacting lineages is that only about one-fifth of the reviewed published papers attempted to disentangle the processes driving codiversification statistically, regardless of the strategy or algorithm used. Most of the published papers (~80%) focused on a specific lineage and discussed the potential role of biotic interactions driving the diversification of each single taxa, with no attempt to compare phylogenies. From a review of previous summary studies on this topic (e.g., Brooks, 1979; Janzen, 1980; Page, 2003; de Vienne et al., 2013; Poisot, 2015; Charleston, 2016; Martínez-Aquino, 2016), it became evident that the concepts used as keywords in our search (such as codiversification and coevolution) or related words (e.g., cospeciation) are defined differently or used interchangeably as also pointed out earlier (Charleston and Perkins, 2006). This may create confusion and has hindered the implementation and validation of a unified statistical approach or the application of these methods of analysis to specific Association types (e.g., commensal associations).

Although they are becoming more popular, cophylogenetic studies of interacting lineages are proportionally fewer than studies that do not compare phylogenies and merely mention concepts such as coevolution. In a similar synoptic study, Poisot (2015) showed that between 1997 and 2012 the ratio between the number of studies addressing cophylogeny analyses and those mentioning coevolution was stable around 0.34. Our review extends this earlier analysis by adding 10 more years of data and shows an increase of the ratio by more than 5 times. This indicates an increasing tendency by researchers to use qualitative methods rather than phylogenetic or cophylogenetic analyses to infer/assume coevolution between/among interacting lineages and to operate over very short (ecological), nonevolutionary, timescales.

The likelihood of obtaining a specific outcome using available cophylogenetic analyses has been tested here by evaluating three main factors: two inherent to the biological system investigated (Association and Ecosystem types) and one pertaining to the statistical method applied. Overall, we retrieved only a few studies (seven published papers) that analyzed commensal associations to uncover the strength of the cophylogenetic associations. A previous review reported that despite commensalism being frequently mentioned in the ecological literature, it has been little studied because of limited understanding of commensal associations (e.g., inconsistent and divergent definitions of the term leading to miscataloging of the associations and lack of empirical evidence) (Mathis and Bronstein, 2020). We speculate that the lack of cophylogenetic studies on commensal associations may be due to the misleading assumption that if no harm or benefit occurs between interacting lineages, then there will be no driving force for cospeciation to happen. On the other hand, a group of commensal bird-associated mites were recently used as model system to infer the historical coevolutionary processes and, despite overall cospeciation signal from both global-fit and event-based methods, the authors recognized that mite-host specificity varies more widely than expected and host switching is common (Matthews et al., 2023). This is a very good illustration that cophylogenetic congruence is not synonymous with and is not always driven by cospeciation (Brooks and McLennan, 1991, 1993, 2002; de Vienne, Giraud, and Shykoff, 2007). We believe that expanding the analyses of cophylogeny into classical commensal study cases may allow explicit tests of the assumption that cospeciation is the only process leading to congruent phylogenies. This erroneous outcome, recently referred as "apparent cospeciation" (Blasco-Costa et al., 2021), has been discussed extensively in other reviews (de Vienne et al., 2013; Charleston, 2016). We also point out that the cophylogenetic structure and the reconstruction of the associations is largely affected by the possible change of the Association type over time, and none of the analytical methods formally take into account this aspect.

By calculating expected frequencies from contingency tables, we provided evidence for the Outcome to be strongly dependent on the Method, and this result is driven by mutualistic and herbivory associations for global-fit methods and parasitic associations for event-based methods. Our meta-analysis yielded a significant number of congruent outcomes among phylogenies of species in mutualistic associations supported by several authors in the reviewed literature (see hypotheses in the Introduction section). However, as suggested by de Vienne et al. (2013), among others, obtaining congruent phylogenies among interacting lineages is not a definitive indication for cospeciation. Indeed, our results also indicate that we cannot confidently expect that phylogenies in mutualistic associations will be congruent when using event-based methods. For host-parasite systems, our analysis showed a confident association to incongruent outcomes especially when using event-based approaches, these results support alternative hypotheses that would have parasites not mirroring the host phylogeny. For parasitic association, the assumption known as Fahrenholz's rule (Fahrenholz, 1913)—that is, the parasite's phylogeny mirrors the host's phylogeny—may have driven more than 50 years of misleading analyses of cophylogeny. For this reason, we expect phylogenetic incongruency among lineages to be much more common than those observed with the available methods. We used a mixed-model approach, which provided evidence for a non-nested structure of the explanatory factors that singularly affect the outcome of specific associations while using two different categories of methods to study cophylogeny. This analysis supported a higher probability for parasites, herbivores, and pollinators to provide incongruent results when compared to their hosts' phylogenies. Although cospeciation is imposed as "Assumption 0" in most of the methods, other processes, such as host switching, that may lead to incongruence between phylogenies, have been extensively discussed (Hoberg et al., 1997; Brooks and McLennan, 2003). Using a discovery-based approach (i.e., with no a priori assumption), implemented in algorithms such as secondary BPA (Brooks and McLennan, 2003) and PACT (Wojcicki and Brooks, 2005), all processes are equally possible. These methods were applied to only a few parasitic associations—for example, the classic case of cospeciation between pocket gophers and lice, which has been reanalyzed using PACT, showing about half of the links between parasite and host are explained by speciation of the parasite after a host switch rather than cospeciation (Brooks et al., 2015).

To obtain a comprehensive understanding of the real effect of "Assumption 0" on the main outcome of cophylogenetic analyses, more datasets from different types of associations need to be tested using algorithms that do not assume one event to be more probable and costly than another or are not founded on the prevailing paradigm of maximum cospeciation.

An alternative paradigm, the Stockholm Paradigm (SP), is formalized on the idea that symbionts do not have to evolve genetic novelties to be able to adapt to a new host, which means that mirroring the host's phylogeny is not the only option (Brooks et al., 2019). Given the opportunity, a symbiont may colonize a new host successfully with no morphological or genetic changes required (according to ecological-fitting theory), eventually resulting in incongruence between phylogenies. Reconstructing the cophylogenetic history of interacting lineages is not merely a reconciliation problem, it is an estimation of the most parsimonious events inferred using a deductive rather than inductive approach (e.g., PACT). A software package for PACT is in preparation (Trivellone, Panassiti, Boeger, and Brooks, in prep.) and will provide an easy-to-use tool to test more phylogenies of interacting lineages.

Moreover, uncovering the processes driving the interactions between lineages also has a broader impact beyond the advancement of knowledge. The episodes of incongruence between phylogenies may be interpreted as extinction, duplication where the parasite speciates while the host does not, or host switching. The SP postulates that many more incongruences than previously thought are expected due to host switching. Those incongruences define specific preexisting capacities of the symbiont to colonize a new host and are phylogenetically conserved. In particular for host-parasite associations, the SP also provides a policy extension (DAMA: Document, Assess, Monitor, Act) that is a workbench that translates the scientific outcomes in action (Brooks et al., 2021; Trivellone et al., 2022). Cophylogenetic analyses, using PACT or similar discovery-based approaches, represent the fundamental tool for the second step in DAMA (Assess). Once all the diversity has been reasonably documented (DAMA—Document), it will inform the phylogenies of interacting lineages, and the cophylogenetic analysis will aid in predicting the extension of the potential host range within an evaluation known as phylogenetic triage (i.e., uncover phylogenetically conservative traits that allow the parasite to colonize a new host). Another tool is available for this step in DAMA, a modeling platform that evaluates the dynamics of host switching through ecological fitting (for a review, see Souza et al., 2022).

To conclude, in our review we highlighted that the method selected may affect the outcome of cophylogenetic analyses, depending on the assumptions applied to a specific type of interacting species used as a study model. Knowing how new associations emerge between pathogens and their hosts is critical for informing a global strategy to anticipate the risk of future disease outbreaks and EIDs. Future research should focus on evaluation of real raw metadata to establish whether deductive versus inductive methods affect the main outcome of the cophylogenetic analysis and the significance of congruence between phylogenies.

Literature Cited

Agosta, S.J. 2006. On ecological fitting, plant-insect associations, herbivore host shifts, and host plant selection. Oikos 114: 556–565. https://doi.org/10.1111/j.2006.0030-1299.15025.x

Agosta, S.J.; Janz, N.; Brooks, D.R. 2010. How specialists can be generalists: resolving the "parasite paradox" and implications for emerging infectious disease. Zoologia (Curitiba) 27: 151–162. https://doi.org/10.1590/S1984-46702010000200001

Aliouat-Denis, C.-M.; Chabé, M.; Demanche, C.; Aliouat, E.M.; Viscogliosi, E.; Guillot, J.; et al. 2008. *Pneumocystis* species, co-evolution and pathogenic power. Infection, Genetics and Evolution 8: 708–726. https://doi.org/10.1016/j.meegid.2008.05.001

AlRyalat, S.A.S.; Malkawi, L.W; Momani, S.M. 2019. Comparing bibliometric analysis using PubMed, Scopus, and Web of Science databases. Journal of Visualized Experiments 152: e58494. https://doi.org/10.3791/58494

Althoff, D.M.; Segraves, K.A.; Johnson, M.T.J. 2014. Testing for coevolutionary diversification: linking pattern with process. Trends in Ecology and Evolution 29: 82–89. https://doi.org/10.1016/j.tree.2013.11.003

Anderson, B.; de Jager, M.L. 2020. Natural selection in mimicry. Biological Reviews 95: 291–304. https://doi.org/10.1111/brv.12564

Anholt, R.R.H. 2020. Chemosensation and evolution of *Drosophila* Host Plant Selection. iScience 23: 100799. https://doi.org/10.1016/j.isci.2019.100799

Araújo, J.P.M.; Hughes, D.P. 2016. Chapter one—Diversity of entomopathogenic fungi: which groups conquered the insect body? In: Genetics and Molecular Biology of Entomopathogenic Fungi, Advances in Genetics. B. Lovett, R.J.S. Leger (eds.). Elsevier/Academic Press, Cambridge, MA. pp. 1–39. https://doi.org/10.1016/bs.adgen.2016.01.001

Arbuckle, K.; de la Vega, R.C.R.; Casewell, N.R. 2017. Coevolution takes the sting out of it: evolutionary biology and mechanisms of toxin resistance in animals. Toxicon 140: 118–131. https://doi.org/10.1016/j.toxicon.2017.10.026

Balbuena, J.A.; Míguez-Lozano, R.; Blasco-Costa, I. 2013. PACo: a novel Procrustes application to cophylogenetic analysis. PLOS ONE 8: e61048. https://doi.org/10.1371/journal.pone.0061048

Balbuena, J.A.; Pérez-Escobar, Ó.A.; Llopis-Belenguer, C.; Blasco-Costa, I. 2020. Random tanglegram partitions (Random TaPas): an Alexandrian approach to the cophylogenetic Gordian knot. Systematic Biology 69: 1212–1230. https://doi.org/10.1093/sysbio/syaa033

Bates, D.; Mächler, M.; Bolker, B.; Walker, S. 2015. Fitting linear mixed-effects models using lme4. Journal of Statistical Software 67: 1–48. https://doi.org/10.18637/jss.v067.i01

Baudet, C. 2021. Coala 1.2.1: COevolution assessment by a likelihood-free approach. My Biosoftware-Bioinformatics Softwares Blog. Accessed October 2022. https://mybiosoftware.com/tag/coala

Baudet, C.; Donati, B.; Sinaimeri, B.; Crescenzi, P.; Gautier, C.; Matias, C.; Sagot, M.-F. 2015. Cophylogeny reconstruction via an approximate Bayesian computation. Systematic Biology 64: 416–431. https://doi.org/10.1093/sysbio/syu129

Beaulieu, J.M. 2017. corHMM: analysis of binary character evolution, version 1.15. R-Forge. https://r-forge.r-project.org/projects/corhmm/

Blasco-Costa, I.; Hayward, A.; Poulin, R.; Balbuena, J.A. 2021. Next-generation cophylogeny: unravelling eco-evolutionary processes. Trends in Ecology and Evolution 36: 907–918.

https://doi.org/10.1016/j.tree.2021.06.006

Boeger, W.A.; Brooks, D.R.; Trivellone, V.; Agosta, S.J.; Hoberg, E.P. 2022. Ecological super-spreaders drive host-range oscillations: Omicron and risk space for emerging infectious disease. Transboundary and Emerging Diseases 69: e1280–e1288. https://doi.org/10.1111/tbed.14557

Bollback, J.P. 2006. SIMMAP: Stochastic character mapping of discrete traits on phylogenies. BMC Bioinformatics 7: 88. https://doi.org/10.1186/1471-2105-7-88

Borcard, D.; Gillet, F.; Legendre, P. 2011. Numerical Ecology with R. Springer, New York. 306 pp.

Boshnakov, G.N.; Putman, C. 2022. rbibutils: Read "Bibtex" Files and Convert between Bibliography Formats. rbibutils 2.2.9 (website). https://geobosh.github.io/rbibutils/

Bronstein, J.L.; Alarcón, R.; Geber, M. 2006. The evolution of plant-insect mutualisms. New Phytologist 172: 412–428. https://doi.org/10.1111/j.1469-8137.2006.01864.x

Brooks, D.R. 1979. Testing the context and extent of host-parasite coevolution. Systematic Zoology 28: 299–307. https://doi.org/10.2307/2412584

Brooks, D.R. 1981. Hennig's parasitological method: a proposed solution. Systematic Biology 30: 229–249. https://doi.org/10.1093/sysbio/30.3.229

Brooks, D.R. 1985. Historical ecology: a new approach to studying the evolution of ecological associations. Annals of the Missouri Botanical Garden 72: 660–680. https://doi.org/10.2307/2399219

Brooks, D.R. 1988. Macroevolutionary comparisons of host and parasite phylogenies. Annual Review of Ecology and Systematics 19: 235–259. https://doi.org/10.1146/annurev.es.19.110188.001315

Brooks, D.R. 1990. Parsimony analysis in historical biogeography and coevolution: methodological and theoretical update. Systematic Biology 39: 14–30. https://doi.org/10.2307/2992205

Brooks, D.R. 2003. The new orthogenesis. Cladistics 19: 443–448. https://doi.org/10.1016/S0748-3007(03)00073-2

Brooks, D.R.; Hoberg, E.P, Boeger, W.A. 2015. In the eye of the cyclops: The classic case of cospeciation and why paradigms are important. Comparative Parasitology 82: 1–8. https://doi.org/10.1654/4724C.1

Brooks, D.R.; Hoberg, E.P.; Boeger, W.A. 2019. The Stockholm Paradigm: Climate Change and Emerging Disease. University of Chicago Press, Chicago.

Brooks, D.R.; Hoberg, E.P.; Boeger, W.A.; Trivellone, V. 2021. Emerging infectious disease: an underappreciated area of strategic concern for food security. Transboundary and Emerging Diseases 69: 254–267. https://doi.org/10.1111/tbed.14009

Brooks, D.R.; McLennan, D.A. 1991. Phylogeny, Ecology, and Behavior: A Research Program in Comparative Biology. University of Chicago Press, Chicago.

Brooks, D.R.; McLennan, D.A. 1993. Parascript: Parasites and the Language of Evolution. Smithsonian Series in Comparative Evolutionary Biology. Smithsonian Institution Press, Washington, D.C.

Brooks, D.R.; McLennan, D.A. 2002. The Nature of Diversity: An Evolutionary Voyage of Discovery. University of Chicago Press, Chicago.

Brooks, D.R.; McLennan, D.A. 2003. Extending phylogenetic studies of coevolution: secondary Brooks parsimony analysis, parasites, and the Great Apes. Cladistics 19: 104–119. https://doi.org/10.1016/S0748-3007(03)00018-5

Burnham, K.P.; Anderson, D.R. 2002. Model Selection and Multimodel Inference: A Practical Information-Theoretic Approach. 2nd ed. Springer, New York.

Charleston, M.; Libeskind-Hadas, R. 2014. Event-based cophylogenetic comparative analysis. In: Modern Phylogenetic Comparative Methods and Their Application in Evolutionary Biology: Concepts and Practice. L.Z. Garamszegi (ed.). Springer, Berlin. pp. 465–480. https://doi.org/10.1007/978-3-662-43550-2_20

Charleston, M.A. 2003. Recent results in cophylogeny mapping. Advances in Parasitology 54: 303–330. https://doi.org/10.1016/s0065-308x(03)54007-6

Charleston, M.A. 2012. TreeMap 3b: A Java program for cophylogeny mapping. Cophylogeny. https://sites.google.com/site/cophylogeny/software

Charleston, M.A. 2016. Cospeciation. In: Encyclopedia of Evolutionary Biology. R.M. Kliman (ed.). Academic Press, Oxford. pp. 381–386. https://doi.org/10.1016/B978-0-12-800049-6.00200-6

Charleston, M.A.; Page, R.D.M. 2002. TreeMap 2: A Macintosh program for cophylogeny mapping. Cophylogeny. https://sites.google.com/site/cophylogeny/software

Charleston, M.A.; Perkins, S.L. 2006. Traversing the tangle: algorithms and applications for cophylogenetic studies. Journal of Biomedical Informatics (Phylogenetic Inferencing: Beyond Biology [special issue]) 39: 62–71. https://doi.org/10.1016/j.jbi.2005.08.006

Clayton, D.H.; Bush, S.E.; Johnson, K.P. 2004. Ecology of congruence: past meets present. Systematic Biology 53: 165–173. https://doi.org/10.1080/10635150490265102

Conow, C.; Fielder, D.; Ovadia, Y.; Libeskind-Hadas, R. 2010. Jane: a new tool for the cophylogeny reconstruction problem. Algorithms for Molecular Biology 5: 16. https://doi.org/10.1186/1748-7188-5-16

Critchlow, D.E.; Pearl, D.K.; Qian, C. 1996. The triples distance for rooted bifurcating phylogenetic trees. Systematic Biology 45: 323–334. https://doi.org/10.1093/sysbio/45.3.323

Darwin, C. 1872. The Origin of Species. 6th ed. John Murray, London.

de Vienne, D.M.; Giraud, T.; Martin, O.C. 2007. A congruence index for testing topological similarity between trees. Bioinformatics 23: 3119–3124. https://doi.org/10.1093/bioinformatics/btm500

de Vienne, D.M.; Giraud, T.; Shykoff, J.A. 2007. When can host shifts produce congruent host and parasite phylogenies? A simulation approach. Journal of Evolutionary Biology 20: 1428–1438. https://doi.org/10.1111/j.1420-9101.2007.01340.x

de Vienne, D.M.; Refrégier, G.; López-Villavicencio, M.; Tellier, A.; Hood, M.E.; Giraud, T. 2013. Cospeciation vs host-shift speciation: methods for testing, evidence from natural associations and relation to coevolution. New Phytologist 198: 347–385. https://doi.org/10.1111/nph.12150

Dismukes, W.; Braga, M.P.; Hembry, D.H.; Heath, T.A.; Landis, M.J. 2022. Cophylogenetic methods to untangle the evolutionary history of ecological interactions. Annual Review of Ecology, Evolution, and Systematics 53: 275–298. https://doi.org/10.1146/annurev-ecolsys-102320-112823

Doña, J.; Johnson, K.P. 2019. Assessing symbiont extinction risk using cophylogenetic data. EcoEvoRxiv preprint. https://doi.org/10.32942/osf.io/ry9zm

Ehrlich, P.R.; Raven, P.H. 1964. Butterflies and plants: a study in coevolution. Evolution 18: 586–608. https://doi.org/10.2307/2406212

Fahrenholz, H. 1913. Ectoparasiten und abstammungslehre. Zoologischer Anzeiger 41: 371–374.

Falagas, M.E.; Pitsouni, E.I.; Malietzis, G.A.; Pappas, G. 2008. Comparison of PubMed, Scopus, Web of Science, and Google Scholar: strengths and weaknesses. The FASEB Journal 22: 338–342. https://doi.org/10.1096/fj.07-9492LSF

Farris, J.S.; Källersjö, M.; Kluge, A.G.; Bult, C. 1995. Testing significance of incongruence. Cladistics 10: 315–319.

Filipiak, A.; Zając, K.; Kübler, D.; Kramarz, P. 2016. Coevolution of host-parasite associations and methods for studying their cophylogeny. Invertebrate Survival Journal 13: 56–65. https://doi.org/10.25431/1824-307X/isj.v13i1.56-65

Friendly, M. 1994. Mosaic displays for multi-way contingency tables. Journal of the American Statistical Association 89: 190–200. https://doi.org/10.1080/01621459.1994.10476460

Gao, C.-H. 2021. ggVennDiagram: A "ggplot2" implement of Venn Diagram. gaospecial/ggVennDiagram. https://github.com/gaospecial/ggVennDiagram

Goldman, N.; Anderson, J.P.; Rodrigo, A.G. 2000. Likelihood-based tests of topologies in phylogenetics. Systematic Biology 49: 652–670. https://doi.org/10.1080/106351500750049752

Goudet, J. 2002. FSTAT, a program to estimate and test gene diversities and fixation indices. Lausanne, Switzerland.

Groussin, M.; Mazel, F.; Alm, E.J. 2020. Co-evolution and co-speciation of host-gut bacteria systems. Cell Host and Microbe 28: 12–22. https://doi.org/10.1016/j.chom.2020.06.013

Guimarães, P.R., Jr.; Guimarães, P. 2006. Improving the analyses of nestedness for large sets of matrices. Environmental Modelling and Software 21: 1512–1513. https://doi.org/10.1016/j.envsoft.2006.04.002

Hadfield, J.D. 2010. MCMC methods for multi-response generalized linear mixed models: the MCMCglmm R package. Journal of Statistical Software 33: 1–22. https://doi.org/10.18637/jss.v033.i02

Hadfield, J.D.; Krasnov, B.R.; Poulin, R.; Nakagawa, S. 2014. A tale of two phylogenies: comparative analyses of ecological interactions. The American Naturalist 183: 174–187. https://doi.org/10.1086/674445

Hagberg, A.A.; Swart, P.J.; Schult, D.A. 2008. Exploring network structure, dynamics, and function using NetworkX. Proceedings of the 7th Python in Science Conference (SciPy 2008). https://www.osti.gov/servlets/purl/960616

Harmon, L.J.; Andreazzi, C.S.; Débarre, F.; Drury, J.; Goldberg, E.E.; Martins, A.B.; et al. 2019. Detecting the macroevolutionary signal of species interactions. Journal of Evolutionary Biology 32: 769–782. https://doi.org/10.1111/jeb.13477

Hartig, F. 2022. DHARMa: residual diagnostics for hierarchical (multi-level/mixed) regression models. https://github.com/florianhartig/DHARMa

Hartmann, F.E.; Rodríguez de la Vega, R.C.; Carpentier, F.; Gladieux, P.; Cornille, A.; Hood, M.E.; Giraud, T. 2019. Understanding adaptation, coevolution, host specialization, and mating system in castrating anther-smut fungi by combining population and comparative genomics. Annual Review of Phytopathology 57: 431–457. https://doi.org/10.1146/annurev-phyto-082718-095947

Hayward, A.; Poulin, R.; Nakagawa, S. 2021. A broadscale analysis of host-symbiont cophylogeny reveals the drivers of phylogenetic congruence. Ecology Letters 24: 1681–1696. https://doi.org/10.1111/ele.13757

Hembry, D.H.; Althoff, D.M. 2016. Diversification and coevolution in brood pollination mutualisms: windows into the role of biotic interactions in generating biological diversity. American Journal of Botany 103: 1783–1792. https://doi.org/10.3732/ajb.1600056

Hembry, D.H.; Yoder, J.B.; Goodman, K.R. 2014. Coevolution and the diversification of life. The American Naturalist 184: 425–438. https://doi.org/10.1086/677928

Hernández-Hernández, T.; Miller, E.C.; Román-Palacios, C.; Wiens, J.J. 2021. Speciation across the Tree of Life. Biological Reviews 96: 1205–1242. https://doi.org/10.1111/brv.12698

Ho, S.Y.W.; Duchêne, S.; Duchêne, D. 2015. Simulating and detecting autocorrelation of molecular evolutionary rates among lineages. Molecular Ecology Resources 15: 688–696. https://doi.org/10.1111/1755-0998.12320

Hoberg, E.P.; Brooks, D.R. 2008. A macroevolutionary mosaic: episodic host-switching, geographical colonization and diversification in complex host-parasite systems. Journal of Biogeography 35: 1533–1550. https://doi.org/10.1111/j.1365-2699.2008.01951.x

Hoberg, E.P.; Brooks, D.R. 2015. Evolution in action: climate change, biodiversity dynamics and emerging infectious disease. Philosophical Transactions of the Royal Society B 370: 20130553. https://doi.org/10.1098/rstb.2013.0553

Hoberg, E.P.; Brooks, D.R.; Siegel-Causey, D. 1997. Host-parasite co-speciation: history, principles, and prospects. In: Host-Parasite Evolution: General Principles and Avian Models. D.H. Clayton, J. Moore (eds.). Oxford University Press, Oxford, UK. pp. 212–235.

Hommola, K.; Smith, J.E.; Qiu, Y.; Gilks, W.R. 2009. A permutation test of host-parasite cospeciation. Molecular Biology and Evolution 26: 1457–1468. https://doi.org/10.1093/molbev/msp062

Hurlbert, S.H. 1984. Pseudoreplication and the design of ecological field experiments. Ecological Monographs 54: 187–211. https://doi.org/10.2307/1942661

Huson, D.H.; Richter, D.C.; Rausch, C.; Dezulian, T.; Franz, M.; Rupp, R. 2007. Dendroscope: An interactive viewer for large phylogenetic trees. BMC Bioinformatics 8: 460. https://doi.org/10.1186/1471-2105-8-460

Huson, D.H.; Scornavacca, C. 2012. Dendroscope 3: An interactive tool for rooted phylogenetic trees and networks. Systematic Biology 61: 1061–1067. https://doi.org/10.1093/sysbio/sys062

Hutchinson, M.C.; Cagua, E.F.; Stouffer, D.B. 2017. Cophylogenetic signal is detectable in pollination interactions across ecological scales. Ecology 98: 2640–2652. https://doi.org/10.1002/ecy.1955

Jackson, A.P. 2004. A reconciliation analysis of host switching in plant-fungal symbioses. Evolution 58: 1909–1923. https://doi.org/10.1111/j.0014-3820.2004.tb00479.x

Jacomy, M.; Venturini, T.; Heymann, S.; Bastian, M. 2014. ForceAtlas2, a continuous graph layout algorithm for handy network visualization designed for the Gephi software. PLOS ONE 9: e98679. https://doi.org/10.1371/journal.pone.0098679

Janzen, D.H. 1980. When is it coevolution? Evolution 34: 611–612. https://doi.org/10.2307/2408229

Jeffreys, H. 1961. Theory of Probability. 3rd ed. Oxford University Press, Oxford, UK.

Kariñho Betancourt, E. 2018. Plant-herbivore interactions and secondary metabolites of plants: ecological and evolutionary perspectives. Botanical Sciences 96: 35–51. https://doi.org/10.17129/botsci.1860

Kellogg, V.L. 1913. Distribution and species-forming of ecto-parasites. The American Naturalist 47, 129–158.

Kishino, H.; Hasegawa, M. 1989. Evaluation of the maximum likelihood estimate of the evolutionary tree topologies from DNA sequence data, and the branching order in hominoidea.

Journal of Molecular Evolution 29: 170–179. https://doi.org/10.1007/BF02100115

Kuhner, M.K.; Yamato, J. 2015. Practical performance of tree comparison metrics. Systematic Biology 64: 205–214. https://doi.org/10.1093/sysbio/syu085

Laine, A.-L. 2009. Role of coevolution in generating biological diversity: spatially divergent selection trajectories. Journal of Experimental Botany 60: 2957–2970. https://doi.org/10.1093/jxb/erp168

Legendre, P.; Desdevises, Y.; Bazin, E. 2002. A statistical test for host-parasite coevolution. Systematic Biology 51: 217–234. https://doi.org/10.1080/10635150252899734

Libeskind-Hadas, R. 2019. Jane. https://www.cs.hmc.edu/~hadas/jane/

Lieberman, B.S. 2001. Applying molecular phylogeography to test paleoecological hypotheses: a case study involving *Amblema plicata* (Mollusca: Unionidae). In: Evolutionary Paleoecology. W.D. Allmon, D.J. Bottjer (eds.). Columbia University Press, New York. pp. 83–103.

Lieberman, B.S. 2003a. Paleobiogeography: the relevance of fossils to biogeography. Annual Review of Ecology, Evolution, and Systematics 34: 51–69. https://doi.org/10.1146/annurev.ecolsys.34.121101.153549

Lieberman, B.S. 2003b. Unifying theory and methodology in biogeography. In: Evolutionary Biology. R.J. Macintyre, M.T. Clegg (eds.). Springer US, Boston, MA. pp. 1–25. https://doi.org/10.1007/978-1-4757-5190-1_1

Lieutier, F.; Bermudez-Torres, K.; Cook, J.; Harris, M.O.; Legal, L.; Sallé, A.; et al. 2017. From plant exploitation to mutualism. Advances in Botanical Research 81: 55–109. https://doi.org/10.1016/bs.abr.2016.10.001

Llaberia-Robledillo, M.; Lucas-Lledó, J.I.; Pérez-Escobar, O.A.; Krasnov, B.R.; Balbuena, J.A. 2022. Rtapas: An R package to assess cophylogenetic signal between two evolutionary histories. bioRxiv preprint. https://doi.org/10.1101/2022.05.17.492291

Maddison, W.P.; Maddison, D.R. 2021. Mesquite: a modular system for evolutionary analysis. http://mesquiteproject.org

Mantel, N. 1967. The detection of disease clustering and a generalized regression approach. Cancer Research 27: 209–220.

Maron, J.L.; Agrawal, A.A.; Schemske, D.W. 2019. Plant-herbivore coevolution and plant speciation. Ecology 100: e02704. https://doi.org/10.1002/ecy.2704

Marquis, R.J.; Salazar, D.; Baer, C.; Reinhardt, J.; Priest, G.; Barnett, K. 2016. Ode to Ehrlich and Raven or how herbivorous insects might drive plant speciation. Ecology 97: 2939–2951. https://doi.org/10.1002/ecy.1534

Marquitti, F.M.D.; Guimarães, P.R., Jr.; Pires, M.M.; Bittencourt, L.F. 2014. MODULAR: software for the autonomous computation of modularity in large network sets. Ecography 37: 221–224. https://doi.org/10.1111/j.1600-0587.2013.00506.x

Martínez-Aquino, A. 2016. Phylogenetic framework for coevolutionary studies: A compass for exploring jungles of tangled trees. Current Zoology 62: 393–403. https://doi.org/10.1093/cz/zow018

Mathis, K.A.; Bronstein, J.L. 2020. Our current understanding of commensalism. Annual Review of Ecology, Evolution, and Systematics 51: 167–189. https://doi.org/10.1146/annurev-ecolsys-011720-040844

Matthews, A.E.; Wijeratne, A.J.; Sweet, A.D.; Hernandes, F.A.; Toews, D.P.L.; Boves, T.J. 2023. Dispersal-limited symbionts exhibit unexpectedly wide variation in host specificity. Systematic Biology: syad014. https://doi.org/10.1093/sysbio/syad014

Mattiucci, S.; Nascetti, G. 2008. Chapter 2, Advances and trends in the molecular systematics of Anisakid nematodes, with implications for their evolutionary ecology and host-parasite co-evolutionary processes. Advances in Parasitology 66: 47–148. https://doi.org/10.1016/S0065-308X(08)00202-9

Medina, M.; Baker, D.M.; Baltrus, D.A.; Bennett, G.M.; Cardini, U.; Correa, A.M.S.; et al. 2022. Grand challenges in coevolution. Frontiers in Ecology and Evolution 9: 618251. https://doi.org/10.3389/fevo.2021.618251

Meier-Kolthoff, J.P.; Auch, A.F.; Huson, D.H.; Göker, M. 2007. COPYCAT: cophylogenetic analysis tool. Bioinformatics 23: 898–900. https://doi.org/10.1093/bioinformatics/btm027

Merkle, D.; Middendorf, M. 2005. Reconstruction of the cophylogenetic history of related phylogenetic trees with divergence timing information. Theory in Biosciences 123: 277–299. https://doi.org/10.1016/j.thbio.2005.01.003

Merkle, D.; Middendorf, M.; Wieseke, N. 2010. A parameter-adaptive dynamic programming approach for inferring cophylogenies. BMC Bioinformatics 11 (Supplement 1): S60. https://doi.org/10.1186/1471-2105-11-S1-S60

Meyer, D.; Zeileis, A.; Hornik, K. 2006. The strucplot framework: visualizing multi-way contingency tables with vcd. Journal of Statatistical Software 17: 1–48. https://doi.org/10.18637/jss.v017.i03

Minh, B.Q.; Hahn, M.W.; Lanfear, R. 2020. New methods to calculate concordance factors for phylogenomic datasets. Molecular Biology and Evolution 37: 2727–2733. https://doi.org/10.1093/molbev/msaa106

Moher, D.; Liberati, A.; Tetzlaff, J.; Altman, D.G. 2009. Preferred reporting items for systematic reviews and meta-analyses: the PRISMA statement. Annals of Internal Medicine 151: 264–269. https://doi.org/10.7326/0003-4819-151-4-200908180-00135

Morris, C.E.; Moury, B. 2019. Revisiting the concept of host range of plant pathogens. Annual Review of Phytopathology 57: 63–90. https://doi.org/10.1146/annurev-phyto-082718-100034

Oksanen, J.; Blanchet, F.G.; Kindt, R.; Legendre, P.; Minchin, P.R.; O'Hara, R.; et al. 2015. Vegan: Community Ecology Package. R package vegan, version 2.2-1 software.

Page, R.D.M. 1994. Parallel phylogenies: reconstructing the history of host-parasite assemblages. Cladistics 10: 155–173. https://doi.org/10.1111/j.1096-0031.1994.tb00170.x

Page, R.D.M. 2003. Tangled Trees: Phylogeny, Cospeciation, and Coevolution. University of Chicago Press, Chicago.

Paradis, E.; Schliep, K. 2019. ape 5.0: an environment for modern phylogenetics and evolutionary analyses in R. Bioinformatics 35: 526–528. https://doi.org/10.1093/bioinformatics/bty633

Patil, I. 2021. Visualizations with statistical details: the "ggstatsplot" approach. JOSS 6: 3167. https://doi.org/10.21105/joss.03167

Penny, D.; Hendy, M.D. 1985. The use of tree comparison metrics. Systematic Biology 34: 75–82. https://doi.org/10.1093/sysbio/34.1.75

Poisot, T. 2015. When is co-phylogeny evidence of coevolution? In: Parasite Diversity and Diversification: Evolutionary Ecology Meets Phylogenetics. S. Morand, B.R. Krasnov, D.T.J. Littlewood (eds.). Cambridge University Press, Cambridge. pp. 420–433. https://doi.org/10.1017/CBO9781139794749.028

Poulin, R. 2011. Evolutionary Ecology of Parasites. 2nd ed. Princeton University Press, Princeton, NJ.

Poulin, R. 2021. The rise of ecological parasitology: twelve landmark advances that changed its history. International Journal for Parasitology 51: 1073–1084. https://doi.org/10.1016/j.ijpara.2021.07.001

R Core Team. 2019. R: a language and environment for statistical computing. R Foundation for Statistical Computing, Vienna, Austria.

Raguso, R.A. 2021. Coevolution as an engine of biodiversity and a cornucopia of ecosystem services. Plants, People, Planet 3: 61–73. https://doi.org/10.1002/ppp3.10127

Robinson, D.F.; Foulds, L.R. 1981. Comparison of phylogenetic trees. Mathematical Biosciences 53: 131–147. https://doi.org/10.1016/0025-5564(81)90043-2

Ronquist, F. 1995. Reconstructing the history of host-parasite associations using generalised parsimony. Cladistics 11: 73–89.

Ronquist, F. 2002. TreeFitter, version 1.2.

Ronquist, F.; Nylin, S. 1990. Process and pattern in the evolution of species associations. Systematic Zoology 39: 323–344.

Ronquist, F.; Teslenko, M.; van der Mark, P.; Ayres, D.L.; Darling, A.; Höhna, S.; et al. 2012. MrBayes 3.2: efficient Bayesian phylogenetic inference and model choice across a large model space. Systematic Biology 61: 539–542. https://doi.org/10.1093/sysbio/sys029

Runghen, R.; Poulin, R.; Monlleó-Borrull, C.; Llopis-Belenguer, C. 2021. Network analysis: ten years shining light on host-parasite interactions. Trends in Parasitology 37: 445–455. https://doi.org/10.1016/j.pt.2021.01.005

Sagoff, M. 2019. When is it co-evolution? A reply to Steen and co-authors. Biology and Philosophy 34: 10. https://doi.org/10.1007/s10539-018-9656-9

Santichaivekin, S.; Mawhorter, R.; Liu, J.; Yang, Q.; Jiang, J.; Wesley, T.; et al. 2022. eMPRess. https://sites.google.com/g.hmc.edu/empress/home

Santichaivekin, S.; Yang, Q.; Liu, J.; Mawhorter, R.; Jiang, J.; Wesley, T.; et al. 2021. eMPRess: a systematic cophylogeny reconciliation tool. Bioinformatics 37: 2481–2482. https://doi.org/10.1093/bioinformatics/btaa978

Schliep, K.P. 2011. phangorn: phylogenetic analysis in R. Bioinformatics 27: 592–593. https://doi.org/10.1093/bioinformatics/btq706

Scornavacca, C.; Zickmann, F.; Huson, D.H. 2011. Tanglegrams for rooted phylogenetic trees and networks. Bioinformatics 27: i248–i256. https://doi.org/10.1093/bioinformatics/btr210

Shimodaira, H. 2002. An approximately unbiased test of phylogenetic tree selection. Systematic Biology 51: 492–508. https://doi.org/10.1080/10635150290069913

Shimodaira, H.; Hasegawa, M. 1999. Multiple comparisons of log-likelihoods with applications to phylogenetic inference. Molecular Biology and Evolution 16: 1114. https://doi.org/10.1093/oxfordjournals.molbev.a026201

SICSG [Swarm Intelligence and Complex Systems Group]. 2022a. CoRe-PA: software for reconstructing cophylogenies. Leipzig University. Accessed October 2022. http://pacosy.informatik.uni-leipzig.de/58-1-Downloads.html

SICSG [Swarm Intelligence and Complex Systems Group]. 2022b. Tarzan: software zur rekonstruktion von kophylogenien. Leipzig University. Accessed October 2022. http://pacosy.informatik.uni-leipzig.de/51-0-Tarzan.html

Sinaimeri, B.; Urbini, L.; Sagot, M.-F.; Matias, C. 2022. Cophylogeny reconstruction allowing for multiple associations through approximate Bayesian computation. arXiv (preprint): 2205.11084. https://doi.org/10.48550/arXiv.2205.11084

Souza, A.T.C.; Araujo, S.B.L.; Boeger, W.A. 2022. The evolutionary dynamics of infectious diseases on an unstable planet: insights from modeling the Stockholm paradigm. MANTER: Journal of Parasite Biodiversity 25. https://doi.org/10.32873/unl.dc.manter25

Stamatakis, A.; Auch, A.F., Meier-Kolthoff, J., Göker, M. 2007. AxPcoords & parallel AxParafit: statistical co-phylogenetic analyses on thousands of taxa. BMC Bioinformatics 8: 405. https://doi.org/10.1186/1471-2105-8-405

Suchan, T.; Alvarez, N. 2015. Fifty years after Ehrlich and Raven, is there support for plant-insect coevolution as a major driver of species diversification? Entomologia Experimentalis et Applicata 157 (special issue): 98–112. https://doi.org/10.1111/eea.12348

Swofford, D.L. 2001. PAUP*: Phylogenetic Analysis Using Parsimony (*and other methods), version 4.0 beta.

Szöllosi, G.J. 2022. ssolo/ALE. Accessed October 2022. https://github.com/ssolo/ALE

Szöllősi, G.J.; Rosikiewicz, W.; Boussau, B.; Tannier, E.; Daubin, V. 2013. Efficient exploration of the space of reconciled gene trees. Systematic Biology 62: 901–912. https://doi.org/10.1093/sysbio/syt054

Szöllősi, G.J.; Rosikiewicz, W.; et al. 2013. ALE program. https://www.slideshare.net/boussau/ale-presentation-36748632

Thompson, J.N. 1982. Interaction and Coevolution. University of Chicago Press, Chicago.

Thompson, J.N. 1994. The Coevolutionary Process. University of Chicago Press, Chicago.

Trivellone, V.; Hoberg, E.P.; Boeger, W.A.; Brooks, D.R. 2022. Food security and emerging infectious disease: risk assessment and risk management. Royal Society Open Science 9: 211687. https://doi.org/10.1098/rsos.211687

Trivellone, V.; Panassiti, B. 2022. A field synopsis, systematic review, and meta-analysis of cophylogenetic studies: what is affecting congruence between phylogenies? MANTER: Journal of Parasite Diversity 24. https://doi.org/10.32873/unl.dc.manter24

Trivellone, V.; Panassiti, B.; Boeger, W.A.; Brooks, D.R. Forthcoming. PACTDis: an R package to phylogenetic analyses of comparing tree by describing distribution with assumption zero. Manuscript in preparation.

Ulrich, W.; Almeida-Neto, M.; Gotelli, N.J. 2009. A consumer's guide to nestedness analysis. Oikos 118: 3–17. https://doi.org/10.1111/j.1600-0706.2008.17053.x

Westgate, M.J. 2019. revtools: An R package to support article screening for evidence synthesis. Research Synthesis Methods 10: 606–614. https://doi.org/10.1002/jrsm.1374

Wojcicki, M.; Brooks, D.R. 2004. Escaping the matrix: a new algorithm for phylogenetic comparative studies of co-evolution. Cladistics 20: 341–361. https://doi.org/10.1111/j.1096-0031.2004.00029.x

Wojcicki, M.; Brooks, D.R. 2005. PACT: an efficient and powerful algorithm for generating area cladograms. Journal of Biogeography 32: 755–774. https://doi.org/10.1111/j.1365-2699.2004.01148.x

Yu, Y.; Harris, A.J.; Blair, C.; He, X. 2015. RASP (Reconstruct Ancestral State in Phylogenies): a tool for historical biogeography. Molecular Phylogenetics and Evolution 87: 46–49. https://doi.org/10.1016/j.ympev.2015.03.008

Yu, Y.; Harris, A.J.; He, X. 2013. RASP (Reconstruct Ancestral State in Phylogenies). Phylogenetics and Evolution Software. Sichuan University. http://mnh.scu.edu.cn/soft/blog/RASP

Zeileis, A.; Meyer, D.; Hornik, K. 2007. Residual-based shadings for visualizing (conditional) independence. Journal of Computational and Graphical Statistics 16: 507–525. https://doi.org/10.1198/106186007X237856

Zohdy, S.; Schwartz, T.S.; Oaks, J.R. 2019. The coevolution effect as a driver of spillover. Trends in Parasitology 35: 399–408. https://doi.org/10.1016/j.pt.2019.03.010

4

Modeling the Stockholm Paradigm: Insights for the Nature and Dynamics of Emerging Infectious Diseases

Angie T. C. Souza, Sabrina B. L. Araujo, and Walter A. Boeger

Abstract

Emerging infectious diseases (EIDs) are, besides a question of food safety and public health, an ecological and evolutionary issue. The recognition of this condition combined with the accumulation of evidence that pathogens are not specialists in their original hosts evidences the need for understanding how the dynamics of interaction between pathogens and hosts occurs. The Stockholm Paradigm (SP) provides the theoretical foundation to understand the dynamics of emergence of diseases and design proactive measures to avoid both the emergence and various reemergence of infectious diseases. In this review, we revisit the models that evaluate several aspects of the proposed dynamics of the SP, including the complex nature of the elements that have been associated with this new framework for the evolution of associations. We integrate the results from these studies into a putative dynamic of infectious diseases, discuss subordinate elements of this dynamic, and provide suggestions on how to integrate these findings into the DAMA (Document, Assess, Monitor, Act) protocol.

Keywords: individual-based model (IBM), agent-based model (ABM), computer modeling, evolution, emergent infectious diseases, DAMA (Document, Assess, Monitor, Act) protocol

Introduction

Among the most worrisome consequences of the changes we are presently subjected to on Earth is the alarming increase in both emergence and reemergence of infectious diseases (EIDs) (Brooks and Ferrao, 2005; Fauci and Morens, 2012). Although many did not recognize it, emergences have accumulated in the recent and distant past, with serious consequences to humans, crops, and livestock (Morens et al., 2004; Brooks and Hoberg, 2007; Fauci and Morens, 2012; Brooks et al., 2014, 2019; Hoberg and Brooks, 2015; Brooks and Boeger, 2019; Trivellone et al., 2022).

We struggle to understand the dynamics of pathogens, having assumed they were a special and unique part of the biosphere. This has not allowed us to anticipate and prevent the emergence of new infirmities in the human-associated ecological network, and we remain greatly dependent on reactive measures following the establishment of a disease (Brooks et al., 2014). However, we are learning that biology is not fragmented in relatively independent systems (e.g., hosts and pathogens; plants and insect; predator and prey) that follow their own rules (Nylin et al., 2018). Since the beginning, life has been linked in a single, vast, and complex network that evolves under the influence of the interactions of its elements and the environment. Darwin was one of the first to recognize this and expressed it in his metaphor of an "entangled bank" (Darwin, 1872). He also recognized that the complex association among the involved actors was driven by independent elements—the *nature of the organism* and the *nature of the conditions*—and that this interaction results in common and universal properties of the entire biological system: evolution and ecology (Brooks and Agosta, 2012; Agosta and Brooks, 2020).

For a long time, we have ignored these most fundamental elements of evolution posed by Darwin, especially for pathogens. Pathogens are usually thought to be specialists to their hosts species and, hence, trapped in a single lineage of hosts (Haldane, 1951; Gioti et al., 2013; Rychener et al., 2017; Scheiner and Mindell, 2019). However, pathogens are resource specialists (Agosta et al., 2010), and specific sets of resources may be widespread among distinct host species. Often what has been assumed to be coadaptation or coevolution actually reflects *ecological fitting* (EF; see Janzen, 1985; Brooks and McLennan, 2002; Agosta, 2006; Agosta and Klemens, 2008). More than a characteristic of antagonistic associations, EF is widely common relative to ecological changes in general (Wilkinson, 2004; Le Roux et al., 2017; Cipollini and Peterson, 2018) and influences the dynamics of ecological networks on this planet. EF is the most basal element of an emerging paradigm, the *Stockholm Paradigm* (SP) (Brooks et al., 2019), a theorical framework that accommodates Darwinian theory into the ecology and evolution of antagonistic associations. This paradigm provides a strong explanatory structure for the understanding of the present emergent infectious diseases crisis (Brooks and Hoberg, 2013).

For the SP, the most significant factors that drive the present crisis are the opportunities generated by changes in species distribution, of both or either hosts and pathogens, and the capacity of pathogens to exploit new hosts (Hoberg and Brooks, 2015; Brooks and Boeger, 2019). On Earth—with unchecked human population growth, climate change, and unmeasurable connectivity associated with human travel and commerce—we have generated a perfect scenario for the emergence of new associations among pathogens and compatible hosts to occur (Gubler, 2010). We introduce species in new areas through socioeconomic interests, force species to move because of habitat loss caused by landscape changes, and even carry them through geographic space both intentionally and unintentionally (Ribeiro Prist et al., 2022). Like numerous Trojan horses, actively or passively translocated host individuals may contain parasites that, once inserted in a new locality, can establish new associations with *compatible* resident species and cause diseases previously unknown (Hulme, 2014) or become hosts of local symbionts (Steward et al., 2022).

The SP recognizes that the interaction between actors that compose host-pathogen networks is complex (Brooks and Boeger, 2019). While the production of new or available empirical data provides opportunities to study complex systems (Patella et al., 2017), computer models simulating biological scenarios provide important insights for understanding the emerging ecological and evolutionary processes (Dieckmann and Doebeli, 1999; Giacomini, 2007). Models can also provide adequate testing for pure theoretical proposals; integrate empirical results and theory; simulate future scenarios; and explore putative solutions to minimize, mitigate, or even avoid the emergence of new antagonistic associations (Giacomini, 2007; Altizer et al., 2013; Christaki, 2015).

One set of models that can handle adequately the simulation of complex interactions that are expected in the biological system is known as the agent-based model (ABM), also known as the individual-based model (IBM) (Dada and Mendes, 2011). In this approach, individuals that compose the system are explicitly modeled. The characteristics of individuals can be freely defined by the modeler along with their behavioral rules of interaction with other individuals and with the environment in which they are inserted. It is from this set of rules of behavior, limitations, and individual needs that the system dynamics emerge (Giacomini, 2007). Thus, in this chapter, we revisit the models that evaluate several aspects of the proposed dynamics of the SP, including the complexity nature of the elements that have been associated with this new framework for the evolution of associations. We integrate the results from these studies into a putative dynamic of diseases, discuss subordinate elements of this dynamics, and provide suggestions on how to integrate these findings into the DAMA protocol.

Synthesis of the Models Developed under the Stockholm Paradigm

Since 2015, the time of publication of the first model that simulated theoretical assumptions associated with the SP (Araujo et al., 2015), several papers have been published that test many elements of this theoretical framework (see Table 4.1). Among other outcomes, subsequent models explored the accumulated evidence that the different elements of the SP represent emergent properties in a complex system that are directly linked to the ability of biological systems (e.g., molecules, species, communities) to realize EF (Janzen, 1985; Agosta, 2006).

Within the framework of the SP, pathogens (and all other consumer species) continuously explore their environments for associated host (resource species for consumers in general) at reach. Successful exploration may result in colonization by EF and, if successful, in the exploitation of new hosts (when the new association persists, and the evolutionary path of the pathogen is subjected to the new selective regime imposed by the new host). These steps compose what is widely called "host switching," a term inadequate to the process of emergence of new symbiotic associations because there is no switch of hosts but rather the expansion of the host repertoire (Braga and Janz, 2021) by the pathogen or parasites (or any consumer species). The interaction between opportunity and compatibility determines the possibility and the extent of exploration and

Table 4.1. Synthesis of models created from the theoretical assumptions associated with SP

Article	Goals	Methods	Main results
Araujo et al. (2015)	To explore the potential of host-switches for a parasite species with variable phenotype amplitudes (expression of the fitness space)	▪ Pathogen individual is characterized by a phenotype value that was exposed to new host resources at each generation. ▪ Hosts do not evolve but are characterized by a carrying capacity and an optimum phenotype value (p_r) imposed on pathogen. ▪ At each pathogen generation, a new host, whose p_r value is randomly defined, is available. ▪ Each individual pathogen has the same probability of dispersing to a new random host. ▪ The model dynamics is composed of the cycles of sexual reproduction, dispersion to new hosts, and selection.	▪ Cyclical changes in the phenotype amplitude—colonization results in reduction and exploitation, in increase. ▪ Colonization of a new host does not require prior evolutionary novelty. ▪ Survival of pathogen populations in suboptimal adaptive regions. ▪ Exploiting host increases the FS and the chance of host-switching to hosts more distant. ▪ Host-switching between hosts representing highly divergent resources occurs by a stepping-stone process.
Braga et al. (2018)	To offer a mechanistic basis for the origins of macroevolutionary patterns of pathogen diversity and host range that emerges from a heterogeneous fitness landscape	▪ Pathogen individuals are characterized by the species identity, a genotype (a binary string whose sum corresponds to individual phenotype). ▪ Hosts do not evolve but are characterized by a carrying capacity and an optimum phenotype value (p_r) imposed on pathogen. ▪ The phylogenetic distance between hosts is represented by the difference between p_r values imposed by each one. ▪ Each individual pathogen has the same probability of dispersing to any other host. ▪ The model dynamics are composed of the cycles of sexual reproduction, dispersion to new hosts and selection. As the model explicitly describes the individuals' genotype, speciation events are also recorded.	▪ Colonization of a new host increases phenotypic variation. ▪ Use of multiple hosts facilitates speciation (divergent selection by including a new fitness peak). ▪ Pathogen's species richness and phenotypic range are mainly affected by "mutation" rate. ▪ Host range negatively affected by distance between hosts. ▪ Phenotypic amplitude was positively correlated with species richness. ▪ Host range oscillates through the time (specialists, generalist).

Article	Goals	Methods	Main results
Feronato et al. (2021)	To explore the significance and the interaction of the reproduction rate, the rate of novelty emergence and the propagule size for the success of colonization of new host species and its consequences to the phenotypic profile evolution of the new population	• Pathogen individuals are characterized by the genotype (a binary string whose sum corresponds to individual phenotype). • The model considers a unique host, characterized by a carrying capacity and an optimum phenotype value (p_r) imposed on pathogens. • At the beginning of simulations, the host is not parasitized, and n pathogens individuals (with phenotype p_0) are allowed to attempt colonization of the host. • Each time step represents a new cycle of asexual reproduction, and selection.	• Maximization of all parameters (evolutionary novelty rate, reproduction rate, and propagule size) results in a synergetic facilitation of the colonization. • The evolutionary novelty rate has the smallest effect on the establishment success in the new host. • Higher evolutionary rates accelerate population growth. • Population size stabilizes (reaches maximum) before phenotypic stabilization. • Even in the absence of evolutionary novelty, and in a suboptimal condition, population size reaches carrying capacity. • Small evolutionary novelty rates ($< 10^{-3}$) result in a smaller phenotypical range, the loss of ancestral phenotypes, and a delay for the population to stabilize around the new optimum imposed by the newly colonized host when compared to larger evolutionary novelty rates.
D'Bastiani et al. (2022)	To understand how host-switching intensity affects parasite evolution	• Pathogen individuals are characterized by their used host species and genetic identity. • Hosts evolve through time without being influenced by the presence of the pathogens (based on empirical phylogenies). • Each host species has the same carrying capacity. • Sexual reproduction. • Continuous host-switching, with probability of success inversely proportional to evolutionary distance between hosts. • Comparison with nine empirical interaction networks using Sackin Index (balance of phylogenetic trees) and beta diversity.	• The model was able to reproduce ecological and evolutionary patterns of the parasites (beta diversity and Sackin index) of all communities analyzed, suggesting that host-switching is determinant in parasite evolution. • Beta diversity is inversely proportional to host-switching intensities, suggesting that this metric can be proxy for host-switching intensity. • The variation in the Sackin index revealed that stochastic host-switching events can change the evolutionary trajectory of parasites.

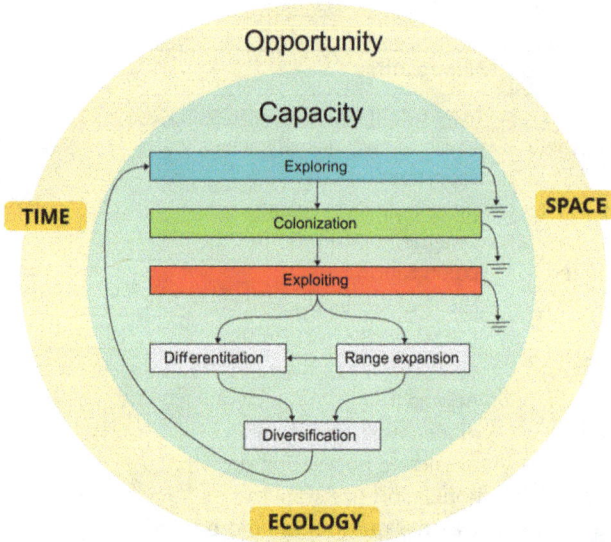

Figure 4.1. Pathogens explore, colonize, and exploit hosts according to their capacity and opportunity (determined by space, time, and ecology).

exploitation (Araujo et al., 2015; Braga et al., 2018; Brooks et al., 2019). Compatibility defines the possibility of realizing EF and is the symmetrical expression of the capacities of the actors in the association (in this case, pathogen and

host). Opportunity is the chance of encounter among potential actors of an association, and thus opportunity is determined by ecology, time, and space (Figure 4.1) (see also Combes, 2001; Araujo et al., 2015). The entire process of host-repertoire expansion triggers emergences or reemergences of new infectious diseases.

The first three models (Araujo et al., 2015; Braga et al., 2018; Feronato et al., 2021) have the following elements in common: (1) they explicitly describe each pathogen individual, characterized by a phenotype (z_i); (2) the resources impose selection pressure on parasite individuals around an optimum phenotype (z_h); and (3) individuals that survive this selection can reproduce, and the offspring inherits the parental phenotype with a probability of incorporating a variation due to the random origin evolutionary novelties (e.g., mutation). These elements essentially follow Darwin's theory of evolution (Darwin, 1872) (Figure 4.2)—surviving organisms reproduce according to their frequency within the parental generation. The *fitness space* (FS) in the model is assumed to be correlated to the phenotypic amplitude of a population in each generation (PA in Figure 4.2b), since the greater the phenotype amplitude, the wider is the range of possible hosts with which the pathogen could interact.

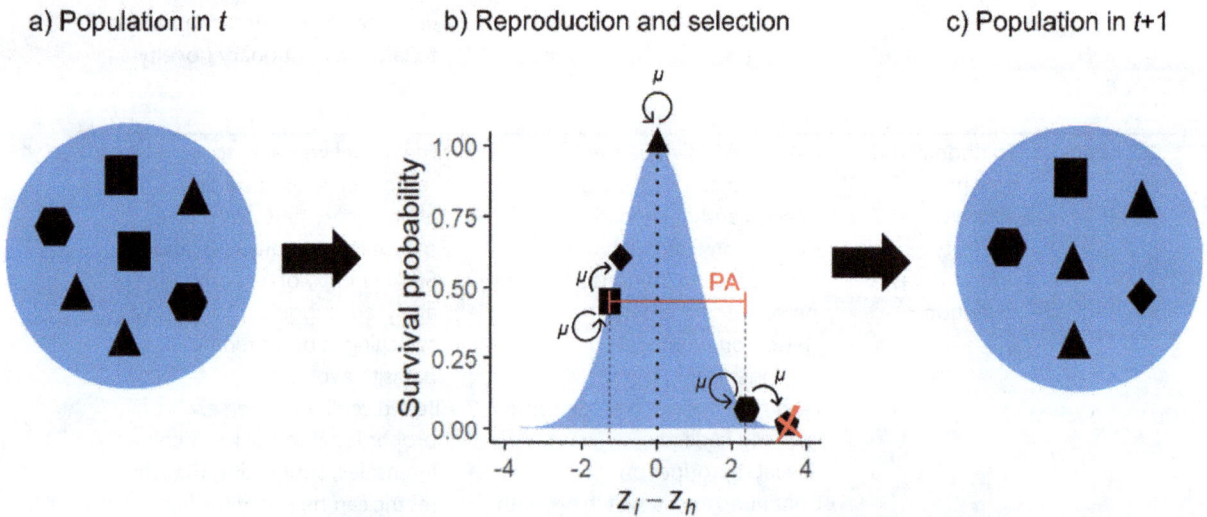

Figure 4.2. Population dynamics common to individual-based models (IBMs) of Araujo et al. (2015), Braga et al. (2018), and Feronato et al. (2021). **(a)** In time t, the parasite population inhabiting a given host (blue circle) is composed of three different phenotypes (triangle, square, and hexagon). **(b)** All phenotypes survive and reproduce in the host; however, the probability of survival decreases with the increase in the distance between the phenotype expressed by the individual (z_i) and the optimal phenotype value imposed by the host (z_h). Reproduction can occur in a sexual (Araujo et al., 2015; Braga et al., 2018) or asexual way (Feronato et al., 2021); the offspring inherits the parental phenotype with a μ probability of incorporating evolutionary novelties, and **(c)** individuals that survive the selection imposed by the host form the population present in $t + 1$. The phenotype amplitude indicated by PA (in graph b) is correlated to the concept of fitness space (FS).

The first model (Araujo et al., 2015) evaluated the relationship between the historical fluctuations of the FS and the potential for pathogens to colonize new hosts by EF. At each time step, a reproductive cycle occurs, and a new host is available to be colonized (see Figure 4.3). A fraction of the pathogen individuals explores the available host and, when colonization is successful, only the evolution of this new population is subsequently recorded. Due to the model dynamics, the phenotypic amplitude (i.e., the FS) can vary and evolve by accumulating evolutionary novelties through time. The simulations showed that (1) successful host colonization does not require "adaptive" evolutionary novelties emerging immediately before colonization, (2) that the FS varies in amplitude (i.e., it oscillates), (3) that pathogen populations can survive for long periods under suboptimal conditions, and (4) that host colonization can occur by a "stepping-stone" process (subsequent colonization of hosts depicting different nature of resources).

Subsequently, Braga et al. (2018) extended the previous model (Araujo et al., 2015) by allowing the evolution of the pathogens in more than one host to coinhabit the same community and by monitoring the evolution of all pathogen populations simultaneously. During one generation, a certain proportion of the pathogens can migrate to a randomly chosen host. When the exploration of the new host results in its colonization by the pathogen, the same lineage of pathogens exploits this host and evolves under different selective regimes. The model describes the pathogen genome and restricts mating to a minimal genotypic similarity. This approach determines the possibility of gene flow among individuals and was used as a proxy to delimit species. When gene flow is reduced between populations of pathogens (imposed by the limits of phenotypic similarity), a speciation event occurs. The model shows that the exploitation of new hosts increases phenotypic and genotypic variation of the pathogen population, which, with reduction in reproductive exchange, may result in speciation of the pathogen, generating host range cycles through time (= oscillations).

The model by Feronato et al. (2021) challenged pathogens to explore new host resources and evaluated the influence of demographic parameters of pathogens (reproduction rate, rate of novelty emergence, and propagule size) on the success of colonization. In the beginning of the simulation, the pathogen population had a single opportunity to colonize a predetermined host resource; following successful colonization, subsequent steps represented new cycles of asexual reproduction and selection. Supporting the model by Araujo et al. (2015) and contrary to the prevailing belief, the rate of novelty emergence (e.g., mutations) depicted a secondary contribution to the success of colonization—even in the absence of emergence of evolutionary novelties, pathogens could survive under suboptimal conditions and reach the carrying capacity imposed by the host.

Finally, motivated by the empirical suggestions that host expansion of a pathogen lineage is common among closely related host species (Braga et al., 2015), D'Bastiani et al. (2021) designed a model to estimate the intensity of host-switching observed in nature and how this parameter affects the phylogenetic history of parasites. In this model, the evolution of a parasite occurs freely along preestablished phylogenies of hosts, with the possibility of migration among hosts at any time. Based on the idea that phylogenetically close hosts present more similar resources (Gilbert and Webb, 2007; Streicker et al., 2010; Imrie et al., 2021), D'Bastiani et al. (2021) assume that the probability of success in the exchange of pathogens between closely related hosts is high and that this probability decays as the hosts diversify and differentiate. The model reproduced the ecological and evolutionary patterns of all nine empirical studies analyzed, suggesting that host-switching is a strong determinant in parasite evolution and diversification. The ecological and evolutionary patterns were measured by the dissimilarity of parasite composition per host species (beta diversity—Baselga, 2010, 2013) and the balance of the phylogenetic tree (Sackin index—Blum and François, 2005), respectively. The variation in the Sackin Index revealed that stochastic host-switching events (leading to host range expansion) can change the evolutionary trajectory of parasites. Beta diversity was inversely proportional to host-switching intensities, suggesting that this metric can represent a proxy for host-switching intensity.

Although the mathematical models presented here enable a good understanding of resource-pathogen dynamics, the code of all these models does not provide a user-friendly interface, restricting many potential users from manipulating the model with their own data sets. Recently, Trivellone and collaborators created an R package, "HostSwitch" (Trivellone et al., 2023), that provides several accessible functions to explore host-switching dynamics. The authors implemented and expanded the original model by Araujo et al. (2015). Users can easily change model parameters and plot the outputs (see Figure 4.3 as an example). They also indicate a method to parameterize the model using three real-world scenarios drawn from selected ecology, agriculture, and parasitology literature. This publication is an effort to facilitate the use of theoretical tools, helping the users build hypotheses of pathogens' evolution.

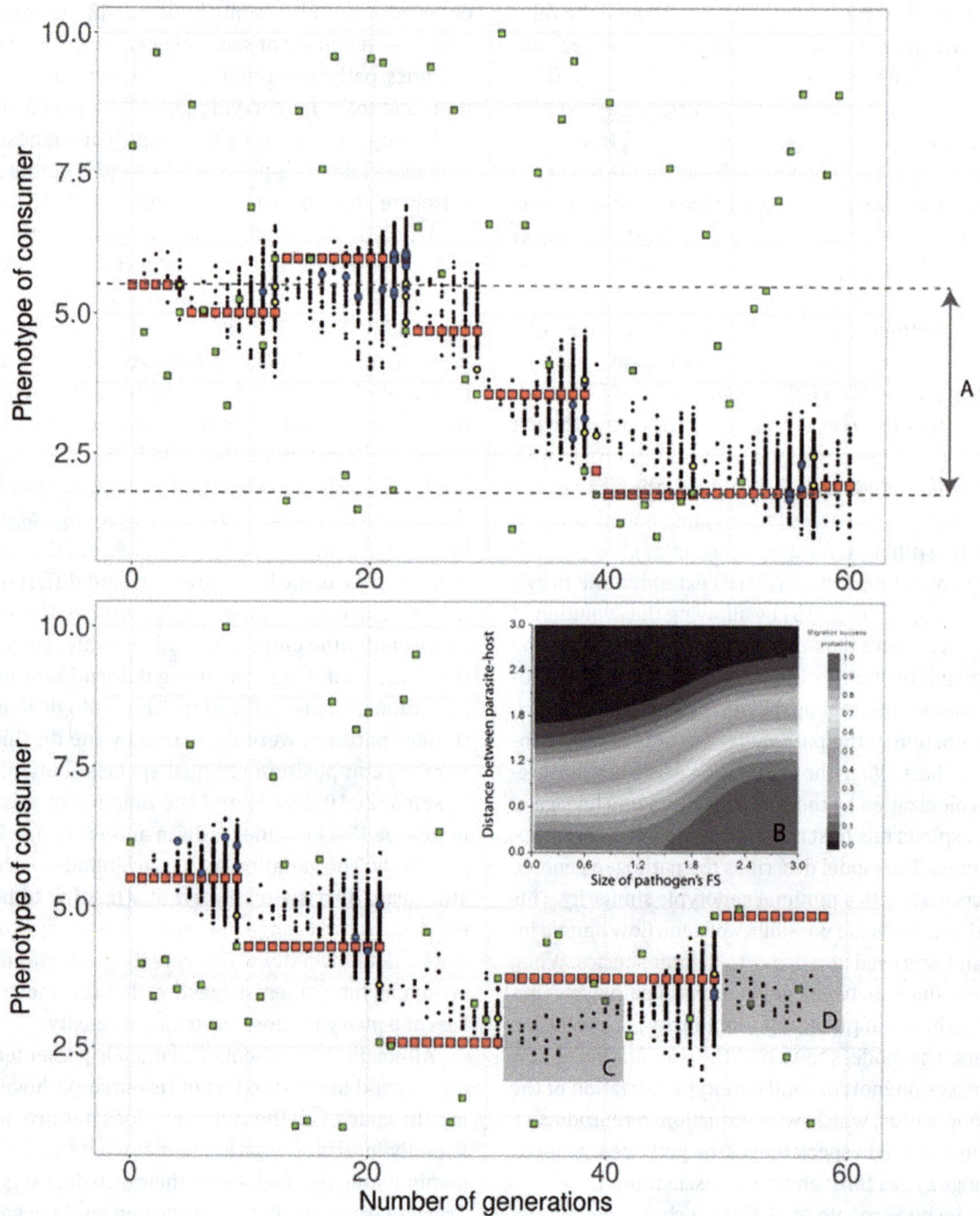

Figure 4.3. Two independent simulations of temporal evolution of the phenotype of the pathogen population (consumer). This graph was generated from "HostSwitch" package (Trivellone et al., 2021) based on the model of Araujo et al. (2015). The green squares represent the optimum phenotype of the pathogen to survive on that specific host resource; red squares are host resources in use at that moment; black dots are the pathogen's phenotype. A = distance between the first and final host in a stepping-stone chain of host expansion; B = mean values of successful colonization of new host resources according to the distance pathogen-host and the size of the pathogen's Fitness Space; C and D = individual parasite phenotypes surviving for many generations of a not-ideal host.

Insights derived from the simulations

Connecting the elements of the Stockholm Paradigm through complexity levels

The series of models developed under the framework of the SP strongly suggest the recognition that the eco-evolutionary dynamics of infectious diseases represent a complex system. Species of such associations are never playing in pairs but in a network of interactions among many other species, something that creates a level of complexity such that its behavior cannot be understood nor predicted easily. The rules of the "game" of interactions may be simple, but complexity comes from interaction among multiple agents. Published models (Araujo et al., 2015; Braga et al., 2018; D'Bastiani et al., 2021) indicate that at least one of the elements of the SP, oscillation, is a putative emergent property of communities in which interactions are driven by EF. Taxon pulse is the interaction between the increased opportunity associated with environmental disruptions, and thus it likely also represents an emergent property in this chain of complexity—an emergent property resulting from the interaction of communities with potential for oscillation and an unstable environment.

By exploring the available capacity represented by *sloppy fitness space* (SFS) (Agosta and Klemens, 2008; Agosta et al., 2010), a pathogen can colonize new hosts, exploiting new elements of a community (i.e., the resources offered by hosts). In the simulation presented in Araujo et al. (2015), some of the individuals in the pathogen population try to colonize a new host, but just a fraction succeeds. Consequently, the FS of the population in this new host is reduced compared to the original host. During exploitation of the new host, the accumulation of evolutionary novelties often resulted in an increase of the FS but presented a distinct nature according to the influence of the new host-associated selection. This oscillatory nature of the simulated FS in Araujo et al. (2015) strongly suggested the emergence of oscillations (Janz and Nylin, 2008) in host repertoire, another fundamental element of the SP (Brooks et al., 2019). To test for the evidence that evolution under EF may generate oscillations as an emergent property, in the following model, the opportunity to colonize variable hosts was constant and all pathogen populations were followed over time. The simulations of Braga et al. (2018) replicated the pattern expected from the hypothesis of oscillation proposed by Janz and Nylin (2008), in which pathogens' lineages oscillate between generalists and specialists through time. Hence, the result supports the observation that host oscillation is an emergent property of

a community of interacting species that change their ecology by EF, and that oscillation does not necessarily require the geographic vector suggested by Janz and Nylin (2008).

The evidence that the SP is composed of elements defined as fundamental (EF) and emergent properties (oscillation and taxon pulse) reveals the flexibility of the many levels of complexity of biological systems. This flexibility is far greater than that expected under the prevailing paradigm of evolutionary theory (i.e., maximum adaptation/specialization). That includes greater than expected flexibility at the metabolic (Khersonsky et al., 2006; Carbonell et al., 2011), cellular (Margulis, 1971; Alison et al., 2002), organism, population (Schradin et al., 2012), and community levels (Wilkinson, 2004; Malcicka et al., 2015; Hui and Richardson, 2018). Hence, this property of life, replicated at all levels of complexity, is certainly a fundamental element of evolution that favors the survival of life on an unstable planet and biosphere (Brooks and Agosta, 2012; Agosta and Brooks, 2020).

This relatively great flexibility of the actors involved in the complex system and the instability of the planet also increases the capacity of pathogens to explore, colonize, and exploit available hosts. That is the fundamental reason for the ongoing emerging infectious disease crisis on a planet greatly interconnected by human activities and under climate change.

The dynamics of infectious diseases under the SP

Modeling has allowed recognition of ecological and evolutionary patterns of pathogens during processes of exploration, colonization, and exploitation of host species in a continuously changing community caused by geographical, geological, climatological, and inherent biological processes (Brooks et al., 2019).

Evolution is, despite anecdotal knowledge, a highly conservative process (Gómez et al., 2010), and this most likely reflects conservatism of resources (of the host) and capacity (of the pathogen). Closely related hosts have a greater possibility of sharing the same characteristics (e.g., biochemistry, physiology, morphology) that are required by pathogens as resources. Correspondingly, closely related species of pathogens likely depict similar capacities to utilize these hosts that share traits (especially those representing the fundamental resources for the maintenance of a pathogen's infrapopulations). Combining these elements, it is evident that the history of any association is about compatibility (and potential compatibility) between the actors involved (Gilbert and Webb, 2007; De Vienne et al., 2009). However, the fit between phylogeny and compatibility is not perfect nor equally effective in determining the extent of the arena of possible host incorporation by pathogens (Gilbert and Webb, 2007) since capacity of the pathogen and the nature of the resource (a host property) can be homoplasious (e.g., subjected to convergent evolution) (Brooks and McLennan, 2002).

The IBM simulations of host repertoire expansion performed by D'Bastiani et al. (2021) assumed host phylogeny as a proxy for pathogens' colonization. They assumed that the closer the phylogenetic relationship between the donor and the recipient host species, the greater the probability of successful host expansion by the simulated parasite species. Simulated relationships resulted in scenarios compatible with empirical studies when host repertoire increase is considered (i.e., host-switching). Moreover, the authors recovered the expected pattern that host-repertoire expansion is higher on a local than regional scale. This supports the conclusion that intense exploration favors new associations. This is an expected result for a group of closely related host species—and the putative intensity of host-repertoire expansion should be smaller for an entire community composed by variably related hosts. The study also supports the conclusion of Braga et al. (2018) that although evolution of pathogens within a community may generate cycles of oscillation in host repertoire (i.e., specialization), and that these tend to stabilize as closely related hosts, that is, those bearing a similar nature of resources are colonized and exploited (Brooks et al., 2019).

This scenario of multihost dynamics is compatible with the accumulated knowledge on the ecology of associations (Nylin et al., 2018; Brooks et al., 2019; Agosta and Brooks, 2020), and it recently became more conspicuous in the ongoing SARS-CoV-2 pandemic (Fenollar et al., 2021; Boeger et al., 2022; Hoberg et al., 2022; Kuchipudi et al., 2022). SARS-CoV-2 further exposed the importance of opportunity, especially those derived from human activities, in the dynamics of antagonistic associations (Hoberg et al., 2022).

Temporal variation in the presence or in the levels of permeability of barriers among communities can result in changes in species distributions and, as a consequence of this variability, large- or small-scale changes can occur in the structure of ecological networks, which offer new opportunities for the emergence of new ecological interactions (Hoberg and Brooks, 2010). This scenario can facilitate and even enable an unmeasurable intensity of change in the opportunity of encounter—in time, space, and ecology—between pathogen and host species. However, species (including actual or potential hosts) in a community are usually not each other's closest relatives and, thus, communities present different combinations of pathogens and resources in both quality and quantity (Figure 4.4.1).

Whenever opportunity exists, pathogens are continuously probing the nature of the resource provided by different host species within a community (Figure 4.4.2). Some explorations (exploratory infections) are successful, resulting in colonization and exploitation (Figure 4.4.3), but most are likely not (Figure 4.4.3—extinction of the red circle in the dark pink host). Expected differences in the success of colonization of new hosts are dependent on a series of factors that influence compatibility among pathogens and hosts, and many of these have been revealed by the simulations generated with IBMs (Araujo et al., 2015; Braga et al., 2018; Feronato et al., 2021).

Both stochastic and deterministic processes are involved in shaping the success of each colonization attempt of compatible hosts (Araujo et al., 2015; Feronato et al., 2021). One of the most significant results of the simulation of Araujo et al. (2015) indicates that even when the phenotype amplitude of a pathogen population is null, colonization of other host species with the same or slightly different compatibility (i.e., hosts presenting different resource quality and/or quality) is still possible (Figure 4.3). That same simulation also suggested that there is an upper limit to the extent of successful colonization—that is, hosts may represent a set of resources too distant from that of the original donor species to be successfully colonized. In this case, exploration occurs but colonization—and hence exploitation—is not achieved, and the pathogen cannot exploit the resource provided by the host (Figure 4.3, inset B).

However, empirical evidence and simulations have recovered a process that makes it possible for pathogens to indirectly colonize hosts bearing relatively distant resources. This is known as the host-repertoire expansion by stepping-stone (Araujo et al., 2015; Braga et al., 2015). In fact, stepping-stone embraces distinct processes, involving either opportunity or capacity. Hosts, intermediary in the chain of transmission, may favor contact between compatible host species. For instance, while bats are important reservoirs for zoonotic viruses (Calderon et al., 2016), the opportunity of contact with humans is limited and transmission often occurs through other hosts that bridge ecological and spatial distances between viruses and humans (Hoberg et al., 2022). There is no logical reason to believe that this type of transmission involves only a single species intermediary in the expansion to new hosts. Given time, especially for microorganisms that depict fast evolutionary rates, stepping-stone may also facilitate colonization of distant host resources through changes in the nature of the *capacity space* (CS) and FS of the pathogen in response to gradual and sequential influence of the selection imposed by hosts intermediary in the process (Figure 4.3) (Araujo et al., 2015; Brooks et al., 2019).

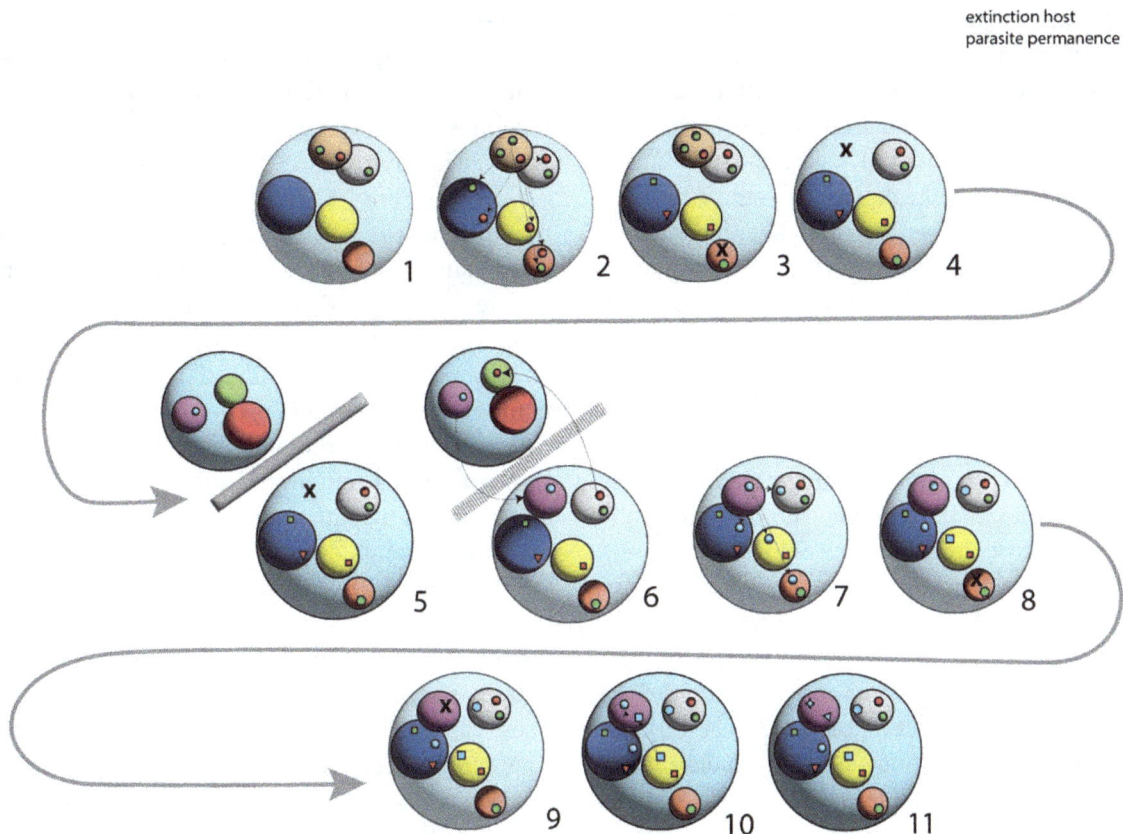

Figure 4.4. The putative dynamics of diseases under the perspective of the Stockholm Paradigm. **(1)** In an isolated community, five host species are present, two of which are in association with pathogens (the orange species with two pathogen lines—green and red—and the ash also parasitized by the green pathogen. **(2)** Still in isolation, pathogens exploit the species available when the opportunity presents itself. **(3)** Exploration may result in successful or nonsuccessful colonization (failure to establish the red pathogen represented by an X). The strains capable of surviving and reproducing in the new hosts eventually differ from their ancestors (pathogens with new shapes represent descendent strains of the ancestral forms with the same color). **(4)** The loss of a host (X) does not imply co-extinction of the strains of pathogens with which it was associated because the same pathogen lineage may be associated with more than one host species. **(5)** Two communities remain isolated by a barrier (gray bar). **(6)** The loss or increase of barrier permeability allows migration of hosts and pathogens between communities. **(7)** A new phase of exploration of new hosts takes place. **(8)** During the period of stability and exploitation of hosts whose association was established, new diversification events occur. **(9)** The loss (X) of pathogens, natural or human mediated, can happen. **(10)** However, retrocolonization from a descending variant (light blue square) present in the population and that has retained the ability to survive in the ancestral host can happen. **(11)** A new period of stability follows. No demography is represented here.

Successful colonization is also strongly associated with inherent and demographic properties and processes of the pathogen attempting to populate the new hosts. Feronato et al. (2021) varied key demographic features—reproductive rate, rate of emergence of evolutionary novelties (e.g., mutations), and propagule pressure—for simulated pathogens and concluded that propagule pressure was the most important in determining the success of colonization of new hosts. Contrary to what is commonly assumed, the rate of emergence of new evolutionary novelties of the pathogen species was shown to be less important to ascertain the success of colonization. However, synergy among these simulated parameters maximizes the colonization and apparently provides explanatory evidence for the observed success of microorganisms in expanding to new hosts. Maximizing the values of the evaluated parameters during simulation results in an unexpected increase in the success of colonization of hosts representing resources of variable compatibility by parasites that are prolific, present high mutation rates, and generate large propagule sizes, such as viruses.

Once the process of colonization of a new host species is successful, the pathogen population may have different outcomes, depending on the heterogeneity of the resources offered by the parasitized hosts (Braga et al., 2018), and it is strongly dependent on the lagload (Smith, 1976) imposed by the new host species. Lagload (Smith, 1976) originates by differences in the nature of the new resource being explored when compared to the donor host—the difference in selective pressure between the original and the newly colonized host species. Change in lagloads may result in the qualitative and quantitative accumulation of evolutionary novelties (mutations for viruses, for instance) in the pathogen (Bashor et al., 2021, 2022). If there are no significant differences between the nature of the resources, the pathogen may not diverge rapidly from its original profile (e.g., genetic, phenotypic) unless demographic processes take place (i.e., intense bottleneck following isolation in the new host). In this case, from the view of the observer, the pathogen has simply expanded its host repertoire (*sensu* Braga and Janz, 2021) (Figure 4.4.3—the green and red circles). However, if the pathogen can be subjected to a sufficiently strong lagload that may impose relocation of the *realized fitness space* (RFS) within the FS (Figure 4.5), the accumulation of evolutionary novelties (e.g., mutations in viruses) can generate new variants or even new species (Figure 4.4.3—red triangle and square, and green square). This scenario is also well represented by the dynamics of emergence of variants of SARS-CoV-2 (Boeger et al., 2022; Kuchipudi et al., 2022). Boeger et al. (2022) suggest that long branches in the phylogeny of selected SARS-CoV-2 sequences of the spike protein is evidence of faster evolutionary rates imposed by a larger lagload that originated from the virus colonization of new mammal host species.

From Figure 4.4.1 to Figure 4.4.4, pathogens are alternating between exploring, colonizing, and exploiting—under stable opportunity within an isolated community. The simulations (Araujo et al., 2015; Braga et al., 2018) strongly suggest that this dynamic results from the cyclic variation of the capacity of the pathogen—that is, variation in FS—a consequence of demographic processes associated with colonization of new host resources. This fluctuation of diversity in the CS of pathogens is registered in species of viruses (Sacristán et al., 2003; Holmes, 2009; Ali and Roossinck, 2010), but the simulations suggest that this may be a common feature in the processes of host repertoire expansion for all other symbiotic species (Moxon and Kussell, 2017; Pérez et al., 2019; Techer et al., 2021).

Among other consequences of host expansion within a community, pathogens with a large host repertoire have a greater chance of survival even when the population of one or some of these hosts become locally extinct (Figure 4.4.4—the X marks the extinction of the orange host population). Pathogens that exploit more than a single species of host may survive the extinction event by persisting in other host species (Figure 4.4.4—the red and green circles survived in the gray host), even if they are marginally fit to the surviving host. That entire process certainly is important to maximize permanence of pathogen species within a community, sometimes at expected low prevalence, often undetected by traditional sampling efforts.

However, we predict that cycles of oscillation within an isolated community stabilize through time. Most likely, pathogen exploration (i.e., probing) of new host species never ceases, but successful colonization decreases in rate through time as compatible hosts become colonized and are exploited by pathogens. Furthermore, evolution has generated enough diversity in the nature of the resource (i.e., hosts) in such magnitude that many hosts are never reached by or exposed to a specific lineage of pathogen, either directly or by a stepping-stone process (despite maximized opportunity) (see, for instance, Braga et al., 2014). These gaps in the nature of the resource within a biological community likely results from differences in the historical pathway of lineages of hosts and consequent historical constraints of the CS of the pathogens and community assemblage and composition. However, no community is perfectly isolated, and even during periods of considerable stability, the dynamics may be resumed through the introduction of new species from other communities (Figure 4.4.6). Hence, the special concern by health authorities with migratory birds, invasive species, human traveling, and species translocations (Pinder et al., 2005; Peeler et al., 2006; Hoberg, 2010; Conn, 2014).

The simulations of Araujo et al. (2015) provide another alarming insight regarding pathogens breaking sanitary barriers. As previously mentioned, this model suggests that pathogens may survive for many generations on hosts that represent only marginally adequate resources (Figure 4.3, C and D; Figure 4.5). This result also suggests that even if evolutionary novelties that favor the exploitation of the new resource never emerge, this does not preclude the continuous use of the host species by a small population of the pathogen (low prevalence and low intensity of infection) (see Figure 4.3, C and D). Under this scenario, pathogen populations may be small because of strong selective pressure within this host, and as a consequence, pathogen detection by sanitary inspections is hampered. However, Feronato et al. (2021) indicated that even in the absence of new evolutionary novelties (e.g., mutations)—hence, without the possibility of generating a more fit pathogen population—the

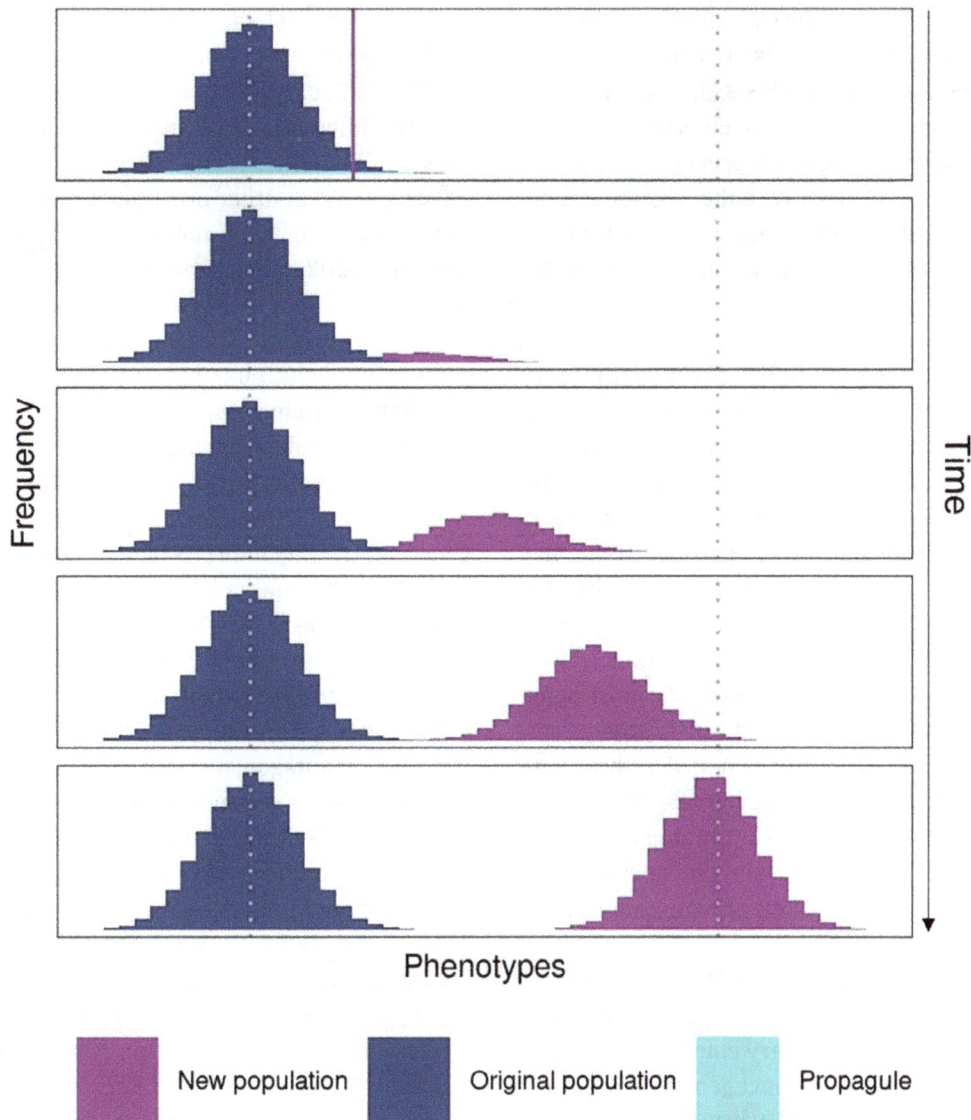

Figure 4.5. Evolution of the phenotypic profile of a population of pathogens after host switch. A portion (propagule) of the original population of pathogens can migrate to a new host. Only a fraction of these individuals survives the new selective pressure, and the mean phenotype of the survivors is identified by the pink vertical line. Over time the new population has an increase in the number of individuals and, because of the new selective pressure imposed by the new host (dotted line), the phenotypic population profile is directed to the optimal value imposed, stabilizing around it.

pathogen population may, with time, reach similar populational size to those with greater fitness.

This same outcome of the simulations of Araujo et al. (2015) also provides a potential explanation for the fallacious conclusion that the emergence of new diseases is associated with the evolution of new genetic strains of the pathogen species. Although a common belief, emergence due to new mutations does not appear to be the case for many EID evaluated by Morse (2001); this study concluded

that most emergences appear to be associated with increasing opportunity. As distribution of fitness of the pathogen strains in the original host is likely not uniform (Figure 4.5), by chance or because of selective differences, marginally fit, low-frequency strains may have a better opportunity to explore and exploit new host species, some of which may bear distinct but exploitable resources by EF. Since the probability of sampling marginal and low-frequency variants of the pathogen in its original host is comparatively smaller, an inadequate sampling scheme may conclude that the pathogen is absent in that host while it is present in larger frequencies in the new host species. This is likely the main reason for the fallacious assumption that emergences of new, or previously unknown pathogens in previously unutilized hosts are necessarily associated with the "right mutation" occurring just at the opportune time, which basically extends the question of Kellogg (1907) to parasites or pathogens in general.

Environmental and ecological disruptions can promote changes in the relative permeability of ecological barriers (Figure 4.4.6), creating or increasing interfaces among communities (or systems)—opportunity for encounters between symbionts previously not in sympatry. This process may increase the probability of contact among previously isolated potential hosts and pathogens that had been maintained in geographic separation in different communities or habitats. New opportunity drives new cycles of exploration, colonization, and exploitation (Figure 4.4.6–4.4.10). Currently, humans are likely the most significant and consistent agent of ecological disruption, transporting pathogens throughout the planet, directly or indirectly (Boeger et al., 2022). Besides humans themselves, inserting populations in literally all biomes on Earth, SARS-CoV-2 is perhaps the most convincing example of this process, and it is clear that we are super-spreaders of diseases (Hoberg et al., 2022).

In recent time, even before the COVID-19 pandemic, we had hints of our great influence on the spreading of diseases—including the contemporary emergence of ZikaV, dengue, and chikungunya in distinct geographical areas. However, the spatial and temporal behavior of SARS-CoV-2 revealed an unexpected dynamic of host use. While the involvement of other mammal species in the epidemiology of SARS-CoV-2 has been reported since the beginning of the COVID-19 pandemic, most researchers ignored it, assuming a more traditional perspective on the evolution of pathogens—that is, that pathogens are highly specialized and incapable of crossing host barriers easily, except when releasing mutations occur (CDC, 2022). The SP predicted (Agosta et al., 2010; Brooks et al., 2014, 2019; Hoberg and Brooks, 2015), and it is recently becoming empirically evident (Fenollar et al., 2021; Bashor et al., 2022; Kuchipudi et al., 2022; Mallapaty, 2022), that nonhuman mammals (at least) likely play a significant role not only as reservoirs for the virus in urban, peri-urban, and wildlife systems but also in the origin of new variants (Boeger et al., 2022; Hoberg et al., 2022). The above-proposed dynamics for SARS-CoV-2 was likely replicated in many regions of the planet, involving a much larger number of species than we presently know (Boeger et al., 2022).

While allopatry is often considered the most common mode of differentiation—the generation of new species or variants (Fitzpatrick et al., 2009)—mathematical models have shown that other processes are also likely to occur (de Aguiar et al., 2009; Yamaguchi and Iwasa, 2017; Princepe et al., 2022). Princepe et al. (2022), for instance, using a neutral model for two islands, demonstrated that migration promotes species diversification through two processes: (1) the founding population has no ability (similarity) to reproduce with resident species/variant, creating a new population with, at least in the beginning, independent evolution; (2) the founding population can reproduce with resident species/variant, introducing new genetic variability in the resident population. In this latter case, sympatric speciation is induced by migration. These results highlight the fact that the contact between variants increases the probability of new variants emerging and should be considered seriously during the management of pandemics. The proposed generalized model for the dynamics of antagonistic associations (Figure 4.4)—such as diseases—are dependent on both historical (time) and spatial (distribution) processes. Unfortunately, because of tradition, we have not approached the problem of epidemiology of diseases by integrating all these elements. This has hampered the way we understand and deal with emerging and reemerging diseases.

For instance, the question whether biodiversity influences amplification or dilution (Clay et al., 2009; Keesing et al., 2010; Ostfeld and Keesing, 2012; Rohr et al., 2019) of pathogens, thus affecting the emergence or reemergence of diseases in humans, has been in discussion for some time now. Ecological factors are often assumed to influence the observed patterns (Luis et al., 2018), but at least part of the answer may be associated with the dynamics of pathogens through time and under the influence of environmental disruptions, as synthesized in Figure 4.4. It is intuitive to recognize that the dilution effect may result from the early process of oscillation when species are exploring hosts within the limits of opportunity and capacity—presenting larger host-range but low levels of parasitism. Otherwise, during exploitation, pathogens specialize and diverge, each lineage occupying now one or a limited number of hosts within the community at higher prevalence levels (see, for instance, Patella et al., 2017), maximizing the number of propagules and amplifying exploration and, hence, the emergence of diseases in species newly introduced in the community (i.e., us, new crop or livestock). Hence, the answer to the dilution/amplification paradox may be eco-evolutionary and needs to be evaluated in this way in the future.

Putting the insights to work

These insights derived from the models and supported by empirical data also provide important elements that should be considered when applying the DAMA protocol (Brooks et al., 2014, 2019; Boeger et al., 2022; Hoberg, 2022; Molnár et al., 2022; Trivellone et al., 2022). For instance, it is not enough to **document** (D of DAMA) the biodiversity of pathogens associated with known host species. As suggested by the models and in consonance with accumulated empirical data, pathogens may reach us and species of our direct (and indirect) interest by several ways, including stepping-stone, recolonization, convergent or plesiomorphic nature of the resource, or simply by historically changing its own capacity to explore new and more distant host species. **Hence, prospective efforts should not be limited to pathogens nor to a group of species closely related to the focal host species (e.g., us) within a community but expanded to all those that may be involved in the previously described processes of colonization.**

The theoretical and empirical evidence that stepping-stone host expansion occurs shows a necessity of a more comprehensive knowledge on the composition of potential host species of a pathogen. Hence, the need to also recognize the composition of potential hosts within a community since these may provide the conditions (either associated to capacity or opportunity) for pathogens to reach focal host species. This is a counterintuitive conclusion contrasting with the proposal that biodiversity constrains the emergence of infectious disease (see Keesing et al., 2010). In fact, it is not the richness of species that may facilitate the emergence of diseases but the composition of phylogenetically close species of hosts in a community that may result in a slow but effective process for pathogen lineages to reach distant host resources by stepping-stone (Braga et al., 2014).

For instance, mammal species living in a same geographic area in the same or close communities may represent elements in the chain toward colonization of ecologically distant mammal hosts. Indeed, the origin of the Omicron variant (and likely of many others—Boeger et al., 2022) is thought by many to have been the result of exploring and exploiting of different host mammals (Wei et al., 2021; Kuchipudi et al., 2022).

Increased capacity to reach new hosts within a community may also be a matter of increasing the pathogen's FS through time (Araujo et al., 2015). For microorganisms, especially those with large mutation rates, this may signify a short period of time as erroneously perceived by us (Manrubia, 2012; Sprouffske et al., 2018). The putative cycle of reduction and increase in FS of such pathogens is expectedly fast, and exploration of available hosts should result in many cases of serial and successful colonizations and exploitations by EF. **Monitoring the ecological and evolutionary dynamics** of change of pathogens is, thus, fundamental.

The recent crisis generated by the SARS-Cov2 virus is an excellent example of the potential that successful encounters between a pathogen and compatible hosts can generate (Boeger et al., 2022). Most likely, access to a susceptible individual triggered colonization of humans followed by a quick spreading throughout the world catalyzed by demographics and connectivity of our species (Hoberg et al., 2022). We highlight two lessons from the SARS-CoV-2 pandemic: (1) humans are not detached from nature; we are at the mercy of ecological and evolutionary processes like any other species in the biosphere; and (2) technology did not allow us to react efficiently when a threat of this magnitude presented itself. This is the first major pandemic in an era of high technology and communication, and even with scientists around the world working to develop ways to minimize its effects, we were unable to save 6.28 million lives (WHO, 2022). However, studies of COVID-19 generated an enormous amount of data that can provide a more comprehensive understanding of the dynamics of diseases (emergent and reemergent) through testing of theoretical developments proposed recently, such as the SP.

Indeed, SARS-CoV-2 revealed how the evolutionary dynamics of the virus influenced the epidemiological characteristics of the disease. Substantial evidence shows that SARS-CoV-2 can expand into other mammal species by EF in spite of minor differences in the nature of the membrane-bound angiotensin-converting enzyme 2 (ACE2) receptor (Damas et al., 2020). Under different lagloads (i.e., selective pressure), the virus population may change, but it likely retains the ability to rapidly recolonize humans (Boeger et al, 2022; Hoberg et al., 2022) supporting the suggestion that rapidly evolving pathogens, such as viruses, can augment their capacity to reach more diverse resources (Araujo et al., 2015; Braga et al., 2018) (and hosts) but still preserve the ability to return to the original host species (Feronato et al., 2021).

Hence, the **document** step of the DAMA protocol **needs to be continuous**, combined with the equally continuous **monitoring** step (the M of DAMA) since the scale of evolution of many pathogens is many times faster than that of their actual and potential host species within a community. **Exploration and exploitation are thought to continuously renew the risk space** (i.e., the sum of all potential pathogens of the focal species in a community) through evolution. In this scenario of variable lagloads (i.e., selective scenarios), pathogens may rapidly change and be recognized, upon return to the focal species, as a new variant or even as a new species (Boeger et al., 2022).

Sampling schemes for documenting and monitoring should use **effective and sensitive sampling protocols** to reveal the total variability of pathogens and its distribution in a community. The simulations have revealed the significance of low-frequency variants of pathogens in the colonization of new host resources (Araujo et al, 2015). These variant pathogens are often greatly concealed within local hosts until opportunity favors colonization of other hosts species, often causing the emergence of new diseases. Finally, monitoring of pathogens and hosts demographics are fundamental due to the evidence that **propagule pressure represents the most influential characteristic of pathogens to accomplish colonization of new host resources** (Feronato et al., 2021).

While **assessing** (Assess = first A of DAMA) the potential of pathogens to cause emergences, all these factors just mentioned need to be considered. These same factors will determine the compatibility and probability of encounter and emergence of new antagonistic associations. Thus, **Assess** is not as simple as analyzing the phylogenetic relationship of unknown pathogens with their known relatives to determine their zoonotic potential—although this is an important part of this process.

Using an analogy to illustrate the potential zoonoses we live with: we are living in a minefield in which new mines are being installed and replaced continuously. We predict that **the risk space for focal species will never reduce, and the pathogen capacity to colonize new hosts will increase over time and as evolution continues.** The DAMA protocol provides the continuous feedback for adjustments of the **Act** element of DAMA.

Models help us understand the dynamics of the diseases based on the elements of the SP. However, they may also represent important assets to provide anticipatory scenarios for specific pathogens under the opportunity provided by environmental and human-related disruptions. Hence, these models and future models may confer predictive capacity that will be key in the design of specific methodology associated with DAMA.

Literature Cited

Agosta, S.J. 2006. On ecological fitting, plant-insect associations, herbivore host shifts, and host plant selection. Oikos 114, 556–565.

Agosta, S.J.; Brooks, D.R. 2020. The Major Metaphors of Evolution: Darwinism Then and Now. Evolutionary Biology—New Perspectives on Its Development series, vol. 2. Springer, Cham, Switzerland. https://doi.org/10.1007/978-3-030-52086-1

Agosta, S.J.; Janz, N.; Brooks, D.R. 2010. How specialists can be generalists: resolving the "parasite paradox" and implications for emerging infectious disease. Zoologia (Curitiba) 27: 151–162. https://doi.org/10.1590/S1984-46702010000200001

Agosta, S.J.; Klemens, J.A. 2008. Ecological fitting by phenotypically flexible genotypes: implications for species associations, community assembly and evolution. Ecology Letters 11: 1123–1134. https://doi.org/10.1111/j.1461-0248.2008.01237.x

Ali, A.; Roossinck, M.J. 2010. Genetic bottlenecks during systemic movement of Cucumber mosaic virus vary in different host plants. Virology 404: 279–283. https://doi.org/10.1016/j.virol.2010.05.017

Alison, M.R.; Poulsom, R.; Forbes, S.; Wright, N.A. 2002. An introduction to stem cells. Journal of Pathology 197: 419–423. https://doi.org/10.1002/path.1187

Altizer, S.; Ostfeld, R.S.; Johnson, P.T.J.; Kutz, S.; Harvell, C.D. 2013. Climate change and infectious diseases: from evidence to a predictive framework. Science 341: 514–519. https://doi.org/10.1126/science.1239401

Araujo, S.B.; Braga, M.P.; Brooks, D.R.; Agosta, S.J.; Hoberg, E.P.; von Hartenthal, F.W.; Boeger, W.A. 2015. Understanding host-switching by ecological fitting. PLOS ONE 10: e0139225. https://doi.org/10.1371/journal.pone.0139225

Baselga, A. 2010. Partitioning the turnover and nestedness components of beta diversity. Global Ecology and Biogeography 19: 134–143. https://doi.org/10.1111/j.1466-8238.2009.00490.x

Baselga, A. 2013. Separating the two components of abundance-based dissimilarity: balanced changes in abundance vs. abundance gradients. Methods in Ecology and Evolution 4: 552–557. https://doi.org/10.1111/2041-210X.12029

Bashor, L.; Gagne, R.B.; Bosco-Lauth, A.; Stenglein, M.; VandeWoude, S. 2022. Rapid evolution of SARS-CoV-2 in domestic cats. Virus Evolution: veac092. https://doi.org/10.1093/ve/veac092

Bashor, L.; Gagne, R.B.; Bosco-Lauth, A.M.; Bowen, R.A.; Stenglein, M.; VandeWoude, S. 2021. SARS-CoV-2 evolution in animals suggests mechanisms for rapid variant selection. Proceedings of the National Academy of Sciences 118: e2105253118. https://doi.org/10.1073/pnas.2105253118

Blum, M.G.B.; François, O. 2005. On statistical tests of phylogenetic tree imbalance: the Sackin and other indices revisited. Mathematical Biosciences 195, 141–153. https://doi.org/10.1016/j.mbs.2005.03.003

Boeger, W.A.; Brooks, D.R.; Trivellone, V.; Agosta, S.; Hoberg, E. 2022. Ecological super-spreaders drive host-range oscillations: Omicron and risk-space for emerging infectious disease. Illinois Experts reprint. https://doi.org/10.22541/au.164342794.41467213/v1

Braga, M.P.; Araujo, S.B.L.; Agosta, S.; Brooks, D.; Hoberg, E.; Nylin, S.; et al. 2018. Host use dynamics in a heterogeneous fitness landscape generates oscillations in host range and diversification. Evolution 72: 1773–1783. https://doi.org/10.1111/evo.13557

Braga, M.P.; Araújo, S.B.L.; Boeger, W.A. 2014. Patterns of interaction between Neotropical freshwater fishes and their gill Monogenoidea (Platyhelminthes). Parasitology Research 113: 481–490. https://doi.org/10.1007/s00436-013-3677-8

Braga, M.P.; Janz, N. 2021. Host repertoires and changing insect-plant interactions. Ecological Entomology 46: 1241–1253. https://doi.org/10.1111/een.13073

Braga, M.P.; Razzolini, E.; Boeger, W.A. 2015. Drivers of parasite sharing among Neotropical freshwater fishes. Journal of Animal Ecology 84: 487–497. https://doi.org/10.1111/1365-2656.12298

Brooks, D.R.; Agosta, S.J. 2012. Children of time: the extended synthesis and major metaphors of evolution. Zoologia (Curitiba) 29: 497–514. https://doi.org/10.1590/S1984-46702012000600002

Brooks, D.R.; Boeger, W.A. 2019. Climate change and emerging infectious diseases: evolutionary complexity in action. Current Opinion in Systems Biology 13: 75–81.

Brooks, D.R.; Ferrao, A.L. 2005. The historical biogeography of co-evolution: emerging infectious diseases are evolutionary accidents waiting to happen. Journal of Biogeography 32: 1291–1299. https://doi.org/10.1111/j.1365-2699.2005.01315.x

Brooks, D.R.; Hoberg, E.P. 2007. How will global climate change affect parasite-host assemblages? Trends in Parasitology 23: 571–574. https://doi.org/10.1016/j.pt.2007.08.016

Brooks, D.R.; Hoberg, E.P. 2013. The emerging infectious diseases crisis and pathogen pollution. In: The Balance of Nature and Human Impact. K. Rohde (ed.). Cambridge University Press, Cambridge, England. 215–230 p. https://doi.org/10.1017/CBO9781139095075.022

Brooks, D.R.; Hoberg, E.P.; Boeger, W.A. 2019. The Stockholm Paradigm: Climate Change and Emerging Disease. University of Chicago Press, Chicago.

Brooks, D.R.; Hoberg, E.P.; Boeger, W.A.; Gardner, S.L.; Galbreath, K.E.; Herczeg, D.; et al. 2014. Finding them before they find us: informatics, parasites, and environments in accelerating climate change. Comparative Parasitology 81: 155–164.

Brooks, D.R.; McLennan, D.A. 2002. The Nature of Diversity: An Evolutionary Voyage of Discovery. University of Chicago Press, Chicago.

Calderon, A.; Guzman, C.; Salazar-Bravo, J.; Figueiredo, L.T.; Mattar, S.; Arrieta, G. 2016. Viral zoonoses that fly with bats: a review. MANTER Journal of Parasite Biodiversity 6. https://doi.org/10.13014/K2BG2KWF

Carbonell, P.; Lecointre, G.; Faulon, J.-L. 2011. Origins of specificity and promiscuity in metabolic networks. Journal of Biological Chemistry 286: 43994–44004. https://doi.org/10.1074/jbc.M111.274050

CDC [Center for Disease Control and Prevention]. 2022. What you should know about COVID-19 and pets. [WWW document]. Accessed April 27, 2022. https://www.cdc.gov/healthypets/covid-19/pets.html

Christaki, E. 2015. New technologies in predicting, preventing and controlling emerging infectious diseases. Virulence 6: 558–565. https://doi.org/10.1080/21505594.2015.1040975

Cipollini, D.; Peterson, D.L. 2018. The potential for host switching via ecological fitting in the emerald ash borer–host plant system. Oecologia 187: 507–519. https://doi.org/10.1007/s00442-018-4089-3

Clay, C.A.; Lehmer, E.M.; Jeor, S.; Dearing, M.D. 2009. Sin Nombre virus and rodent species diversity: a test of the dilution and amplification hypotheses. PLOS ONE 4: e6467. https://doi.org/10.1371/journal.pone.0006467

Combes, C. 2001. Parasitism: The Ecology and Evolution of Intimate Interactions. University of Chicago Press, Chicago.

Conn, D.B. 2014. Aquatic invasive species and emerging infectious disease threats: a One Health perspective. Aquatic Invasions 9: 383–390. https://doi.org/10.3391/ai.2014.9.3.12

Dada, J.O.; Mendes, P. 2011. Multi-scale modelling and simulation in systems biology. Integrative Biology 3: 86–96. https://doi.org/10.1039/c0ib00075b

Damas, J.; Hughes, G.M.; Keough, K.C.; Painter, C.A.; Persky, N.S.; Corbo, M.; et al. 2020. Broad host range of SARS-CoV-2 predicted by comparative and structural analysis of ACE2 in vertebrates. Proceedings of the National Academy of Sciences 117: 22311–22322. https://doi.org/10.1073/pnas.2010146117

Darwin, C. 1872. The Origin of Species. 6th ed. John Murray, London.

D'Bastiani, E.; Princepe, D.; Marquitti, F.M.D.; Boeger, W.A.; Campião, K.M.; Araujo, S.L.B. 2021. Effect of host-switching on the eco-evolutionary patterns of parasites. bioRxiv preprint. https://doi.org/10.1101/2021.11.27.470149

de Aguiar, M.A.M.; Baranger, M.; Baptestini, E.M.; Kaufman, L.; Bar-Yam, Y. 2009. Global patterns of speciation and diversity. Nature 460: 384–387. https://doi.org/10.1038/nature08168

de Vienne, D.M.; Hood, M.E.; Giraud, T. 2009. Phylogenetic determinants of potential host shifts in fungal pathogens. Journal of Evolutionary Biology 22: 2532–2541. https://doi.org/10.1111/j.1420-9101.2009.01878.x

Dieckmann, U.; Doebeli, M. 1999. On the origin of species by sympatric speciation. Nature 400: 354–357. https://doi.org/10.1038/22521

Fauci, A.S.; Morens, D.M. 2012. The perpetual challenge of infectious diseases. New England Journal of Medicine 366: 454–461. https://doi.org/10.1056/NEJM1108296

Fenollar, F.; Mediannikov, O.; Maurin, M.; Devaux, C.; Colson, P.; Levasseur, A.; et al. 2021. Mink, SARS-CoV-2, and the human-animal interface. Frontiers in Microbiology 12: 745. https://doi.org/10.3389/fmicb.2021.663815

Feronato, S.G.; Araujo, S.; Boeger, W.A. 2021. 'Accidents waiting to happen'—insights from a simple model on the emergence of infectious agents in new hosts. Transboundary and Emerging Diseases 69: 1727–1738. https://doi.org/10.1111/tbed.14146

Fitzpatrick, B.M.; Fordyce, J.A.; Gavrilets, S. 2009. Pattern, process and geographic modes of speciation. Journal of Evolutionary Biology 22: 2342–2347. https://doi.org/10.1111/j.1420-9101.2009.01833.x

Giacomini, H.C. 2007. Sete motivações teóricas para o uso da modelagem baseada no indivíduo em ecologia [Seven theoretical reasons for using individual-based modeling in ecology]. Acta Amazonica 37: 431–445. https://doi.org/10.1590/S0044-59672007000300015

Gilbert, G.S.; Webb, C.O. 2007. Phylogenetic signal in plant pathogen–host range. Proceedings of the National Academy of Sciences 104: 4979–4983. https://doi.org/10.1073/pnas.0607968104

Gioti, A.; Stajich, J.E.; Johannesson, H. 2013. Neurospora and the dead-end hypothesis: genomic consequences of selfing in the model genus. Evolution 67: 3600–3616. https://doi.org/10.1111/evo.12206

Gómez, J.M.; Verdú, M.; Perfectti, F. 2010. Ecological interactions are evolutionarily conserved across the entire tree of life. Nature 465: 918–921. https://doi.org/10.1038/nature09113

Gubler, D.J. 2010. The global threat of emergent/re-emergent vector-borne diseases. In: Vector Biology, Ecology and Control. P.W. Atkinson (ed.). Springer Netherlands, Dordrecht. 39–62 p. https://doi.org/10.1007/978-90-481-2458-9_4

Haldane, J.B.S. 1951. Everything Has a History. Routledge/Taylor & Francis Group, London.

Hoberg, E. 2022. The DAMA protocol, an introduction: finding pathogens before they find us. MANTER Journal of Parasite Biodiversity 21. https://doi.org/10.32873/unl.dc.manter21

Hoberg, E.P. 2010. Invasive processes, mosaics and the structure of helminth parasite faunas. Revue Scientifique et Technique—Office International des Épizooties 29: 255–272. https://doi.org/10.20506/rst.29.2.1972

Hoberg, E.P.; Boeger, W.A.; Brooks, D.R.; Trivellone, V.; Agosta, S.J. 2022. Stepping-stones and mediators of pandemic expansion—a context for humans as ecological super-spreaders. MANTER Journal of Parasite Biodiversity 18. https://doi.org/10.32873/unl.dc.manter18

Hoberg, E.P.; Brooks, D.R. 2010. Beyond vicariance: integrating taxon pulses, ecological fitting, and oscillation in evolution and historical biogeography. In: The Geography of Host-Parasite Interactions. S. Morand and B. Krasnov (eds.). Oxford University Press, Oxford, UK. 7–20 p.

Hoberg, E.P.; Brooks, D.R. 2015. Evolution in action: climate change, biodiversity dynamics and emerging infectious disease. Philosophical Transactions of the Royal Society B—Biological Sciences 370: 20130553. https://doi.org/10.1098/rstb.2013.0553

Holmes, E.C. 2009. The Evolution and Emergence of RNA Viruses. Oxford Series in Ecology and Evolution. Oxford University Press, Oxford, UK.

Hui, C.; Richardson, D.M. 2018. How to Invade an ecological network. Trends in Ecology and Evolution 34: 121–131. https://doi.org/10.1016/j.tree.2018.11.003

Hulme, P.E. 2014. Invasive species challenge the global response to emerging diseases. Trends in Parasitology 30: 267–270. https://doi.org/10.1016/j.pt.2014.03.005

Imrie, R.M.; Roberts, K.E.; Longdon, B. 2021. Between virus correlations in the outcome of infection across host species: evidence of virus by host species interactions. Evolution Letters 5: 472–483. https://doi.org/10.1002/evl3.247

Janz, N.; Nylin, S. 2008. The oscillation hypothesis of host-plant range and speciation. In: Specialization, Speciation, and Radiation: The Evolutionary Biology of Herbivorous Insects. K. Tilmon (ed.). University of California Press. 203–215 p. https://doi.org/10.1525/california/9780520251328.001.0001

Janzen, D.H., 1985. On ecological fitting. Oikos 45: 308–310. https://doi.org/10.2307/3565565

Keesing, F.; Belden, L.K.; Daszak, P.; Dobson, A.; Harvell, C.D.; Holt, R.D.; et al. 2010. Impacts of biodiversity on the emergence and transmission of infectious diseases. Nature 468: 647–652.

Kellogg, V. 1907. Darwinism Today. Holt, New York.

Khersonsky, O.; Roodveldt, C.; Tawfik, D.S. 2006. Enzyme promiscuity: evolutionary and mechanistic aspects. Current Opinion in Chemical Biology (Analytical Techniques/Mechanisms special issue) 10: 498–508. https://doi.org/10.1016/j.cbpa.2006.08.011

Kuchipudi, S.V.; Surendran-Nair, M.; Ruden, R.M.; Yon, M.; Nissly, R.H.; Vandegrift, K.J.; et al. 2022. Multiple spillovers from humans and onward transmission of SARS-CoV-2 in white-tailed deer. Proceedings of the National Academy of Sciences 119: e2121644119. https://doi.org/10.1073/pnas.2121644119

Le Roux, J.J.; Hui, C.; Keet, J.-H.; Ellis, A.G. 2017. Co-introduction vs ecological fitting as pathways to the establishment of effective mutualisms during biological invasions. New Phytologist 215: 1354–1360. https://doi.org/10.1111/nph.14593

Luis, A.D.; Kuenzi, A.J.; Mills, J.N. 2018. Species diversity concurrently dilutes and amplifies transmission in a zoonotic host-pathogen system through competing mechanisms. Proceedings of the National Academy of Sciences 115: 7979–7984. https://doi.org/10.1073/pnas.1807106115

Malcicka, M.; Agosta, S.J.; Harvey, J.A. 2015. Multi level ecological fitting: indirect life cycles are not a barrier to host switching and invasion. Global Change Biology 21: 3210–3218. https://doi.org/10.1111/gcb.12928

Mallapaty, S. 2022. How sneezing hamsters sparked a COVID outbreak in Hong Kong. Nature (News, February 4, 2022). https://doi.org/10.1038/d41586-022-00322-0

Manrubia, S.C. 2012. Modelling viral evolution and adaptation: challenges and rewards. Current Opinion in Virology 2: 531–537. https://doi.org/10.1016/j.coviro.2012.06.006

Margulis, L. 1971. Symbiosis and evolution. Scientific American 225: 48–57. https://doi.org/10.1038/scientificamerican0871-48

Molnár, O.; Knickel, M.; Marizzi, C. 2022. Taking action: turning evolutionary theory into preventive policies. MANTER Journal of Parasite Diversity 28. https://doi.org/10.32873/unl.dc.manter28

Morens, D.M.; Folkers, G.K.; Fauci, A.S. 2004. The challenge of emerging and re-emerging infectious diseases. Nature 430: 242–249.

Morse, S.S. 2001. Factors in the emergence of infectious diseases. In: Plagues and Politics. A.T. Price-Smith (ed.). Global Issues series. Palgrave Macmillan, London. 8–26 p. https://doi.org/10.1057/9780230524248_2

Moxon, R.; Kussell, E. 2017. The impact of bottlenecks on microbial survival, adaptation, and phenotypic switching in host-pathogen interactions. Evolution 71: 2803–2816. https://doi.org/10.1111/evo.13370

Nylin, S.; Agosta, S.; Bensch, S.; Boeger, W.A.; Braga, M.P.; Brooks, D.R.; et al. 2018. Embracing colonizations: a new paradigm for species association dynamics. Trends in Ecology and Evolution 33: 4–14. https://doi.org/10.1016/j.tree.2017.10.005

Ostfeld, R.S.; Keesing, F. 2012. Effects of host diversity on infectious disease. Annual Review of Ecology, Evolution, and Systematics 43: 157–182. https://doi.org/10.1146/annurev-ecolsys-102710-145022

Patella, L.; Brooks, D.R.; Boeger, W.A. 2017. Phylogeny and ecology illuminate the evolution of associations under the Stockholm Paradigm: Aglaiogyrodactylus spp. (Platyhelminthes, Monogenoidea, Gyrodactylidae) and species of Loricariidae (Actinopterygii, Siluriformes). Vie et Milieu 67: 91–102.

Peeler, E.; Thrush, M.; Paisley, L.; Rodgers, C. 2006. An assessment of the risk of spreading the fish parasite Gyrodactylus salaris to uninfected territories in the European Union with the movement of live Atlantic salmon (Salmo salar) from coastal waters. Aquaculture 258, 187–197. https://doi.org/10.1016/j.aquaculture.2005.07.042

Pérez, S.D.; Grummer, J.A.; Fernandes-Santos, R.C.; José, C.T., Medici, E.P., Marcili, A. 2019. Phylogenetics, patterns of genetic variation and population dynamics of Trypanosoma terrestris support both coevolution and ecological host-fitting as processes driving trypanosome evolution. Parasites & Vectors 12: 473. https://doi.org/10.1186/s13071-019-3726-y

Pinder, A.C.; Gozlan, R.E.; Britton, J.R. 2005. Dispersal of the invasive topmouth gudgeon, Pseudorasbora parva, in the UK: a vector for an emergent infectious disease. Fisheries Management and Ecology 12: 411–414. https://doi.org/10.1111/j.1365-2400.2005.00466.x

Princepe, D.; Czarnobai, S.; Pradella, T.M.; Caetano, R.A.; Marquitti, F.M.D.; de Aguiar, M.A.M.; Araujo, S.B.L. 2022. Diversity patterns and speciation processes in a two-island system with continuous migration. Evolution 76: 2260–2271. https://doi.org/10.1111/evo.14603

Ribeiro Prist, P.; Reverberi Tambosi, L.; Filipe Mucci, L.; Pinter, A.; Pereira de Souza, R.; de Lara Muylaert, R.; Rhodes, J.R., et al. 2022. Roads and forest edges facilitate yellow fever virus dispersion. Journal of Applied Ecology 59: 4–17. https://doi.org/10.1111/1365-2664.14031

Rohr, J.R.; Civitello, D.J.; Halliday, F.W.; Hudson, P.J.; Lafferty, K.D.; Wood, C.L.; Mordecai, E.A. 2019. Towards common ground in the biodiversity-disease debate. Nature Ecology and Evolution 4: 24–33. https://doi.org/10.1038/s41559-019-1060-6

Rychener, L.; In-Albon, S.; Djordjevic, S.P.; Chowdhury, P.R.; Nicholson, P.; Ziech, R.E.; et al. 2017. Clostridium chauvoei, an evolutionary dead-end pathogen. Frontiers in Microbiology 8: 1–13. https://doi.org/10.3389/fmicb.2017.01054

Sacristán, S.; Malpica, J.M.; Fraile, A.; García-Arenal, F. 2003. Estimation of population bottlenecks during systemic movement of Tobacco mosaic virus in tobacco plants. Journal of Virology 77: 9906–9911. https://doi.org/10.1128/jvi.77.18.9906-9911.2003

Scheiner, S.M.; Mindell, D.P. (eds.). 2019. The Theory of Evolution: Principles, Concepts, and Assumptions. University of Chicago Press, Chicago. https://doi.org/10.7208/chicago/9780226671338.001.0001

Schradin, C.; Lindholm, A.K.; Johannesen, J.; Schoepf, I.; Yuen, C.-H.; König, B.; Pillay, N. 2012. Social flexibility and social evolution in mammals: a case study of the African striped mouse (*Rhabdomys pumilio*). Molecular Ecology 21: 541–553. https://doi.org/10.1111/j.1365-294X.2011.05256.x

Smith, J.M. 1976. What determines the rate of evolution? American Naturalist 110: 331–338. https://doi.org/10.1086/283071

Sprouffske, K.; Aguilar-Rodríguez, J.; Sniegowski, P.; Wagner, A. 2018. High mutation rates limit evolutionary adaptation in *Escherichia coli.* PLOS Genetics 14: e1007324. https://doi.org/10.1371/journal.pgen.1007324

Steward, R.A.; Epanchin-Niell, R.S.; Boggs, C.L. 2022. Novel host unmasks heritable variation in plant preference within an insect population. Evolution 76: 2634–2648. https://doi.org/10.1111/evo.14608

Streicker, D.G.; Turmelle, A.S.; Vonhof, M.J.; Kuzmin, I.V.; McCracken, G.F.; Rupprecht, C.E. 2010. Host phylogeny constrains cross-species emergence and establishment of rabies virus in bats. Science 329: 676–679. https://doi.org/10.1126/science.1188836

Techer, M.; Roberts, J.; Cartwright, R.; Mikheyev, A. 2021. The first steps toward a global pandemic: reconstructing the demographic history of parasite host switches in its native range. Research Square preprint. https://doi.org/10.21203/rs.3.rs-196900/v1

Trivellone, V.; Araujo, S.B.L.; Panassiti B. 2021. HostSwitch: Simulate the Extent of Host Switching by Consumers. 12 pp. https://cran.r-project.org/web/packages/HostSwitch/HostSwitch.pdf

Trivellone, V.; Araujo, S.B.L.; Panassiti, B. 2023. HostSwitch: an R package to simulate the extent of host-switching by a consumer. The R Journal 14: 179–194. https://doi.org/10.32614/RJ-2023-005

Trivellone, V.; Hoberg, E.P.; Boeger, W.A.; Brooks, D.R. 2022. Food security and emerging infectious disease: risk assessment and risk management. Royal Society Open Science 9: 211687. https://doi.org/10.1098/rsos.211687

Wei, C.; Shan, K.-J.; Wang, W.; Zhang, S.; Huan, Q.; Qian, W. 2021. Evidence for a mouse origin of the SARS-CoV-2 Omicron variant. Journal of Genetics and Genomics. https://doi.org/10.1016/j.jgg.2021.12.003

WHO [World Health Organization]. 2022. WHO Coronavirus (COVID-19) Dashboard [WWW document]. Accessed January 19, 2022. https://covid19.who.int

Wilkinson, D.M. 2004. The parable of Green Mountain: Ascension Island, ecosystem construction and ecological fitting. Journal of Biogeography 31: 1–4. https://doi.org/10.1046/j.0305-0270.2003.01010.x

Yamaguchi, R.; Iwasa, Y. 2017. Parapatric speciation in three islands: dynamics of geographical configuration of allele sharing. Royal Society Open Science 4: 160819. https://doi.org/10.1098/rsos.160819

Section II

Putting Evolution to Work

5

Prevent-Prepare-Palliate: The 3P Framework—Integrating the DAMA Protocol into Global Public Health Systems

Orsolya Molnár, Eric P. Hoberg, Valeria Trivellone, Gábor Földvári, and Daniel R. Brooks

Abstract

The COVID-19 pandemic is the latest example of the profound socioeconomic impact of the emerging infectious disease (EID) crisis. Current health security measures are based on a failed evolutionary paradigm that presumes EID is rare and cannot be predicted because emergence requires the prior evolution of novel genetic capacities for colonizing a new host. Consequently, crisis response through preparation for previously emerged diseases and palliation following outbreaks have been the only health security options, which have become unsustainably expensive and unsuccessful. The Stockholm Paradigm (SP) is an alternative evolutionary framework that suggests host changes are the result of changing conditions that bring pathogens into contact with susceptible hosts, with novel genetic variants arising in the new host after infection. Host changes leading to EID can be predicted because preexisting capacities for colonizing new hosts are highly specific and phylogenetically conservative. This makes EID prevention through limiting exposure to susceptible hosts possible. The DAMA (Document, Assess, Monitor, Act) protocol is a policy extension of the SP that can both prevent and mitigate EID by enhancing traditional efforts through adding early warning signs and predicting transmission dynamics. Prevention, preparation, and palliation compose the 3P framework, a comprehensive plan for reducing the socioeconomic impact of EID.

Keywords: emerging infectious diseases, Stockholm Paradigm, DAMA protocol, prevention strategies, public health, global health security

The Emerging Infectious Diseases Crisis

The last fifty years have seen some of the most significant technological and scientific advancements in history. This period also coincides with the crisis of emerging infectious diseases (EIDs) (Brooks and Ferrao, 2005; Rohde, 2013). Technological advancements produce unanticipated ecological disruptions through (1) human intrusion into natural ecosystems and (2) creation of artificial habitats suitable for rapid spread of particular pathogens, sometimes called industrial pathology (Breiman, 1996; Foster et al., 2021). Since the 1970s, more than 40 EIDs and more than 1,100 epidemics have been documented in humans. Among those were the 2003 SARS outbreak in Toronto, which had the potential of pandemic development and also revealed how epidemics foster and amplify social tensions between ethnic minorities even within developed countries and high-income populations, in this case involving the marginalization of the Asian-Canadian community (Jacobs, 2007). The 2015 Zika outbreak was one of a series of emergences of this disease that had swept through Africa, Asia, and the Americas, affecting those with limited access to health care services, especially women deprived of reproductive health services (Plourde and Bloch, 2016). Concurrently, an Ebola epidemic disproportionately affected disadvantaged populations across Liberia, Sierra Leone, and Guinea (Burkle and Burkle, 2015), which lacked public health and physical infrastructures.

EIDs affecting livestock and crops also have substantial economic and social impacts (Brooks et al., 2022; Trivellone et al., 2022). Avian influenza has a mortality rate in domestic poultry of 90–100% within 48 hours of infection (Centers for Disease Control and Prevention, 2022). The 2014–15 avian influenza pandemic affected more than 45 million birds in the US alone, resulting in poultry export bans across 75 countries and the doubling of egg prices within a few weeks (Newton and Kuethe, 2015). The 2018–19 African swine fever (ASF) pandemic resulted in the culling of nearly 20% of Vietnam's pig population, representing almost 6 million animals. China suffered an economic loss of US$141 billion by September 2019 directly due to the bans on international trade and led to the collapse of half of what was the world's pork export market prior to the outbreak (FAO, 2019). Among those emerging in crops, coconut lethal yellowing (LY) is a fatal disease of several species of palms (Martinez and Roberts, 1967) that has severe repercussions on local cultures for which coconuts provide economic security (Gurr et al., 2016). LY destroyed 95% of the coconut palms in a single region in Mexico in only two decades, millions of coconut trees that provided a livelihood for more than 30,000 rural families in Nigeria by 2010, and almost 99% of tall palms and 72% of dwarfs in West Africa by 2006 (Datt et al., 2020).

Prior to the COVID pandemic, treatment costs and production losses as a result of EIDs of all kinds reached US$1 trillion per year globally (see Brooks et al. 2019). Apart from short-term impacts, EIDs produce long-lasting negative social effects, mostly affecting marginalized and/or low-income communities (Leach and Dry, 2010). In addition to more than 200 million confirmed cases and more than 4.3 million deaths, the first six months of the COVID-19 pandemic catalyzed the greatest unemployment rate and economic deflation since World War II (ILO, 2021). Household expenditures were elevated because of medical expenses, and incomes dropped by as much as 50% (WHO and World Bank, 2019). Healthcare infrastructure was overwhelmed, leading to decreased accessibility for low-income, vulnerable groups, and at the same time, food insecurity rose and access to education was reduced (Blacke and Wadhwa, 2020).

The EID crisis calls for a reevaluation of current disease management approaches that focus only on managing EIDs after they have emerged (Evans, 2010; Hadler et al., 2015; Apari et al., 2019). Health services are constantly and unpleasantly surprised by each apparently unexpected EID, and these services are then forced into expensive and time-consuming crisis response. Global health systems are backed into a corner, trying to plan protective measures against opponents they know nothing about until they announce themselves.

Approaches to Coping with Emerging Infectious Disease

Disease has been a constant challenge for modern humanity since the early Holocene, when human settlement created highly dense populations that lived in close proximity with domesticated animals. Infections differing in virulence, transmission dynamics, and persistence have influenced human history, from the emergence of sporadic cases of mysterious symptoms to the decimation of human populations of entire regions (see review in Lakoff, 2017). Humanity has therefore always tried to establish an appropriate response to combat diseases, designing actions based on the often limited knowledge that was available.

Palliate: Medicate, Vaccinate

The oldest tradition for combating diseases is palliation, treating and curing those infected and alleviating signs and symptoms of an infection, with the immediate aim of improving patients' life quality and the longer-range aim of reducing mortality and morbidity. Hygienic restrictions and regulations were among the first palliative measures first introduced in health care facilities by pioneering physicians such as Ignaz Semmelweiss (Bowden et al., 2003; Lane et al., 2010). The medical tools involved in palliation compose two major functional groups: (1) Medication refers to interfering with pathological physiological and/or biochemical pathways of patients or reproducing pathogens within. Although certain medicaments are available for use as prophylaxis for a limited period of time, the majority of application is linked to a prior development of a disease. (2) Vaccination, on the other hand, aims to reduce the morbidity, mortality, and/or incidence number of a known communicable disease by triggering a mild immune response through administering whole or partial pathogens as an antigen (Bowden et al., 2003).

Although they originated before pathogenic agents were known, today's medical practices still include multiple treatment options that target the physiological pathway responsible for the symptoms and not the pathogen itself. Palliation to a large extent has continued as the backbone of combating infectious disease because most infectious diseases are not deemed predictable and thus not preventable.

Global inequality in access to palliative medicine led to the founding of organizations such as Médecins Sans Frontières (MSF, August 18, 2021), an international network that brings medical professionals to treat diseases such as malaria, yellow fever, dengue, hepatitis, or cholera in developing countries and develops educational programs to improve health care infrastructure. The Bill and Melinda Gates Foundation supports research initiatives and medical intervention programs that target inequality, poverty, and health care program improvement or establishment in developing countries (Bill & Melinda Gates Foundation, August 18, 2021).

Once humans discovered that particular disease signs and symptoms were caused by infection with specific pathogens, palliation could be made easier by maintaining high standards of hygiene in health care settings and preparing for infectious diseases by having appropriate health care facilities as well as medications and vaccines on hand. Also, understanding the transmission dynamics of particular infectious pathogens allowed clinicians and public health workers to prepare by anticipating where and at what time of year certain infectious diseases would occur. Ironically, assumptions about seasonality contributed to notions about apparent disappearance of pathogens at local to regional scales when diseases were not observed in circulation. These notions persist, despite J.R. Audy's groundbreaking observation in 1958 that distribution of a pathogen is always broader than the disease it causes (Audy, 1958). These measures set the foundation for modern public health initiatives and infrastructure to prepare for infectious disease outbreaks.

Prepare: Stockpile and Eradicate

The late 20th century saw large-scale outbreaks of infectious diseases that affected social structure, economic processes, and industrial production. Starting with the 1976 influenza epidemic, followed by an outbreak of both seasonal (in 2003) and avian influenza (in 2005) that affected poultry production, disease management became an issue of national security (Lakoff, 2017). Medical intervention then needed to be planned using risk assessment metrics and national distribution networks, sparking close collaboration between public health and other sectors such as the pharmaceutical industry, military, and government agencies and the development of new methods, such as scenario-based exercises to estimate the severity of the threat and improve "response strategies" (Johns Hopkins Center for Civilian Biodefense Studies et al., 2001). Intersectoral collaborations led to international alliances (e.g., International Health Regulations (WHO, 2005)), providing further insight into the behavior of known pathogens. Monitoring stations and surveillance networks established in tropical ecosystems fed information into global databases (e.g., Global Influenza Surveillance and Response System (GISRS) (GISRS, August 17, 2021), Global Outbreak Alert and Response

Network (GOARN) (GOARN, August 17, 2021), Global Early Warning System (GLEWS) (FAO et al., 2006)) to spot early onset and instigate timely crises response.

Collaborative networks of veterinarians, physicians, and public health experts have been established since the early 1900s, but the EID crisis gave impetus to creating global initiatives. The One Health initiative (AVMA, 2008) was established in 2006 to target vectors and detect reservoirs of known pathogens. Partly in response to the 2003 avian flu pandemic (Burns et al., 2008), the PREDICT project was launched in 2009 as part of USAID's Emerging Pandemic Threats Program to anticipate future pandemics. The Wildlife (Preservation) Trust renamed itself the EcoHealth Alliance (EcoHealth Alliance, September 13, 2021) and joined forces with the PREDICT project in 2010 to study suspected vectors in areas highly affected by malaria, yellow fever, hepatitis, and dengue. Efforts to prepare for EID now include activities as diverse as studies of zoonoses, campaigns for wildlife vaccination, and distribution of insecticides and mosquito nets. Programs aimed at local communities involve establishing education programs, distributing insecticides, procuring equipment to treat the sick in local health care facilities, and initiating mass vaccination programs to mitigate effects.

Once the transmission dynamics of particular pathogens were known, preparation to cope with infectious disease began to include plans to modify the landscape in ways that would disrupt pathogen transmission, mitigating the severity of outbreaks.

Why Palliate and Prepare Are Not Sufficient

Despite growing awareness of the severity of the EID crisis and efforts to prepare better for coping with EIDs, the new millennium saw a concerning increase in the number of previously unknown pathogens that had no previous record of infecting humans (e.g., SARS, MERS, Hepatitis D, COVID-19) and previously known infections either manifesting novel symptoms (e.g., Zika, West Nile virus) or developing resistance to applied treatment (e.g., resistant malaria, MRSA). Each emergence triggered augmented crisis response, which quickly became a proxy for measuring health security of a given nation. Indices such as the Global Health Security Index (Cameron et al., 2019) or the Epidemic Preparedness Index (Oppenheim et al., 2019) were introduced to evaluate how efficiently a nation's public health infrastructure can respond to emergent disease outbreaks—and stop a potential pandemic. Although global estimates all agreed that Western

European and North American countries were far better equipped to mitigate damages of a potential pandemic, COVID-19 proved prediction wrong. During the time it took for preparedness measures to take effect, the disease had exacted a devastating toll on the affected countries, with the largest losses in both economic value and human life linked to those thought to be best prepared. Preparation alone had two substantial soft spots: (1) all preparations were based on the assumption that any outbreaks would be caused by known pathogens and (2) EIDs are emerging far more rapidly than expected.

Preparatory programs target known pathogens and, when appropriate, known vectors in areas where outbreaks have previously occurred. For outbreaks not covered by preparatory programs, specimens are often not collected intensively or extensively, so taxonomy is often unclear. Also, when collected specimens are not stored under conditions that allow for long-term preservation, material is not available for further research, such as phylogenetic comparative studies (Colella et al., 2021). Crisis response works well in preparation for reemergence of known pathogens, but it is not prevention. Crisis response was designed to combat diseases we have previously been exposed to and have information about from previous outbreaks. The ongoing SARS-CoV-2 pandemic has led to calls for developing a way of preventing the next emergence, but most publications on the subject suggest strengthening, increasing, and improving our response strategies (Khoo and Lantos, 2020; Naguib et al., 2020; WHO, 2020; DeSalvo et al., 2021; Stenseth et al., 2021). However, when faced with a novel pathogen, crisis response always lags behind the sweeping epidemic (Audy, 1958; Brooks et al., 2020). With EID increasing at an unprecedented rate since the 1970s (WHO, 2007), global health security demands that we acknowledge the limitations of crisis response and add another pillar of disease control: prevention.

The main reason prevention has not been a focus of disease management is the way most scientists perceive relationships between pathogens and their hosts. The traditional paradigm assumes that pathogen attributes are strongly selected, resulting in specialized associations with a narrow range (often assumed to be a single species) of hosts. As a result of such specialization, pathogens are assumed to lose their ability to utilize (infect) novel hosts, and any new colonization (emergent disease) must be preceded by the pathogen evolving new capacities (Parrish and Kawaoka, 2005). The assumption is that, given that such new capacities evolve rarely and at random with respect to any particular potential host, emergence is assumed to be rare and unpredictable, thus prevention is in vain. This traditional paradigm, the core of the standard model for pathogens and disease, fails on three counts: (1) it claims pathogens are tightly co-adapted to a restricted range of hosts, based on assumptions lacking empirical support; (2) it claims that EIDs ought to be rare, whereas they are common, as evidenced by the current EID crisis and by phylogenetic studies showing that host colonization has been common throughout evolutionary history; and (3) it claims that host switching occurs at random with respect to environmental perturbations such as climate change, when biogeographical studies show that colonization events in many pathogen clades cluster around climate change perturbations. This contradiction between the traditional paradigm and empirical observations in phylogenetic and real time has been dubbed the *parasite paradox* (Agosta et al., 2010). A new evolutionary framework of host-pathogen associations, the Stockholm Paradigm, resolves this paradox.

The Stockholm Paradigm

The Stockholm Paradigm (SP) (Brooks et al., 2014; Hoberg and Brooks, 2015; Brooks et al., 2019) is based on two Darwinian principles: (1) **Evolutionary outcomes of interacting species are always local.** All organisms that a pathogen could potentially infect (potential hosts) form its *fundamental fitness space*, while those that the pathogen has actually colonized represent its *realized fitness space*. Coevolutionary processes between the pathogen and a host in a certain geographic locality involve only the realized fitness space and have no effect on other potential hosts in other locations. As a consequence, the smaller the pathogen's realized fitness space is relative to the fundamental fitness space, the higher its potential for colonizing a new host without the need for newly evolved capacities. This potential is referred to as *ecological fitting* (Janzen, 1985). (2) **Evolution is conservative.** All organisms exploit particular resources of their environment to survive, which requires certain specialized traits. Given these traits are phylogenetically conservative, the same pathogen will be able to utilize and colonize distantly related host species, and, conversely, a novel host will serve as a competent resource for multiple pathogens. Coronaviruses (e.g., SARS, SARS-CoV-2) use receptors distributed across a wide range of Mammalia (Mahdy et al., 2020; Dicken et al., 2021; Lytras et al., 2021), while human red blood cells are targets of various bacterial (*Bartonella bacilliformis*) as well as eukaryotic (*Plasmodium* spp., *Babesia* spp., *Toxoplasma gondii*) pathogens (McCullough, 2014).

Changes in geographical distribution or ecological structure will lead to novel species encounters and provide pathogens with novel opportunities to colonize novel, suitable hosts and increase their realized fitness space (Agosta et al., 2010). Considering genetic variation within the original population, certain low-frequency variants in the original host may be a "better fit" with and thus start rapidly proliferating in the new host. This "stepping-stone dynamic" is a common antecedent of emergences (Araujo et al., 2015; Braga et al., 2015), such as in the 2003 SARS, 2012 MERS, and 2020 SARS-CoV-2 outbreaks (Morens and Fauci, 2020).

Taken together, emergence is not due to the appearance of novel genetic capacities; rather, it is due to the pathogen's taking advantage of new opportunities using preexisting capacities. Pathogens have faced multiple host-range changes throughout their evolutionary history, and each of these events presented them with new suitable hosts they then colonized. The link between climate change and the EID crisis is therefore straightforward and surprisingly simple: changing environmental conditions create movement among species, bringing pathogens in contact with susceptible hosts (Hoberg et al., 2012; Brooks et al., 2019). Pathogens expand their host range and geographic distribution, producing new diversity in these new settings (Hoberg and Brooks, 2015), which sets the stage for the next set of emergences, given novel opportunities. Emerging diseases are therefore a built-in feature of evolutionary diversification.

Mobilizing Evolution for Public Health Prevention Efforts

Public health policy in itself represents a highly interconnected and complex field that aims to derive clear directions for policy-making from a variety of scientific, social, and economic predictions (Berger et al., 2019). Nevertheless, such a systemic view is necessary to design anticipatory regulations and prevent negative, unanticipated consequences (Vianna Franco et al., 2022). Policies are often driven by outdated paradigms, which are unable to define the appropriate outcome or have limited understanding of the issue at hand (Sheng and Zeng, 2021). If we understand the organic nature of infectious disease and the often neglected importance of context-dependent processes (Bozzuto et al., 2020), managing EIDs can move from preparation to anticipation and even prevention of outbreaks. The concept of planetary health is one of the key underlying paradigms driving changes in not only policy making but redefining educational curricula (Guzman et al., 2021).

From a scientific perspective, the SP is a novel approach that replaces the prevailing paradigm of host-parasite co-evolution. The key differences are not only in the evolutionary processes but in that evolutionary processes described by the SP result in highly context-dependent outcomes, and therefore allow for anticipatory action. The bad news is that pathogens switch to different hosts frequently during conditions of environmental perturbations, such as climate change and globalized trade and travel; therefore, EIDs will continue for an indefinite period of time. The good news is that the conservative nature of genetic capacities makes emergence predictable. The prevailing paradigm used for planning public health policies has linked diseases to the host (and vector) in which they were first found and identified and tries to eliminate that presumed closed cycle (PAHO and WHO, 2008; van den Berg et al., 2012; Cucunubá et al., 2018). Focus is limited to these two target populations, despite overwhelming evidence that any pathogen observed in a host species will be present in numerous sympatric species, even ones distantly related to the original (Parrish et al., 2015; Olivero et al., 2017; Cahan, 2020; Fagre et al., 2021). From a public health perspective, we track the disease to track the pathogen, whereas "the distribution of a pathogen is wider than the disease caused by it, and the latter cannot be understood without understanding the former as a whole" (Audy, 1958). Therefore, even the most organized efforts to eliminate a pathogen from the vector or the host will be faced with an array of reservoirs from which the same pathogen can easily cross over to the newly treated host population, thereby diminishing control efforts in the long term.

The SP offers a way to integrate an evolutionary framework into current disease management strategies to concentrate efforts on preventing emergence. By determining fundamental fitness space as well as observed realized fitness space, we can assign risk to pathogen populations before their emergence. Risk space is maximized along habitat interfaces, where interconnectedness is also increased by climate shifts and human intrusion (Araujo et al., 2015; Brooks et al., 2022). While the ongoing pandemic has all our attention and resources bound by crisis response, the risk spaces for novel EIDs grow by the second. Warming global temperatures and human modified landscapes extend habitable areas for mosquito species such as *Aedes aegypti*, *Aedes albopictus*, or *Anopheles maculipennis*, some of which are competent vectors of malaria, dengue, or yellow fever (Khasnis and Nettleman, 2005; Suzán et al., 2015). Tropical diseases such as chikungunya (Weaver and Lecuit, 2015), dengue (Brady et al., 2012), Zika (Brady and Hay,

2019), and malaria (New map shows, 2018) are constantly reported in Europe, with a large proportion of cases related to travel. Crop diseases such as wheat stem rust (*Puccinia graminis* f. sp. *tritici*) are emerging in crops (Saunders et al., 2019), while the emergent African swine fever virus continues to colonize both swine and wild boar stocks on three continents (Gallardo et al., 2015). Such interfaces and interconnected systems should be primary targets of intensive monitoring with efforts concentrating on tracking pathogens in known reservoirs and alternate vectors. Early interventions should then produce public health, food safety, and wildlife management policies that minimize encounter between reservoirs and susceptible communities, thereby preventing or mitigating impacts of emergence. In the following section, we describe a four-step prevention protocol that stems from the SP.

Prevent: Predict and Act

Managing risk for EIDs relies on identifying targets for control measures before the onset of a crisis. Strategies thus far have primarily focused on crisis response but have not provided actionable information in advance of an outbreak (Fallah et al., 2015; Daszak et al., 2020; Chatterjee et al., 2021). However, with the accelerating rate of novel emergences, finding them before they find us will be more cost effective than palliation or preparation (Brooks et al., 2019; Brooks et al., 2020; Vianna Franco et al., 2022). Phylogenetically conservative traits make the risk space of EIDs large but also make their behavior predictable in novel settings.

The DAMA protocol—Document, Assess, Monitor, Act—is a comprehensive policy plan stemming directly from the SP (Brooks et al., 2014; Brooks et al., 2019) (Figure 5.1). It aims to shift the focus of efforts from crisis response (palliation and preparation) to preventing disease outbreaks and facilitating communication between stakeholders in science, health security, and policy making. As well, when prevention is not possible, the DAMA protocol can aid preparation and early response efforts by mitigating the impact of outbreaks.

It is not possible to cope with pathogens without knowing what they are, where they occur, and which potential host species they are likely to infect. ***Documenting***

Figure 5.1. A schematic representation of the DAMA protocol. Keywords next to the circle in normal font represent the overall aim of the respective protocol phase; keywords in bold type describe the methods used to achieve that aim.

pathogens actually or potentially residing in a given region is thus fundamental; strategic inventory feeding into archives for specimens and information is essential (Dunnum et al., 2018). Taking advantage of the evolutionary and ecological context DAMA provides, we focus our search on reservoir hosts. We know that disease-causing organisms of humans, crops, and livestock reside in at least one host that is not diseased. Such reservoirs are often known or suspected, allowing us to focus on a manageable subset of all the species within an area. When reservoirs live in habitats adjacent to human communities or their crops and livestock, disease transmission occurs in the interface between reservoir and human habitats. African swine fever was transmitted from wild boars living around animal breeding facilities to domestic pigs, avian flu spreads through encounters between wild migratory birds and domestic poultry, SARS spread from bats to humans (Cyranoski, 2017). Furthermore, the transmission of pathogens is highly specific for each disease-causing organism—some are transmitted in food, some in water, some by contact between infected hosts or surfaces that have been in contact with infected hosts. Many are transmitted by vectors such as mosquitoes and ticks. Finally, in addition to the information collected during scientific research, local and traditional ecological knowledge from people living in areas at risk is also a significant resource, calling for the establishment of science-society collaboration (Marizzi et al., 2018; Brooks et al., 2019).

Assessing the risk of documented pathogens is a three-step process. The first step is *phylogenetic triage*, which reveals if the species is: (1) known to cause disease in another place, (2) most closely related to nonpathogens, or (3) a close relative of a known pathogen. In the first case, findings should be reported to health authorities; in the second, representative specimens of these species are archived for future reference. In the third case, species are subjected to the second step of risk assessment, *phylogenetic assessment*, using what is known about the close relatives of the novel pathogen to determine possible reservoirs, their mode of transmission, and their microhabitat preferences. Finally, population modeling is done to gather information on population genetics, focusing on the rare genotypes with higher potential to emerge in a new host.

Monitoring pathogens of potential risk begins with detailed mapping of their distribution in areas where they have already been observed as well as searches for them in areas predicted to be suitable. Changes in pathogen populations lead to regular reassessment, while surveillance provides actionable information about changes in risk

space before the onset of an epidemic. We are looking for *change*—in geographic distribution, in host range, in transmission dynamics, in geographic variation, in what is there, and in early signs of arrival of anticipated pathogens. A context for comparisons to detect change over time is linked to archival resources and baselines.

Monitoring activities must be rapidly translated into effective **action**. Interventions mainly manifested as policy modifications in the sectors of food safety, wildlife management, veterinary medicine, and public health and education target wildlands, urban/peri-urban, and agricultural landscapes (Trivellone et al., 2022) and interfaces between them. Policies must be generated using historical data stored in natural history biorepositories to accurately evaluate pathogenic and spreading potential of the species in focus. Expertise involved must consider global patterns of trade and travel as well as local environments and community circumstances to maximize feasibility of novel policy developments. Prevention relies on developing interdisciplinary and transdisciplinary networks to determine effective courses of action and coordinate their implementation.

The Prevent-Prepare-Palliate (3P) framework

As demonstrated by the past two decades, one of the biggest threats to modern humanity is the EID crisis, dealing unanticipated damage across all socioeconomic landscapes. Gathering information on symptoms, distribution, and mode of transmission is crucial but not sufficient for avoiding novel emergences. Understanding the novel paradigm describing the ease with which pathogens switch to susceptible host, health security must also shift to a paradigm placing Prevention in the heart of public health. The addition of a preventative element to responses to health threats leads to a tripartite framework of Prevent-Prepare-Palliate (3P) for effective action against EIDs (Figure 5.2).

Similar to the existing framework of One Health, the 3P initiative emphasizes the need for efficient communication and collaborative task forces formed among different sectors and disciplines (Tenorio, 2022). Given the multitude of expertise involved in addressing any emerging infectious threat, methods to establish working collaboration require transdisciplinary approaches (Alonso Aguirre et al., 2019)

In the following sections, we describe different methods for establishing and facilitating communication and information sharing between elements of the framework and within processes of prevention measures.

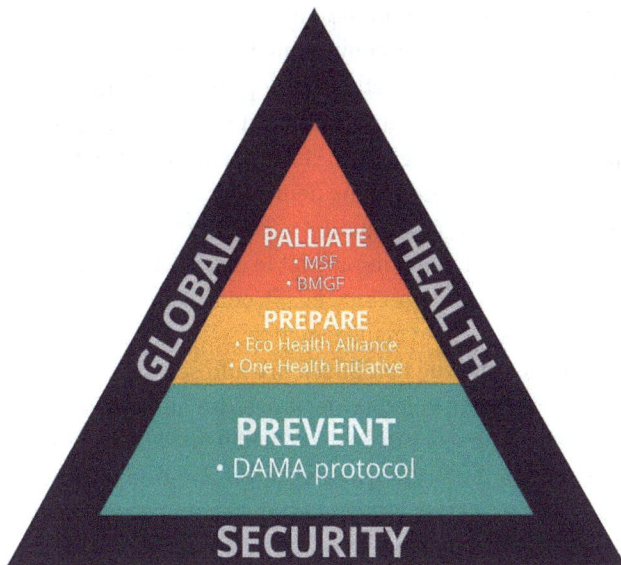

Figure 5.2. Visual representation of the 3P (Prevent-Prepare-Palliate) framework of global health security facing the threat of newly emerging infectious diseases, with examples of organizations and initiatives implementing the Prepare and Palliate elements.

Communication among the components of the 3P framework

Communication needs to be established between elements of the framework to mutually increase impact. Palliation creates vast knowledge and data on the physiological symptoms caused by particular pathogens, distinguishing between strains and subspecies. The channels to disseminate this information to actors of the Prepare element have already been established and provide basis for planning preparatory efforts in case of a resurgence of a known pathogen. However, it is also indispensable knowledge for the Prevent element for assessing the risk of a close relative of a known pathogen. Manifestations, treatment options, morbidity, and mortality of a close relative are all taken into consideration when assigning risk to a potential pathogen. The EpiPulse initiative (European Centre for Disease Control and Prevention, 2021) is an exemplary fusion of multiple, distinct surveillance systems for collecting and sharing information on pathogens and facilitating interdisciplinary and cross-sector collaborations.

Preparatory efforts have been substantially improved by adopting the OneHealth approach and thereby including not only medical but also veterinary, agricultural, and wildlife experts in preparing for the onset of a particular disease. Information on reservoirs, vectors, and their distribution are all highly beneficial during phylogenetic

assessment of potential pathogens to accurately reveal their range of potential host species. If we expand our target range from the current known pathogens to their close relatives, this tool will prove extremely beneficial in finding them before they find us. An existing hub that allows data entry from diverse sectors' pathogen data stores is the International Nucleotide Sequence Database Collaboration (INSDC) (INSDC, March 22, 2022), an open access data platform under the OneHealth approach (Timme et al., 2020). Information, however, must always be linked back directly to specimens of pathogens and hosts in museums, a process that has not been uniform nor assumed.

Finally, information on high-risk potential pathogens identified by the DAMA protocol will be openly available to preparatory and palliative initiatives, thereby substantially mitigating damages in case of an outbreak.

Communication within the DAMA protocol

The DAMA protocol is designed to combine the techniques of both fundamental and applied research and drive efficient policy making that builds on existing infrastructures and local conditions. The key to such a comprehensive protocol involving stakeholders from multiple different sectors is establishing efficient communication strategies for each of the four phases (Figure 5.1).

When documenting and assessing pathogens, it is necessary to have globally available linked databases on previously identified and assessed parasitic or microbial organisms and their known host species. To establish baseline conditions of pathogens and all their hosts and reservoirs, specimen sourcing has to be followed by permanent archiving. Implementing standardized entry protocols; collaborative, directed collection efforts; and globally accessible databases provides opportunity for detecting pathogenic microbes in their reservoirs before host switching can result in human cases. Establishing baseline conditions is then followed by periodic resampling to monitor changes in distribution, genetic composition, and host range. Although calls for creating global repositories and archives have been made previously (Brooks and Hoberg, 2000), large-scale collection efforts have failed to permanently archive their specimens in biorepositories (Kelly et al., 2017; Grange et al., 2021), depriving further studies from the valuable data that were collected at high monetary cost.

Colella et al. (2021) recently proposed that biorepositories, such as those mentioned in the Global Museum initiative (Bakker et al., 2020), present a possible solution for archiving and cataloging potential pathogens along with their reservoirs to build cohesive informatics resources that describe diversity (Dunnum et al., 2018). Biorepositories include in their collections substantial numbers of individual vertebrates and invertebrates that can be and often are screened for the potential pathogens that may occur in archived specimens. Parasites or pathogens detected in these archival collections can then be used as alerts and as starting points for additional field-based studies.

Traditional disease surveillance is either limited by jurisdictional boundaries or comes at a significantly higher cost for larger-scale monitoring (Palmer et al., 2017), both of which limit public health security. The DAMA protocol relies heavily on effective working relationships with local communities, involving members in not only collecting data but also gathering traditional insights and observations regarding reservoirs. Establishing communication channels and training programs between susceptible communities and public health authorities will foster trust in science and allow for a close collaboration with those affected when planning and implementing intervention measures and novel policies. Grassroots-level science has been successfully employed in tracking mosquito and tick vectors (Palmer et al., 2017; Földvári et al., 2022), monitoring avian flu in urban environments (Marizzi et al., 2018; Szekeres et al., 2019), and surveying wildlife health (Lawson et al., 2015), creating the dual benefit of access to larger data and increased awareness in communities at high risk of emergence. Such initiatives are also designed to become bases for long-term science-practice collaboration in their respective localities.

Finally, designing effective *action* in light of information gathered in the previous three phases warrants the close cooperation of various governmental, municipal, scientific, and local stakeholders. A promising approach is adopting the innovative Living Lab methodology (Herrera, 2017; Steen and van Bueren, 2017; Huang and Thomas, 2021) into the context of disease management by establishing multi-actor consensus groups that consist of various regional stakeholders impacted by an emerging disease. The Living Lab method aims to increase the efficiency of implementing novel achievements into real-life contexts by inviting all stakeholders into a collaborative task force (Veeckman et al., 2013). Consensus groups consist of scientific personnel as well as public health policy experts, regional governmental officials, and members of the exposed populations. The aim of this method is for all actors to jointly analyze the issue at hand; identify the implications an emerging pathogen has for residential, legal, healthcare, and scientific landscapes; and then cooperate in designing an efficient and highly adoptable intervention plan,

which takes into account not only wider scientific context but also regional policy environments and local feasibility. Although Living Lab methodology presents multiple challenges in terms of implementation, it has been successfully applied in monitoring health states of elderly patients (Kim et al., 2020) and addressing neurological issues (Richardson et al., 2021). The apparent suitability of this approach for the study of noncommunicable diseases strongly suggests the methodology could be introduced into the context of infectious disease management as well.

The fact that the health-science and scientific community has thus far failed to decrease the number of novel emergences appears to be explained mainly by the lack of effective communication among inter- and trans-disciplinary systems. Cutting-edge research in studies of evolution of pathogens is not contextualized in veterinary science, agriculture, or public health, and implications of evolutionary processes are not considered during policy making.

Disseminating the importance of reducing exposure, monitoring early warning signs, and planning interventions to prevent, not merely contain, outbreaks is the only way to strengthen global health security.

Conclusions

EIDs affect every layer of modern society on local to global scales (Kapiriri and Ross, 2020). Reducing the impacts of EID through prevention rather than crisis response is the key to a sustainable future for humanity. The emergence of another pandemic is not a question of if but when and where. The DAMA protocol is necessary for addressing the EID crisis and requires the interdisciplinary collaboration of experts in healthcare policy, epidemiology, and pathogen-host evolution. Efforts must focus on maintaining the guiding principle of anticipatory action while also integrating the novel approach into existing frameworks and programs. DAMA can help us buy time and save resources in the global effort to cope with the EID crisis. The Prevent-Prepare-Palliate framework clarifies aims and goals of health security initiatives, allowing existing infrastructure to focus on mitigating the impacts of EID that have already emerged, while preventing new EID and mitigating the impacts of those that cannot be prevented. Global climate change and globalized trade and travel create new challenges, and our efforts to cope can and must adapt.

We know what to do, we have the tools and the expertise, and the next step is to implement the DAMA protocol before the next pandemic. Global health security relies on a complex network of institutions and organizations operating in various policy environments and accommodating diverse scientific inputs. Therefore, the emphasis must now be on establishing cross-sector communication channels by (1) creating globally accessible databanks that accommodate input from different scientific fields and public sectors and can be used by disease management to assess risk of potential pathogens, (2) involving grassroots-level contributors in research studies to facilitate communication and trust between public health and susceptible communities, and (3) employing cutting-edge approaches such as the Living Lab method to include all stakeholders in generating solutions to the EID threat.

Literature Cited

Agosta, S. J.; Janz, N.; Brooks, D. R. 2010. How specialists can be generalists: resolving the and "parasite paradox" and implications for emerging infectious disease. Zoologia 27: 151–162. https://doi.org/10.1590/S1984-46702010000200001

Alonso Aguirre, A., Basu, N., Kahn, L.H.; Morin, X.K.; Echaubard, P.; Wilcox, B.A.; Beasley, V.R. 2019. Transdisciplinary and social-ecological health frameworks—novel approaches to emerging parasitic and vector-borne diseases. Parasite Epidemiology and Control 4: e00084. https://doi.org/10.1016/j.parepi.2019.e00084

Apari, P.; Bajer, K.; Brooks, D.R.; Molnár, O. 2019. Hiding in plain sight: an evolutionary approach to the South American Zika outbreak and its future consequences. Zoologia 36: 1–7. https://doi.org/10.3897/zoologia.36.e36272

Araujo, S.B.L.; Braga, M.P.; Brooks, D.R.; Agosta, S.J.; Hoberg, E.P.; von Hartenthal, F.W.; Boeger, W.A. 2015. Understanding host-switching by ecological fitting. PLOS ONE 10: e0139225. https://doi.org/10.1371/journal.pone.0139225

Audy, J.R. 1958. The localization of disease with special reference to the zoonoses. Transactions of the Royal Society of Tropical Medicine and Hygiene 52: 308-334. https://doi.org/10.1016/0035-9203(58)90045-2

AVMA [American Veterinary Medical Association]. 2008. One Health: A New Professional Imperative. One Health Initiative Task Force : Final Report, July 15, 2008. https://www.avma.org/resources-tools/reports/one-health-ohitf-final-report-2008

Bakker, F.T.; Antonelli, A.; Clarke, J.A.; Cook, J.A.; Edwards, S.V.; Ericson, P.G.P.; et al. 2020. The Global Museum: natural history collections and the future of evolutionary science and public education. PeerJ 8: e8225. https://doi.org/10.7717/peerj.8225

Berger, K.M.; Wood, J.L.N.; Jenkins, B.; Olsen, J.; Morse, S.S.; Gresham, L.; et al. 2019. Policy and science for global health security: shaping the course of international health.

Tropical Medicine and Infectious Disease 4: 60. https://doi.org/10.3390/tropicalmed4020060

Bill & Melinda Gates Foundation. n.d. Bill & Melinda Gates Foundation (website). Accessed August 18, 2021. https://www.gatesfoundation.org/

Blake, P.; Wadhwa, D. 2020. 2020 Year in Review: The Impact of COVID-19 in 12 Charts. Accessed October 12, 2021. https://blogs.worldbank.org/voices/2020-year-review-impact-covid-19-12-charts

Bowden, M.E.; Crow, A.B.; Sullivan, T. 2003. Pharmaceutical Achievers: The Human Face of Pharmaceutical Research. Chemical Heritage Foundation, Philadelphia. 214 pp.

Bozzuto, C.; Schmidt, B.R.; Canessa, S. 2020. Active responses to outbreaks of infectious wildlife diseases: objectives, strategies and constraints determine feasibility and success. Proceedings of the Royal Society B: Biological Sciences, 287: 20202475. https://doi.org/10.1098/rspb.2020.2475

Brady, O.J.; Gething, P.W.; Bhatt, S.; Messina, J.P.; Brownstein, J.S.; Hoen, A.G.; et al. 2012. Refining the global spatial limits of dengue virus transmission by evidence-based consensus. PLOS Neglected Tropical Diseases 6: e1760. https://doi.org/10.1371/journal.pntd.0001760

Brady, O.J.; Hay, S.I. 2019. The first local cases of Zika virus in Europe. The Lancet 394: 1991–1992. https://doi.org/10.1016/S0140-6736(19)32790-4

Braga, M.P.; Razzolini, E.; Boeger, W.A. 2015. Drivers of parasite sharing among neotropical freshwater fishes. Journal of Animal Ecology 84: 487–497. https://doi.org/10.1111/1365-2656.12298

Breiman, R.F. 1996. Impact of technology on the emergence of infectious diseases. Epidemiologic Reviews 18: 4–9. https://doi.org/10.1093/oxfordjournals.epirev.a01791

Brooks, D.R.; Ferrao, A.L. 2005. The historical biogeography of co-evolution: emerging infectious diseases are evolutionary accidents waiting to happen. Journal of Biogeography 32: 1291–1299. https://doi.org/10.1111/j.1365-2699.2005.01315.x

Brooks, D.R.; Hoberg, E.P. 2000. Triage for the biosphere: the need and rationale for taxonomic inventories and phylogenetic studies of parasites. Comparative Parasitology 67: 1–25.

Brooks, D.R.; Hoberg, E.P.; Boeger, W.A. 2019. The Stockholm Paradigm: Climate Change and Emerging Disease. University of Chicago Press, Chicago.

Brooks, D.R.; Hoberg, E.P.; Boeger, W.A.; Gardner S.L.; Araujo, S.B.L.; Bajer, K.; et al. 2020. Before the pandemic ends: making sure this never happens again. World Complexity Science Academy Journal 1: 8. https://doi.org/10.46473/wcsaj27240606/15-05-2020-0002//full/html

Brooks, D.R.; Hoberg, E.P.; Boeger, W.A.; Gardner, S.L.; Galbreath, K.E.; Herczeg, D.; et al. 2014. Finding them before they find us: informatics, parasites, and environments in accelerating climate change. Comparative Parasitology 81: 155–164. https://doi.org/10.1654/4724b.1

Brooks, D.R.; Hoberg, E.P.; Boeger, W.A.; Trivellone, V. 2022. Emerging infectious disease: an underappreciated area of strategic concern for food security. Transboundary and Emerging Diseases 69: 254–267. https://doi.org/10.1111/tbed.14009

Burkle, F.M., Jr.; Burkle, C.M. 2015. Triage management, survival, and the law in the age of Ebola. Disaster Medicine and Public Health Preparedness 9: 38–43. https://doi.org/10.1017/dmp.2014.117

Burns, A.; van der Mensbrugghe, D.; Timmer, H. 2008. Evaluating the Economic Consequences of Avian Influenza. The World Bank [working paper]. Accessed September 24, 2022. https://documents.worldbank.org/en/publication/documents-reports/documentdetail/977141468158986545/evaluating-the-economic-consequences-of-avian-influenza

Cahan, E. 2020 Aug 18. COVID-19 hits U.S. mink farms after ripping through Europe. Science. https://doi.org/10.1126/science.abe3870

Cameron, E.E.; Nuzzo, J.B.; Bell, J.A.; Nalabandian, M.; O'Brien, J.; League, A.; et al. 2019. Global Health Security Index: Building Collective Action and Accountability. Accessed March 15, 2023. https://www.ghsindex.org/wp-content/uploads/2020/04/2019-Global-Health-Security-Index.pdf

Centers for Disease Control and Prevention. 2022. Avian Influenza in Birds. Accessed November 2, 2021. https://www.cdc.gov/flu/avianflu/avian-in-birds.htm

Chatterjee, P.; Nair, P.; Chersich, M.; Terefe, Y.; Chauhan, A.S.; Quesada, F.; Simpson, G. 2021. One Health, "Disease X" and the challenge of "unknown" unknowns. Indian Journal of Medical Research 153: 264–271. https://doi.org/10.4103/ijmr.IJMR_601_21

Colella, J.P.; Bates, J.; Burneo, S.F.; Camacho, M.A.; Bonilla, C.C.; Constable, I.; et al. 2021. Leveraging natural history biorepositories as a global, decentralized, pathogen surveillance network. PLOS Pathogens 17: e1009583. https://doi.org/10.1371/journal.ppat.1009583

Cucunubá, Z.M.; Nouvellet, P.; Peterson, J.K.; Bartsch, S.M.; Lee, B.Y.; Dobson, A.P.; Basáñez, M.-G. 2018. Complementary paths to Chagas disease elimination: the impact of combining vector control with etiological treatment. Clinical Infectious Diseases 66: S293–S300. https://doi.org/10.1093/cid/ciy006

Cyranoski, D. 2017. Bat cave solves mystery of deadly SARS virus—and suggests new outbreak could occur. Nature 552: 15–16. https://doi.org/10.1038/d41586-017-07766-9

Daszak, P.; Olival, K.J.; Li, H. 2020. A strategy to prevent future epidemics similar to the 2019-nCoV outbreak. Biosafety and Health 2: 6–8. https://doi.org/10.1016/j.bsheal.2020.01.003

Datt, N.; Gosai, R.C.; Raviwasa, K.; Timote, V. 2020. Key transboundary plant pests of coconut [Cocos nucifera] in the Pacific Island Countries—a biosecurity perspective. Plant Pathology & Quarantine 10: 152–171. https://doi.org/10.5943/ppq/10/1/17

DeSalvo, K.; Hughes, B.; Bassett, M.; Benjamin, G.; Fraser, M.; Galea, S.; et al. 2021. Public health COVID-19 impact assessment: lessons learned and compelling needs. NAM Perspectives. https://doi.org/10.31478/202104c

Dicken, S.J.; Murray, M.J.; Thorne, L.G.; Reuschl, A.-K.; Forrest, C.; Ganeshalingham, M.; et al. 2021. Characterisation of B.1.1.7 and Pangolin coronavirus spike provides insights on the evolutionary trajectory of SARS-CoV-2. bioRxiv preprint. https://doi.org/10.1101/2021.03.22.436468

Dunnum, J.L.; McLean, B.S.; Dowler, R.C.; Systematic Collections Committee of the American Society of Mammalogists. 2018. Mammal collections of the Western Hemisphere: a survey and directory of collections. Journal of Mammalogy 99: 1307–1322. https://doi.org/10.1093/jmammal/gyy151

EcoHealth Alliance. n.d. Scientific Research and Pandemic Prevention. Accessed September 13, 2021. https://www.ecohealthalliance.org/

European Centre for Disease Prevention and Control. 2021. EpiPulse—the European surveillance portal for infectious diseases. Accessed March 22, 2022. https://www.ecdc.europa.eu/en/publications-data/epipulse-european-surveillance-portal-infectious-diseases

Evans, M.R. 2010. The swine flu scam? Journal of Public Health 32: 296–297. https://doi.org/10.1093/pubmed/fdq059

Fagre, A.C.; Lewis, J.; Miller, M.R.; Mossel, E.C.; Lutwama, J.J.; et al. 2021. Subgenomic flavivirus RNA (sfRNA) associated with Asian lineage Zika virus identified in three species of Ugandan bats (family Pteropodidae). Scientific Reports 11: 8370. https://doi.org/10.1038/s41598-021-87816-5

Fallah, M.; Skrip, L.A.; d'Harcourt, E.; Galvani, A.P. 2015. Strategies to prevent future Ebola epidemics. The Lancet 386: 131. https://doi.org/10.1016/S0140-6736(15)61233-8

FAO [Food and Agriculture Organization of the United Nations]. 2019. Food Outlook: Biannual Report on Global Food Markets, November 2019. Rome.

FAO; OiE; WHO [Food and Agriculture Organization of the United Nations; World Organisation for Animal Health; World Health Organization]. 2006. Global Early Warning and Response System for Major Animal Diseases, Including Zoonoses (GLEWS). https://www.woah.org/fileadmin/Home/eng/About_us/docs/pdf/GLEWS_Tripartite-Finalversion010206.pdf

Földvári, G.; Szabó, É.; Tóth, G.E.; Lanszki, Z.; Zana, B.; Varga, Z.; Kemenesi, G. 2022. Emergence of Hyalomma marginatum and Hyalomma rufipes adults revealed by citizen science tick monitoring in Hungary. Transboundary and Emerging Diseases 69: e2240–e2248. https://doi.org/10.1111/tbed.14563

Foster, J.B.; Clark, B.; Holleman, H. 2021. Capital and the ecology of disease. Monthly Review 73: 1–23. https://doi.org/10.14452/MR-073-02-2021-06_1

Gallardo, M.C.; de la Torre Reoyo, A.; Fernández-Pinero, J.; Iglesias, I.; Muñoz, M.J.; Arias, M.L. 2015. African swine fever: a global view of the current challenge. Porcine Health Management 1: 21. https://doi.org/10.1186/s40813-015-0013-y

GISRS [Global Influenza Surveillance and Response System]. n.d. GISRS (website). World Health Organization. Accessed August 17, 2021. https://www.who.int/initiatives/global-influenza-surveillance-and-response-system

GOARN [Global Outbreak Alert and Response Network]. n.d. GOARN [website]. World Health Organization. Accessed August 17, 2021. https://extranet.who.int/goarn/

Grange, Z.L.; Goldstein, T.; Johnson, C.K.; Anthony, S.; Gilardi, K.; Daszak, P.; et al. 2021. Ranking the risk of animal-to-human spillover for newly discovered viruses. Proceedings of the National Academy of Sciences USA 118: e2002324118. https://doi.org/10.1073/pnas.2002324118

Gurr, G.M.; Johnson, A.C.; Ash, G.J.; Wilson, B.A.L.; Ero, M.M.; Pilotti, C.A.; et al. 2016. Coconut lethal yellowing diseases: a phytoplasma threat to palms of global economic and social significance. Frontiers in Plant Science 7: 1521. https://doi.org/10.3389/fpls.2016.01521

Guzmán, C.A.F.; Aguirre, A.A.; Astle, B.; Barros, E.; Bayles, B.; Chimbari, M.; et al. 2021. A framework to guide planetary health education. Lancet Planetary Health 5: e253–e255. https://doi.org/10.1016/S2542-5196(21)00110-8

Hadler, J.L.; Patel, D.; Nasci, R.S.; Petersen, L.R.; Hughes, J.M.; Bradley, K.; et al. 2015. Assessment of arbovirus surveillance 13 years after introduction of West Nile virus, United States. Emerging Infectious Diseases 21: 1159–1166. https://doi.org/10.3201/eid2107.140858

Herrera, N.R. 2017. The emergence of living lab methods. In: Living Labs: Design and Assessment of Sustainable Living. D.V. Keyson, O. Guerra-Santin, D. Lockton (eds.). Springer, Berlin/Heidelberg.

Hoberg, E.P.; Brooks, D.R. 2015. Evolution in action: climate change, biodiversity dynamics and emerging infectious disease. Philosophical Transactions of the Royal Society B: Biological Sciences 370: 20130553. https://doi.org/10.1098/rstb.2013.0553

Hoberg, E.P.; Galbreath, K.E.; Cook, J.A.; Kutz, S.J.; Polley, L. 2012. Northern host-parasite assemblages: history and biogeography on the borderlands of episodic climate and environmental transition. In: Advances in Parasitology 79. D. Rollinson, S.I. Hay (eds.). Elsevier, Amsterdam. 1–97 p. https://doi.org/10.1016/B978-0-12-398457-9.00001-9

Huang, J.H.; Thomas, E. 2021. A review of living lab research and methods for user involvement. Technology Innovation Management Review 11: 88–107. http://doi.org/10.22215/timreview/1467

ILO [International Labour Organization]. 2021. World Employment and Social Outlook: Trends 2021. International Labour Office, Geneva. 161 pp. https://www.ilo.org/global/research/global-reports/weso/2021/WCMS_795453/lang--en/index.htm

INSDC [International Nucleotide Sequence Database Collaboration]. n.d. INSDC (website). Accessed March 22, 2022. https://www.insdc.org/

Jacobs, L.A. 2007. Rights and quarantine during the SARS global health crisis: differentiated legal consciousness in Hong Kong, Shanghai, and Toronto. Law and Society Review 41: 511–552. https://doi.org/10.1111/j.1540-5893.2007.00313.x

Janzen, D.H. 1985. On ecological fitting. Oikos 45: 308–310. https://doi.org/10.2307/3565565

Johns Hopkins Center for Civilian Biodefense Studies, Center for Strategic and International Studies, ANSER Institute for Homeland Security, and National Memorial Institute for the Prevention of Terrorism. 2001. Dark Winter: Bioterrorism Exercise, Andrews Air Force Base, June 22–23, 2001. Final Script, Explanatory Note to the Exercise Script, p. 2. Accessed https://carterheavyindustries.files.wordpress.com/2020/05/dark-winter-script.pdf

Kapiriri, L.; Ross, A. 2020. The politics of disease epidemics: a comparative analysis of the SARS, Zika, and Ebola outbreaks. Global Social Welfare 7: 33–45. https://doi.org/10.1007/s40609-018-0123-y

Kelly, T.R.; Karesh, W.B.; Johnson, C.K.; Gilardi, K.V.K.; Anthony, S.J.; Goldstein, T.; et al. 2017. One Health proof of concept: bringing a transdisciplinary approach to surveillance for zoonotic viruses at the human–wild animal interface. Preventive Veterinary Medicine 137 (Part B): 112–118. https://doi.org/10.1016/j.prevetmed.2016.11.023

Khasnis, A.A.; Nettleman, M.D. 2005. Global warming and infectious disease. Archives of Medical Research 36: 689–696. https://doi.org/10.1016/j.arcmed.2005.03.041

Khoo, E.J.; Lantos, J.D. 2020. Lessons learned from the COVID-19 pandemic. Acta Paediatrica 109: 1323–1325. https://doi.org/10.1111/APA.15307

Kim, J.; Kim, Y.L.; Jang, H.; Cho, M.; Lee, M.; Kim, J.; Lee, H.; et al. 2020. Living labs for health: an integrative literature review. European Journal of Public Health 30: 55–63. https://doi.org/10.1093/eurpub/ckz105

Lakoff, A. 2017. Unprepared: Global Health in a Time of Emergency. University of California Press, Oakland. 240 p.

Lane, H.J.; Blum, N.; Fee, E. 2010. Oliver Wendell Holmes (1809–1894) and Ignaz Philipp Semmelweis (1818–1865): preventing the transmission of puerperal fever. American Journal of Public Health 100: 1008–1009. https://doi.org/10.2105/AJPH.2009.185363

Lawson, B.; Petrovan, S.O.; Cunningham, A.A. 2015. Citizen science and wildlife disease surveillance. EcoHealth 12: 693–702. https://doi.org/10.1007/s10393-015-1054-z

Leach, M.; Dry, S. 2010. Epidemics: Science, Governance and Social Justice. Routledge, Oxfordshire, England. 320 pp.

Lytras, S.; Xia, W.; Hughes, J.; Jiang, X.; Robertson, D.L. 2021. The animal origin of SARS-CoV-2. Science 373: 968–970. https://doi.org/10.1126/science.abh0117

Mahdy, M.A.A.; Younis, W.; Ewaida, Z. 2020. An overview of SARS-CoV-2 and animal infection. Frontiers in Veterinary Science 7: 596391. https://doi.org/10.3389/fvets.2020.596391

Marizzi, C.; Florio, A.; Lee, M.; Khalfan, M.; Ghiban, C.; Nash, B.; et al. 2018. DNA barcoding Brooklyn (New York): a first assessment of biodiversity in Marine Park by citizen scientists. PLOS ONE 13: e0199015. https://doi.org/10.1371/journal.pone.0199015

Martinez, A.P.; Roberts, D.A. 1967. Lethal yellowing of coconuts in Florida. Proceedings of the Florida State Horticultural Society 80: 432.

McCullough, J. 2014. RBCs as targets of infection. Hematology 2014: 404–409. https://doi.org/10.1182/asheducation-2014.1.404

Morens, D.M.; Fauci, A.S. 2020. Emerging pandemic diseases: how we got to COVID-19. Cell 182: 1077–1092. https://doi.org/10.1016/j.cell.2020.08.021

MSF [Médecins Sans Frontières]. n.d. MSF (website). Accessed August 18, 2021. https://www.msf.org/

Naguib, M.M.; Ellström, P.; Järhult, J.D.; Lundkvist, Å.; Olsen, B. 2020. Towards pandemic preparedness beyond COVID-19. Lancet Microbe 1: e185–e186. https://doi.org/10.1016/S2666-5247(20)30088-4

New map shows the presence of Anopheles maculipennis s.l. mosquitoes in Europe. 2018. European Centre for Disease Prevention and Control. Accessed November 29, 2021. https://www.ecdc.europa.eu/en/news-events/new-map-shows-presence-anopheles-maculipennis-sl-mosquitoes-europe

Newton, J.; Kuethe, T. 2015. Economic implications of the 2014–2015 bird flu. farmdoc daily 5: 104, Department of Agricultural and Consumer Economics, University of Illinois at Urbana-Champaign.

Olivero, J.; Fa, J.E.; Real, R.; Farfán, M.Á.; Márquez, A.L.; Vargas, J.M.; et al. 2017. Mammalian biogeography and the Ebola virus in Africa. Mammal Review 47: 24–37. https://doi.org/10.1111/mam.12074

Oppenheim, B.; Gallivan, M.; Madhav, N.K.; Brown, N.; Serhiyenko, V.; Wolfe, N.D.; Ayscue, P. 2019. Assessing global preparedness for the next pandemic: development and application of an Epidemic Preparedness Index. BMJ Global Health 4: e001157. https://doi.org/10.1136/bmjgh-2018-001157

PAHO; WHO [Pan American Health Organization and World Health Organization]. 2008. Resolution CD48.R8: Integrated Vector Management: A Comprehensive Response to Vector-borne Diseases. 48th Directing Council, 60th Session of the Regional Committee, Washington, DC, USA, September 29–October 3, 2008. 3 pp. https://www3.paho.org/english/gov/cd/cd48.r8-e.pdf

Palmer, J.R.B.; Oltra, A., Collantes, F.; Delgado, J.A.; Lucientes, J.; Delacour, S.; et al. 2017. Citizen science provides a reliable and scalable tool to track disease-carrying mosquitoes. Nature Communications 8: 916. https://doi.org/10.1038/s41467-017-00914-9

Parrish, C.R.; Kawaoka, Y. 2005. The origins of new pandemic viruses: the acquisition of new host ranges by canine parvovirus and influenza A viruses. Annual Review of Microbiology 59: 553–586. https://doi.org/10.1146/annurev.micro.59.030804.121059

Parrish, C.R.; Murcia, P.R.; Holmes, E.C. 2015. Influenza virus reservoirs and intermediate hosts: dogs, horses, and new possibilities for influenza virus exposure of humans. Journal of Virology 89: 2990–2994. https://doi.org/10.1128/jvi.03146-14

Plourde, A.R.; Bloch, E.M. 2016. A literature review of Zika virus. Emerging Infectious Diseases 22: 1185–1192. https://doi.org/10.3201/eid2207.151990

Richardson, S.; Sinha, A.; Vahia, I.; Dawson, W.; Kaye, J.; et al. 2021. Brain health living labs. American Journal of Geriatric Psychiatry 29: 698–703. https://doi.org/10.1016/j.jagp.2020.11.010

Rohde, K., ed. 2013. The Balance of Nature and Human Impact. Cambridge University Press, Cambridge, England.

Saunders, D.G.O.; Pretorius, Z.A.; Hovmøller, M.S. 2019. Tackling the re-emergence of wheat stem rust in Western Europe. Communications Biology 2: 51. https://doi.org/10.1038/s42003-019-0294-9

Sheng, A.; Geng, X. 2021. How paradigm blindness leads to bad policy. Project Syndicate. Accessed March 14, 2023. www.project-syndicate.org/commentary/linear-mechanical-paradigm-creating-and-exacerbating-crises-by-andrew-sheng-and-xiao-geng-2021-08

Steen, K.; van Bueren, E. 2017. Urban Living Labs: A Living Lab Way of Working. Amsterdam Institute for Advanced Metropolitan Solutions, Delft University of Technology. 95 pp. https://research.tudelft.nl/en/publications/urban-living-labs-a-living-lab-way-of-working

Stenseth, N.C.; Dharmarajan, G.; Li, R.; Shi, Z.L.; Yang, R.; Gao, G.F. 2021. Lessons learnt from the COVID-19 pandemic. Frontiers in Public Health 9: 694705. https://doi.org/10.3389/fpubh.2021.694705

Suzán, G.; García-Peña, G.E.; Castro-Arellano, I.; Rico, O.; Rubio, A.V.; Tolsá, M.J.; et al. 2015. Metacommunity and phylogenetic structure determine wildlife and zoonotic infectious disease patterns in time and space. Ecology and Evolution 5: 865–873. https://doi.org/10.1002/ece3.1404

Szekeres, S.; Docters van Leeuwen, A.; Tóth, E.; Majoros, G.; Sprong, H.; Földvári, G. 2019. Road-killed mammals provide insight into tick-borne bacterial pathogen communities within urban habitats. Transboundary and Emerging Diseases 66: 277–286. https://doi.org/10.1111/tbed.13019

Tenorio, J.C.B. 2022. Discussing One Health: veterinary public health, health communication, and collaborations and partnerships. Advances in Animal and Veterinary Sciences 10: 852–857. https://doi.org/10.17582/JOURNAL.AAVS/2022/10.4.852.857

Timme, R.E.; Wolfgang, W.J.; Balkey, M.; Gubbala Venkata, S.L.; Randolph, R.; Allard, M.; Strain, E. 2020. Optimizing open data to support One Health: best practices to ensure interoperability of genomic data from bacterial pathogens. One Health Outlook 2: 20. https://doi.org/10.1186/s42522-020-00026-3

Trivellone, V.; Hoberg, E.P.; Boeger, W.A.; Brooks, D.R. 2022. Food security and emerging infectious disease: risk assessment and risk management. Royal Society Open Science 9: 211687. https://doi.org/10.1098/rsos.211687

van den Berg, H.; Zaim, M.; Singh Yadav, R.; Soares, A.; Ameneshewa, B.; Mnzava, A.; et al. 2012. Global trends in the use of insecticides to control vector-borne diseases. Environmental Health Perspectives 120: 577–582. https://doi.org/10.1289/ehp.1104340

Veeckman, C.; Schuurman, D.; Leminen, S.; Westerlund, M. 2013. Linking living lab characteristics and their outcomes: towards a conceptual framework. Technology Innovation Management Review 3: 6–15. https://doi.org/10.22215/timreview748

Vianna Franco, M.P.; Molnár, O.; Dorninger, C.; Laciny, A.; Treven, M.; Weger, J.; et al. 2022. Diversity regained: precautionary approaches to COVID-19 as a phenomenon of the total environment. Science of the Total Environment 825: 154029. https://doi.org/10.1016/j.scitotenv.2022.154029

Weaver, S.C.; Lecuit, M. 2015. Chikungunya virus and the global spread of a mosquito-borne disease. New England Journal of Medicine 372: 1231–1239. https://doi.org/10.1056/nejmra1406035

WHO [World Health Organization]. 2020. The COVID-19 Pandemic: Lessons Learned for the WHO European Region: A Living Document, 15 September 2020. https://apps.who.int/iris/handle/10665/334385

WHO [World Health Organization]. 2007. The World Health Report 2007: A Safer Future: Global Public Health Security in the 21st Century. WHO Press, Geneva, Switzerland. 72 pp. https://www.who.int/publications/i/item/9789241563444

WHO [World Health Organization]. 2005. International Health Regulations (2005). 3rd ed. WHO Press, Geneva, Switzerland. 74 pp. https://www.who.int/publications/i/item/9789241580496

WHO; World Bank [World Health Organization and International Bank for Reconstruction and Development/The World Bank]. 2019. Global Monitoring Report on Financial Protection in Health 2019. 50 pp. https://www.who.int/publications/i/item/9789240003958

6

The DAMA Protocol: Anticipating to Prevent and Mitigate Emerging Infectious Diseases

Eric P. Hoberg, Walter A. Boeger, Orsolya Molnár, Gabor Földvári, Scott L. Gardner, Alicia Juarrero, Vitaliy Kharchenko, Eloy Ortíz, Wolfgang Preiser, Valeria Trivellone, and Daniel R. Brooks

Abstract

Emerging infectious diseases (EIDs) pose a growing threat to technological humanity. Strategies for coping with EIDs have been mostly focused on preparing for recurrences of known pathogens and reactive crisis response for all outbreaks. The DAMA protocol (Document, Assess, Monitor, Act), the policy extension of the Stockholm Paradigm (SP), adds a dimension of prevention to coping with EIDs.

Keywords: emerging infectious disease (EID), Stockholm Paradigm (SP), Audy space, DAMA protocol (Document, Assess, Monitor, Act), EID prevention and mitigation, international cooperation

Introduction

The Stockholm Paradigm (SP) suggests we can anticipate much about how pathogens will behave when they arrive in a new place or colonize a new host, based on their preexisting and evolutionarily conservative capacities. This holds promise that we ought to be able to find them before they announce themselves, preventing at least some EID outbreaks and mitigating the impacts of outbreaks that cannot be prevented (Brooks et al., 2014, 2019; Hoberg and Brooks, 2015; Wille et al., 2021; Vora et al., 2022; Estrada-Peña et al., 2023). The rationale of the DAMA protocol (Document, Assess, Monitor, Act) is to provide an operational framework by which this can be achieved as timely and economically as possible. The ultimate goal is to reduce the cost of EID to sustainable levels.

DAMA is an integrative protocol that combines efforts to strategically document the distribution of complex assemblages of pathogens and their hosts across the biosphere, in the context of dynamic environmental interfaces that provide opportunities for pathogen exchange and emergence. Movement of habitats and animals (a breakdown in ecological isolation) catalyzed by climate change and broader anthropogenic trajectories of environmental disruption provide the landscape of opportunity for emergence of pathogens. Evolutionarily and ecologically conserved capacities for exploitation of host-based resources allow pathogens to persist in one place, or among a particular spectrum of hosts, and knowing about this can provide insights for predicting outcomes of persistence and emergence in novel conditions and across changing ecological interfaces.

EID unfolds in what we are referring to as *Audy space*, an intricate arena circumscribed by the biosphere, encompassing the interfaces within and across managed, agricultural, and wildland ecosystems, where the biological complexity of viruses, bacteria, plants, fungi, animals (including humans), and environments are in continuous or frequent contact (Figure 6.1).

Audy Space

Fitness space in general terms is fairly clear cut (Agosta, 2022; Agosta and Brooks, 2020). *Fundamental fitness space* (FFS) encompasses every place on the planet where a given species could have positive fitness (i.e., it can survive and reproduce, even if marginally), while *realized fitness space* (RFS) refers to the subset of fundamental fitness space in which given species actually exist at any given time. Relative to EID, RFS *for pathogens* is more complicated than the generality of RFS for all species. J.R. Audy (1958) noted more than half a century ago that a pathogen's distribution is not

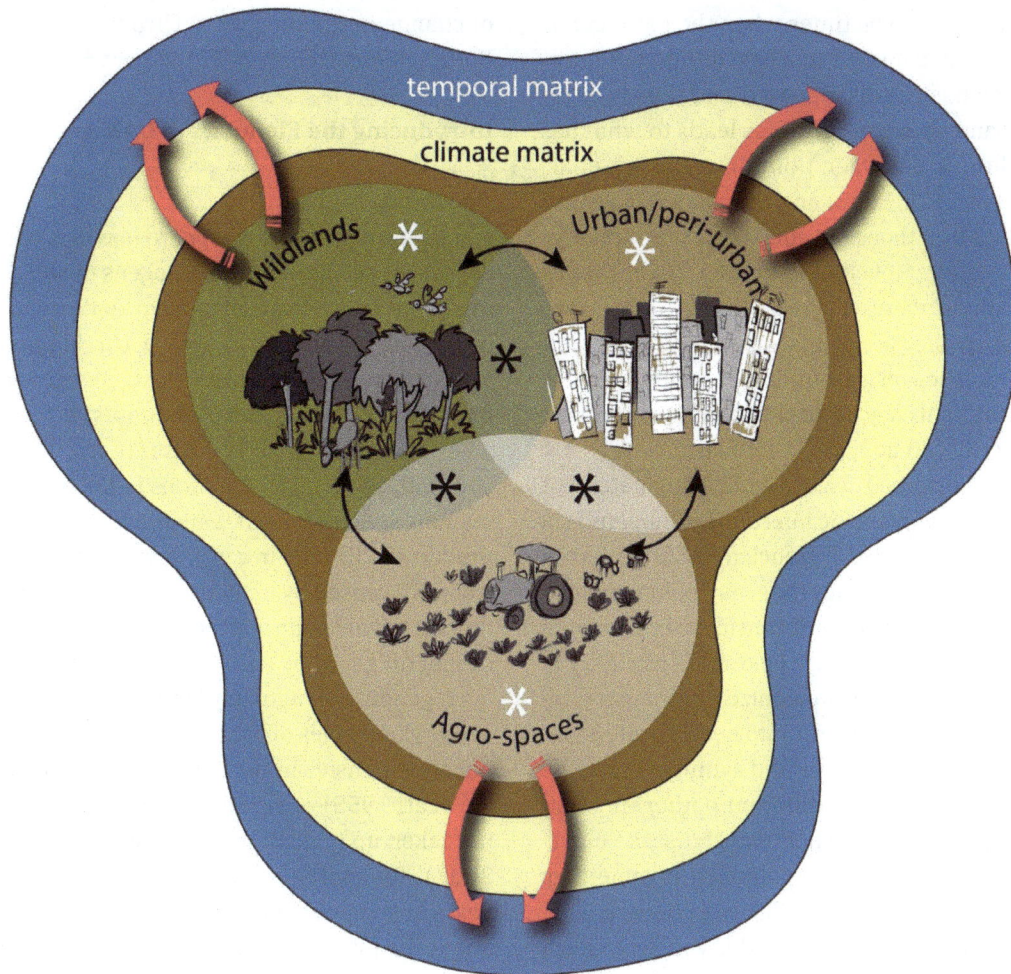

Figure 6.1. Audy space—A complex ecological arena for interfaces, pathogen distribution, and circulation. This generalized representation of Audy space and the operational arena for DAMA (Document, Assess, Monitor, Act) shows a strategic focus for documentation and action regions with the potential for intervention within a matrix that links space, climate, and time. Three major landscapes—wildlands, urban/peri-urban, and agro-spaces—and diverse habitat interfaces (*black asterisks*) represent risk spaces for occurrence of EIDs (*bidirectional black arrows*). Connectivity within and among the landscapes (*red arrows*) is dynamic (temporal matrix) and includes passive (e.g., climate responses) and active pathways (e.g., globalization, land use). In a proactive capacity, interventions are most appropriate and effective within landscapes (*white asterisks*) and among interfaces (*black asterisks*), which we define as *intervention space*, in order to prevent triggers for pathogen expansion. An interconnected scientific community can identify and communicate specific targets for actions to be undertaken. Taking action, however, requires cooperation among an often-siloed scientific community, public institutions, and governmental and nongovernmental organizations and agencies to develop, receive, and translate knowledge, emphasizing participation at local and community levels with insights for global humanity. (Modified from Trivellone, Hoberg, et al., 2022)

homogeneous throughout its geographic range, and he also noted that the geographic and host range of a pathogen always exceeds that of disease caused by it. This transforms RFS for pathogens into what we are calling Audy space.

In Audy space, pathogens that may cause disease are quite common, but actual disease outbreaks may be rare. Some host populations are less tolerant than others, and there may be some pathogen variants that are inherently more pathogenic than others, while other variants of a pathogen are more or less pathogenic in different hosts (Brooks et al., 2019; Brooks, Boeger, et al., 2022). Transmission dynamics may be subtly different in changing environmental conditions or novel host-exposure circumstances, either dampening or amplifying pathogen emergence. All this means there may be a lag between the arrival of a pathogen in a geographic area and a subsequent disease outbreak. There may also be times when the pathogen is present but not causing disease in a host or hosts of interest; the pathogen has not disappeared and thus may re-emerge rapidly and unexpectedly. This leads to what has been termed chronic pathogen pollution (Daszak et al., 2000; Cunningham et al., 2003) and the economically unsustainable "death by a thousand cuts" costs of cumulative small emergences (e.g., Hoberg and Brooks, 2015; Brooks et al., 2019; Brooks, Hoberg, et al., 2022).

Hosts—whether they are plants, humans, or other animals, wild or domestic—may be infected but not diseased, and health professionals pay no attention to asymptomatic hosts (a corollary of "do no harm"). Influenza-D virus, for example, causes widespread disease among ungulates, but antibodies suggest asymptomatic infections in humans (Liu et al., 2020; Leibler et al., 2022; Doucleff, 2023). Infected individuals may be diseased, but they may belong to host species whose disease status is not considered relevant to health care clinicians, and thus are ignored (a corollary of the belief that pathogens rarely colonize other hosts). Infected individuals of species relevant to health professionals may be diseased, but the signs and symptoms may be similar to those produced by a different pathogen. Clinicians who are not trained to expect novel disease emergence may misdiagnose the causal pathogen as something with which the clinician is more familiar (e.g., Hoberg, Trivellone, et al. 2022; Hoberg et al., 2023, this volume), which is generally symptomatic of *anchoring* or *confirmation bias* (e.g., see Aguirre et al., 2019). Further, geographically widespread pathogens distributed among an assemblage of wild and domestic plant and animal hosts provide another example of humans as facilitators of ecological super-spreading events, such as the globalization of pathogens in crops and livestock throughout the world and, most recently, the global distribution of SARS-CoV-2 in terrestrial ecosystems (Boeger et al., 2022; Hoberg, Boeger, Brooks, et al., 2022). The COVID pandemic and the secondary emergence of the Omicron variant of SARS-CoV-2 were associated with all three shortcomings (Boeger et al., 2022; Hoberg, Boeger, Brooks, et al., 2022).

Audy space is complex and dynamic, but the inherited components of host resource requirements, transmission dynamics, and microhabitat preference are highly specific and phylogenetically conservative (Agosta, 2023, this volume). This niche space definition of pathogens (Audy space) makes DAMA operationally feasible. DAMA is an integrative and iterative process strategically designed to reveal the complex interactions of pathogen and host assemblages and emergent disease unfolding in the context of change in the biosphere (Brooks et al., 2014; Hoberg, Boeger, Molnár, et al., 2022) (Figure 6.2).

Introducing the Elements of DAMA

Document

Humans are part of a biosphere in which perhaps as many as 50% of all species are pathogens of some sort. We have identified only a fraction of them. For example, the current cataloged diversity of viruses occurring in mammals and water birds (the vertebrate hosts most commonly identified in the origins of novel zoonoses) is less than 2,000 of an estimated 1.7 million viruses (Carroll et al., 2018; Carlson et al., 2019). The functions in the biosphere of the remaining species are not known, and thus their potential for producing EID remain a mystery. We cannot prevent or mitigate EID caused by pathogens we know nothing about. We do not, however, have to remain ignorant and continually caught off guard by each emergence of disease.

A generation ago, implementing a comprehensive inventory of species of pathogens was a daunting challenge—hugely expensive in time, money, and personnel (see Danald et al., 1955)—and for the most part, the challenge was not taken up (e.g., Brooks and Hoberg, 2000, 2001, 2006, 2007). Technological advances made in the intervening years have allowed us to generate large amounts of data about the presence of pathogens in any given geographic locality for relatively little money and in a short time. DNA technology, and RNA technology in particular, can allow rapid and economical documentation of the diversity and distribution of all manner of pathogens in all manner of hosts living in wildlands and managed landscapes (e.g., USAID, 2014, 2016; Joly et al., 2016; Cook et al., 2017; Dunnum et al., 2017; Colella et al., 2021; Wille et al., 2021; Albery et al., 2022; Harvey and Holmes, 2022; Hoberg, Trivellone, et al., 2022; Trivellone, 2022; Trivellone, Cao, et al., 2022; Trivellone, Hoberg, et al., 2022).

Taxonomic inventories document the identity of pathogens, where to find them (host identity and tissue tropism, georeferenced localities), and as much information about their natural history as possible. When specimens and associated data are properly maintained in archival collections, these data become fundamental and permanent biodiversity knowledge sources that contribute to establishment of baselines, thus providing essential insights that can be shared and compared among specialists desiring to use those data and other resources to prevent and mitigate EID (Dunnum et al., 2017; Colella et al., 2021; Hoberg, Boeger, Molnár, et al., 2022; Hoberg, Trivellone, et al., 2022).

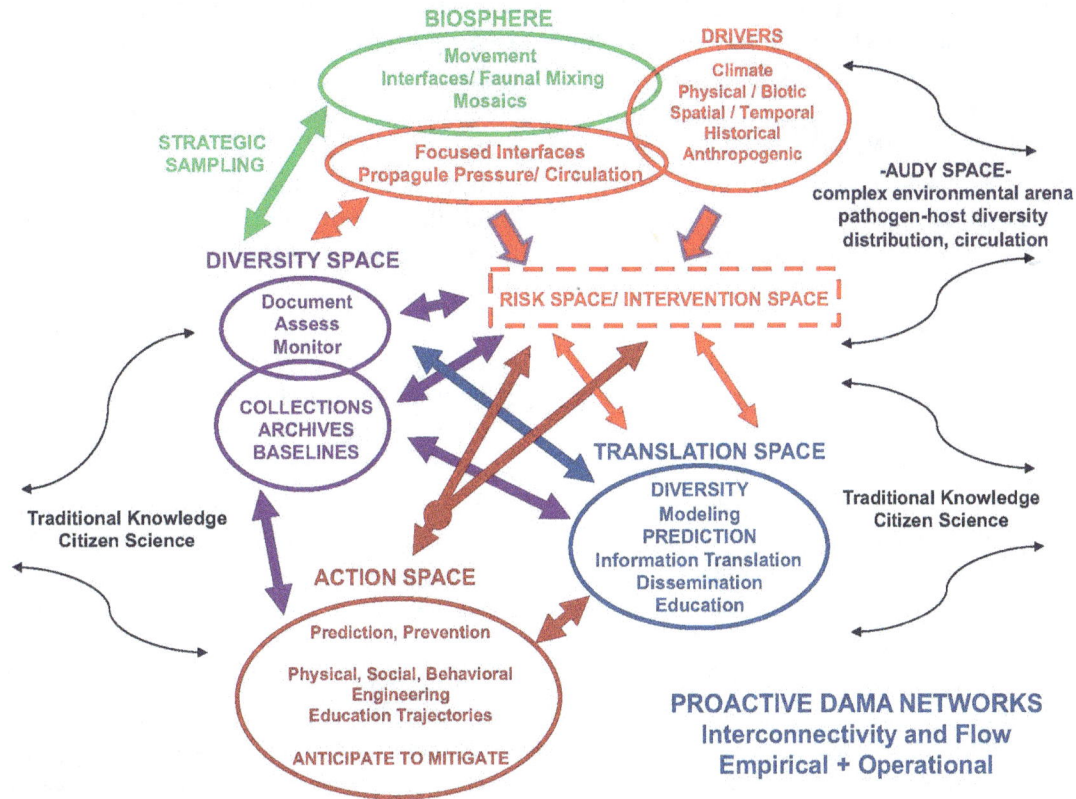

Figure 6.2. DAMA (Document, Assess, Monitor, Act)—an operational network demonstrating connectivity for identifying potentially emergent pathogens by developing and translating essential information for action networks. DAMA networks are an interconnected model that links *empirical observations* within a biosphere umbrella of Audy space through an *operational protocol* that defines connections and feedbacks for development and application of information and elucidation of Risk Space and Intervention Space. We emphasize that EIDs are fundamentally a biosphere phenomenon. At the core of the DAMA protocol, faunal structure and assembly are explored through Strategic Sampling, revealing the limits of Diversity Space. Document—documentation with accumulated, authoritatively identified, pathogen/host specimens archived in biological collections—becomes the basis for Assess and Monitor. Informatics, which consolidate phylogenetic and ecological assessment, feed into Translation Space and downstream to Action Space, establishing a pathway for Intervention. The Stockholm Paradigm (SP) and an operational DAMA link the conceptual and empirical arena of Audy space for the potential to anticipate and mitigate EID.

Assess

Assessment is the process of establishing or estimating risk relative to observed diversity for micro- and macroparasites. Assessment flows from documentation and establishes the primary trajectory for phylogenetic and ecological estimation of risk space, for targeting strategic monitoring, and to focus the resources for translation to actionable information (Figure 6.2) (Hoberg, Boeger, Molnár, et al., 2022; Janik et al., 2023; Brooks, Hoberg, et al., 2023, this volume). In the course of any effort to document pathogen diversity, an enormous number of species that are not known to cause disease in humans or animals or plants of socioeconomic significance to humans will be encountered.

If we had time, we would document every single species we discover and painstakingly study it to determine how these species function in the biosphere. Given the immediacy and magnitude of the EID threat, we cannot afford that luxury. We must make rapid decisions about which species need immediate attention. The precautionary principle gives us a mandate to identify and investigate pathogens that represent disease risks before an outbreak occurs. The primary goals of Assess are to winnow out known and potential disease-causing pathogens from expeditionary survey and inventory data before these pathogens break out and to use modern methods of phylogenetic comparative studies to provide the geographic and ecological information necessary to effectively monitor patterns of changing distribution and circulation across temporal and spatial scales.

Monitor

The goal of monitoring in DAMA is to detect, in as close to real time as possible, changes in host range, geographic distribution, transmission dynamics, and genetic variation of pathogens assessed as high-risk potential that might indicate increased chances of an outbreak.

DAMA replaces "minimum sampling effort" (Mayor et al., 2019) with "maximum focus sampling effort" (Brooks et al., 2014, 2019; Hoberg, Boeger, Molnár, et al., 2022), exerting most survey and inventory activity in strategically exploring reservoir hosts (where pathogens may exist without showing symptoms in their hosts) and habitat interfaces (where humans or nonhuman animal and plant hosts of socioeconomic importance are focused spatially and temporally). The sampling should be concentrated in areas which appear to facilitate the opportunity for exposure and pathogen exchange among infected reservoir hosts and new unexposed hosts. Dispersed pathogen circulation is the background (the generality of Audy space). In contrast, compatibility and the focused dynamics maximizing propagule pressure for pathogens constitute the nascent islands of colonization and emergence unfolding across landscape interfaces; this may also be a function of threshold population density for host assemblages driving transmission and persistence (reviewed in Anderson, 1991).

Monitoring efforts should focus on (1) discovery of reservoir hosts that are infected but not diseased (Audy, 1958) and (2) elucidation of the dynamics for pathogen transmission within core habitats and across boundaries and interfaces on the margins of wildland, urban/peri-urban, and agroscape environments of Audy space (Figure 6.1) (Tinnin, 2003; Hoberg, Boeger, Molnár, 2022; Hoberg, Trivellone, et al., 2022; Hoberg et al., 2023, this volume). Monitoring at local landscape through regional scales is necessary to decipher the complex mosaics and interconnections exhibited by pathogens and hosts on an environmental matrix determined by habitat heterogeneity, biological associations, anthropogenic forcing, and climate oscillations (e.g., Tinnin, 2003; Parmenter and Glass, 2022). Connectivity within those spaces is fluid with pathways influencing spatial and temporal islands of pathogen and disease distribution and persistence.

We need to account for movement and demographic changes in pathogens and pathogen variants within the leading edges of geographic expansion waves as well as subsequent isolation events (Hoberg and Brooks, 2015;

Laaksonen et al., 2015; Hoberg et al., 2017; Kafle et al., 2020). Recognizing and documenting change across slices of time, transcending geographic scale, is essential if we wish to reveal pathogen distribution and diversity (taxonomic and genetic). This is the realm of geographic and taxonomic oscillations and the dynamics of taxon pulses, within the SP, with alternating trends for exploration modes (in expansion) and exploitation modes (in stasis or isolation) (Agosta, 2022; Boeger et al., 2022; Hoberg, Boeger, Brooks, et al., 2022). Considerations of the depth and scope of sampling are needed to reveal occurrences of rare pathogens or pathogen variants. More broadly this reflects outcomes for pathogen mosaics and islands of disease across broader backgrounds of pathogen distribution described in Audy space. Integral to monitoring is periodic resampling and reassessment through integrating surveillance data with initial assessments, applying ecological niche and ecological network models to anticipate pathogen response to changing biotic and abiotic factors, including anthropogenic activities (Botero et al., 2019; Colella et al., 2021; Gardner et al., 2021; Dursahinhan et al., 2023).

Monitoring programs determine if pathogens assessed to be of high risk but not yet causing disease begin to show changes associated with disease outbreaks. When this happens, monitoring programs can alert public officials and the general public to emerging threats. The goal is to create a flow of "observation to data to analysis to projection to decision," a loop operating as close to real time as possible, beginning with those that are most at risk (Ortiz and Juarrero, 2022; Ortiz and Juarrero, 2023, this volume) (Figure 6.2).

Act

Molnár et al. (Molnár, Hoberg, et al., 2022; Molnár, Knickel, et al. 2022; Molnár, Hoberg, et al., 2023, this volume; Molnár, Knickel, et al., 2023, this volume) summarized multiple components of activities associated with making Act effective within the powerful 3P (Prevent-Prepare-Palliate) framework. Most important is that as soon as EID threats have been identified and appropriate public agencies and agents notified, there must be prompt and decisive actions (Molnár, Knickel, et al., 2022; Molnár, Knickel, et al., 2023, this volume). Depending on unfolding events, that may be action by public health, wildlife, agriculture, or veterinary officials to prevent an outbreak if monitoring shows imminent signs; action to prepare to mitigate the impact of pathogens assessed to be of high risk that escape prevention; or action to engage in palliation for those cases in which prevention and preparation failed, and a pandemic occurs.

For action to be effective, information must circulate easily among scientists, clinicians, public officials, and the public (Trivellone, Hoberg, et al., 2022) (Figure 6.2). Barriers to accessibility for critical insights and actions for EID must be eliminated. DAMA initiatives cannot function effectively unless information is shared freely among all concerned parties. The EID crisis is a global concern for security and human well-being such that Act recommendations must be shared with relevant governmental and nongovernmental organizations that can spread an alert as widely as is necessary and take appropriate action.

Public outreach and public involvement is also essential. People need reliable information that provides choices for behavioral change to accommodate EID. Knowing the signs and symptoms of emergent disease is fundamental. Knowing how pathogens are transmitted is critical. Knowing your local environment, checking yourself and your children for ticks, emptying a birdbath, caring properly for companion animals and garden plants, altering dawn and dusk activity patterns, and maintaining access to clean water and unspoiled food can limit chains of exposure and infection. A cost-effective way to do this is through the science-based curriculum of local schools and public natural history museums.

A Call for Cooperation

EIDs are an unsustainably expensive global reality, and humans are the world's primary ecological super-spreaders of pathogens. We must assume these changes are now a permanent trajectory in the biosphere. Preventing as many outbreaks as possible and mitigating the impacts of those that do occur is an existential global concern as well. Effective risk management for EID requires actionable information before a crisis appears (Trivellone, Hoberg, et al., 2022). This can be achieved only if we think that we can prevent EIDs or greatly mitigate their impact. Coping with changes of that magnitude requires the cooperation of many people within and among countries and on an unprecedented scale. In order to cope with the onrushing multiple threats associated with climate change—of which EID is only one—we must integrate diverse human activities on multiple scales, and everyone must contribute (Brooks et al., 2014, 2019; Hoberg, Boeger, Molnár, et al., 2022; Mora et al., 2022).

A number of global programs associated with One Health, Eco-Health, Planetary Health, and Global Health (e.g., PREDICT) and intergovernmental bodies such as the World Health Organization have incorporated some basic elements consistent with a DAMA framework (e.g., Horton and Lo, 2015; Whitmee et al., 2015; Daszak et al., 2020; Daszak and Olival, 2020; Kelly et al., 2020; Alimi et al., 2021; Chatterjee et al., 2021; Ellwanger et al., 2021; Carlson et al., 2022). They have not been cooperative to any great degree. They have not been based in a conceptual framework indicating that prevention is possible. They have not included support for training and employing additional taxonomists, nor have they been concerned with supporting permanent archival collections and centers for data analysis needed for proper assessment, monitoring, and reassessment (Hoberg, Trivellone, et al., 2022; Hoberg et al., 2023, this volume). They have not been associated with any particular action plan because mostly they have been chasing pathogens that have already emerged, leaving business as usual as the only option.

Like the blind men encountering the elephant, each proposal for addressing EID has recognized a component of the larger challenge but has remained incomplete in defining the problem. Often overlapping and seemingly competing explanations border on single explanations for EID with limited proposals for solutions and a pathway forward beyond increasingly rapid response (Plowright et al., 2017; Brooks et al., 2019; Hoberg, Boeger, Molnár, et al., 2022). Climate forcing and the downstream influence on diversity, biotic movement, and dynamics at interfaces remain the overall determinant for the environmental opportunity and the potential for EID (Hoberg et al., 2017; Brooks et al., 2019). Missing from this array of initiatives, and where DAMA differs, is in an explicit evolutionary foundation with synthesis and an integrative process across field-based collections, archives and actionable information to establish insights that reflect a historical and contemporary context for EID (Figure 6.2). We endorse a broad perspective with increasingly transboundary and interdisciplinary cooperation bringing evolution, ecology, biodiversity, environmental health, and medicine to the forefront to address complex challenges for humanity and the global landscape (Hoberg et al., 2015; Whitmee et al., 2015; Brooks et al., 2019; Molnár, Hoberg, et al., 2022; Molnár, Knickel, et al., 2022; Trivellone, Hoberg, et al., 2022) (Figure 6.3). The core of DAMA can be the basis for cultural transformation in the disease-ecology community (Brooks et al., 2014; Brooks, Hoberg, et al., 2022; Hoberg, Boeger, Molnár, et al., 2022; Hoberg, Trivellone, et al., 2022; Brooks, Boeger, et al., 2023, this volume; Wallace-Wells, 2023).

Business as usual is not working, and the EID crisis is accelerating. It is time for a new approach, based in the SP and DAMA and characterized by global cooperation among

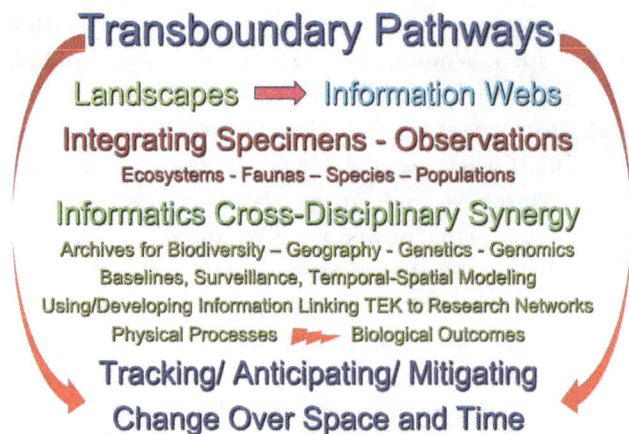

Figure 6.3. Transboundary pathways of DAMA. A generalized scheme for information flow across Document, Assess, and Monitor as the primary interacting focus for Act, which supports prevention through anticipation and mitigation.

different groups of specialists, each uniquely capable of contributing an essential element of coping with the crisis (Brooks et al., 2019; Hoberg, Boeger, Molnár, et al., 2022; Hoberg, Trivellone, et al., 2022).

Conclusions

DAMA samples and explores critical components of the biosphere, and as a process DAMA is taxonomically, temporally, and spatially scalable, geographically portable and maximally flexible. Conceptual foundations from the SP drive operational components of DAMA for explorations of diversity in Audy space (Hoberg, Boeger, Molnár, et al., 2022; Brooks, Boeger, et al., 2023, this volume; Brooks, Hoberg, et al., 2023, this volume; Janik et al., 2023) with identification of environmental change, interfaces, and anticipation for outcomes of movement leading to faunal or floral mixing and mosaics as the basis for pathogen emergence and disease. SP and DAMA emphasize that anticipation/prediction is attainable and is a capability that must be pursued in a precolonization arena prior to emergence (e.g., Boeger et al., 2022; Hoberg, Boeger, Brooks, et al., 2022; Hoberg, Boeger, Molnár, et al., 2022; Molnár, Hoberg, et al., 2022; Molnár, Knickel, et al., 2022). Prediction occurs under a biosphere umbrella—a view posed in transboundary linkages by strategic collections, archives, phylogeny, and translated information (Figure 6.3). Collections and archives with assessment are the operational/empirical cores from which actionable information flows.

The DAMA protocol is one proposal within a broader framework for buying time in an arena of accelerating change and must be at the forefront of policies aimed at reducing the socioeconomic impact of threat multipliers driven by climate change and related perturbation (Brooks, 2022; Vasbinder and Sim, 2022). If we hope to take effective action against EID, DAMA must be the first priority in the 3P infrastructure. We must acknowledge that every EID represents a failure to prevent an outbreak, and every pandemic represents a failure to mitigate an outbreak that occurs. If colonizing new hosts is based on preexisting specific and phylogenetically conservative capacities, outbreaks can be largely anticipated and *prevented*. Once a new host is colonized—that is, once an outbreak occurs—the impact of the outbreak can be mitigated to the extent that *preparation* for anticipated outbreaks has been made and the mitigating actions are undertaken quickly. The longer a pathogen exists and is isolated in a new host, the greater the likelihood that new variants will evolve and emerge (e.g., Boeger et al., 2022). Moreover, those new variants will be unpredictable, as demonstrated by the recurring events for emergence of COVID-19 in the past 3 years, leaving us with no option except unsustainably expensive measures for *palliation* in a rapidly changing arena for hosts and pathogens.

Implementing DAMA will be expensive, but projected costs of interminable episodes of crisis response are unsustainable (Brooks et al., 2019; Brooks, Hoberg, et al., 2022; Hoberg, Boeger, Molnár, et al., 2022; Trivellone, Hoberg, et al., 2022; Vora et al., 2022; Brooks, Boeger, et al., 2023, this volume; Wallace-Wells, 2023).

Literature Cited

Agosta, S.J. 2022. The Stockholm paradigm explains the dynamics of Darwin's entangled bank, including emerging infectious disease. MANTER: Journal of Parasite Biodiversity Number 27. https://doi.org/10.32873/unl.dc.manter27

Agosta, S.J. 2023. The Stockholm Paradigm explains the eco-evolutionary dynamics of the biosphere in a changing world, including emerging infectious disease. In: An Evolutionary Pathway for Coping with Emerging Infectious Disease. S.L. Gardner, D.R. Brooks, W.A. Boeger, E.P. Hoberg (eds.). Zea Books, Lincoln, NE.

Agosta, S.J.; Brooks, D.R. 2020. The Major Metaphors of Evolution: Darwinism Then and Now. Springer International, Chan, Switzerland. 273 p. https://doi.org/10.1007/978-3-030-52086-1

Aguirre, L.E.; Cheung, T.; Iorio, M.; Mueller, M. 2019. Anchoring bias, Lyme disease, and the diagnosis conundrum. Cureus 11: e4300. https://doi.org/10.7759/cureus.4300

Albery, G.F.; Carlson, C.J.; Cohen, L.E.; Eskew, E.A. Gibb, R.; Ryan, S.J.; Sweeny, A.R.; Becker, D.J. 2022. Urban-adapted mammal species have more known pathogens. Nature Ecology and Evolution 6: 794–801. https://doi.org/10.1038/s41559-022-01723-0

Alimi, Y.; Bernstein, A.; Epstein, J.; Espinal, M.; Kakkar, M.; Kochevar, D.; Werneck, G. 2021. Report of the Scientific Task Force on Preventing Pandemics. Harvard Global Health Institute and the Center for Climate, Health, and the Global Environment at Harvard T.H. Chan School of Public Health. 36 pp.

Anderson, R.M. 1991. Discussion: the Kermack-McKendrick epidemic threshold theorem. Bulletin of Mathematical Biology 53: 3–32. https://doi.org/10.1016/S0092-8240(05)80039-4

Audy, J.R. 1958. The localization of disease with special reference to the zoonoses. Transactions of the Royal Society of Tropical Medicine and Hygiene 52: 308–334. https://doi.org/10.1016/0035-9203(58)90045-2

Boeger, W.A.; Brooks, D.R.; Trivellone, V.; Agosta, S.J.; Hoberg, E.P. 2022. Ecological super-spreaders drive host-range oscillations: Omicron and risk space for emerging infectious disease. Transboundary and Emerging Diseases 69: e1280–e1288. https://doi.org/10.1111/tbed.14557

Botero-Cañola, S.; Dursahinhan, A.T.; Racz, S.E.; Lowe, P.V.; Ubelaker, J.E.; Gardner, S.L. 2019. The ecological niche of Echinococcus multilocularis in North America: understanding biotic and abiotic determinants of parasite distribution with new records in New Mexico and Maryland, United States. Therya 10: 91–102.

Brooks, D.R. 2022. What are emerging infections diseases? In: Buying Time for Climate Action: Exploring Ways around Stumbling Blocks. J.W. Vasbinder, J.Y.H. Sim (eds.). World Scientific Publishing, Singapore. 43–49 p.

Brooks, D.R.; Boeger, W.A.; Hoberg, E.P. 2022. The Stockholm paradigm: lessons for the emerging infectious disease crisis. MANTER: Journal of Parasite Biodiversity 22. https://doi.org/10.32873/unl.dc.manter22

Brooks, D.R.; Boeger, W.A.; Hoberg, E.P. 2023. The Stockholm Paradigm: the conceptual platform for coping with the emerging infectious disease crisis. In: An Evolutionary Pathway for Coping with Emerging Infectious Disease. S.L. Gardner, D.R. Brooks, W.A. Boeger, E.P. Hoberg (eds.). Zea Books, Lincoln, NE.

Brooks, D.R.; Hoberg, E.P. 2000. Triage for the biosphere: the need and rationale for taxonomic inventories and phylogenetic studies of parasites. Comparative Parasitology 67: 1–25.

Brooks, D.R.; Hoberg, E.P. 2001. Parasite systematics in the 21st century: opportunities and obstacles. Trends in Parasitology 17: 273–275. https://doi.org/10.1016/s1471-4922(01)01894-3

Brooks, D.R.; Hoberg, E.P. 2006. Systematics and emerging infectious diseases: from management to solution. Journal of Parasitology 92: 426–429. https://doi.org/10.1645/GE-711R.1

Brooks, D.R.; Hoberg, E.P. 2007. How will global climate change affect parasite-host assemblages? Trends in Parasitology 23: 571–574. https://doi.org/10.1016/j.pt.2007.08.016

Brooks, D.R.; Hoberg, E.P.; Boeger, W.A. 2019. The Stockholm Paradigm: Climate Change and Emerging Disease. University of Chicago Press, Chicago. 423 pp.

Brooks, D.R.; Hoberg, E.P.; Boeger, W.; Gardner, S.L.; Galbreath, K.E.; Herczeg, D.; et al. 2014. Finding them before they find us: informatics, parasites, and environments in accelerating climate change. Comparative Parasitology 81: 155–164. https://doi.org/10.1654/4724b.1

Brooks, D.R.; Hoberg, E.P.; Boeger, W.A.; Trivellone, V. 2022. Emerging infectious disease: an underappreciated area of strategic concern for food security. Transboundary and Emerging Diseases 69: 254–267. https://doi.org/10.1111/tbed.14009

Brooks, D.R.; Hoberg, E.P.; Boeger, W.A.; Trivellone, V. 2023. Assess: using evolution to save time and resources. In: An Evolutionary Pathway for Coping with Emerging Infectious Disease. S.L. Gardner, D.R. Brooks, W.A. Boeger, E.P. Hoberg (eds.). Zea Books, Lincoln, NE.

Carlson, C.J.; Boyce, M.R.; Dunne, M.; Graeden, E.; Lin, J.; Abdellatif, Y.O.; et al. 2022. The World Health Organization's disease outbreak news: a retrospective database. Preprint version posted March 23, 2022, in medRxiv. https://doi.org/10.1101/2022.03.22.22272790

Carlson, C.J.; Zipfel, C.M.; Garnier, R.; Bansal, S. 2019. Global estimates of mammalian viral diversity accounting for host sharing. Nature Ecology Evolution 3: 1070–1075. https://doi.org/10.1038/s41559-019-0910-6

Carroll, D.; Daszak, P.; Wolfe N.D.; Gao, G.F.; Morel, C.M.; Morzaria, S.; et al. 2018. The Global Virome Project. Science 359: 872–874. https://doi.org/10.1126/science.aap7463

Chatterjee, P.; Nair, P.; Chersich, M.; Terefe, Y.; Chauhan, A.S.; Quesada, F.; Simpson, G. 2021 One Health, "disease X" and the challenge of "unknown" unknowns. Indian Journal Medical Research 153: 264–271. https://doi.org/10.4103/ijmr.IJMR_601_21

Colella, J.P.; Bates, J.; Burneo, S.F.; Camacho, M.A.; Carrion Bonilla, C.; Constable, I.; et al. 2021. Leveraging natural history biorepositories as a global, decentralized pathogen surveillance network. PLOS Pathogens 17: e1009583. https://doi.org/10.1371/journal.ppat.1009583

Cook, J.A.; Galbreath, K.E.; Bell, K.C.; Campbell, M.L.; Carrière, S.; Colella, J.P.; et al. 2017. The Beringian Coevolution Project: holistic collections of mammals and associated parasites reveal novel perspectives on evolutionary and environmental change in the North. Arctic Science 3: 585–617. https://doi.org/10.1139/as-2016-0042

Cunningham, A.A.; Daszak, P.; Rodríguez, J.P. 2003. Pathogen pollution: defining a parasitological threat to biodiversity conservation. Journal of Parasitology 89: S78–S83.

Danald, G.E.; Woodbuiry, A.M.; Newcombe, C.L.; Rees, D.M.; Gebhardt, L.P.; Dorrell, W.W. 1955. Symposium on Ecology of Disease Transmission in Native Animals. Dugway Proving Ground, Dugway, UT, April 6–8, 1955. Ecological Research, University of Utah, and US Army Chemical Corps. 112 pp.

Daszak, P.; Amuasi, J.; das Neves, C.G.; Hayman, D.; Kuiken, T.; Roche, B.; et al. 2020. IPBES Workshop on Biodiversity and Pandemics: Workshop Report. Intergovernmental Science-Policy Platform on Biodiversity and Ecosystem Services, Bonn, Germany. Retrieved from https://ipbes.net/pandemics

Daszak, P.; Cunningham, A.A.; Hyatt, A.D. 2000. Emerging infectious diseases of wildlife—threats to biodiversity and human health. Science 287: 443–449. https://doi.org/science.287.5452.443

Daszak, P.; Olival, K.J.; Li, H. 2020. A strategy to prevent future epidemics similar to the 2019-nCoV outbreak. Biosafety and Health 2: 6–8. https://doi.org/10.1016/j.bsheal.2020.01.003

Doucleff, M. 2023. A new flu is spilling over from cows to people in the U.S. How worried should we be? National Public Radio, Morning Edition. 29 March 2023.

Dunnum, J.L.; Yanagihara, R.; Johnson, K.M.; Armien, B.; Batsaikhan, N.; Morgan, L.; Cook, J.A. 2017. Biospecimen repositories and integrated databases as critical infrastructure for pathogen discovery and pathobiology research. PLOS Neglected Tropical Diseases 11: e0005133. https://doi.org/10.1371/journal.pntd.0005133

Dursahinhan, A.T.; Botero-Cañola, S; Gardner, S.L. 2023. Intercontinental comparisons of subterranean host-parasite communities using bipartite network analyses. Parasitology 150: 446–454. https://doi.org/10.1017/S0031182023000148

Ellwanger, J.H.; da Veiga, A.B.G.; Kaminski, V.D.L.; Valverde-Villegas, J.M.; de Freitas, A.W.Q.; Chies, J.A.B. 2021 Control and prevention of infectious diseases from a One Health perspective. Genetics Molecular Biology 44: e20200256. https://doi.org/10.1590/1678-4685-GMB-2020-0256

Estrada-Peña, A.; Guglielmone, A.A.; Nava, S. 2023. Worldwide host associations of the tick genus *Ixodes* suggest relationships based on environmental sharing rather than co-phylogenetic events. Parasites and Vectors 16: 75. https://doi.org/10.1186/s13071-022-05641-9

Gardner, S.L.; Botero-Cañola, S.; Aliaga-Rossel, E.; Dursahinhan, A.T.; Salazar-Bravo, J.A. 2021. Conservation status and natural history of *Ctenomys*, tuco-tucos in Bolivia. Therya 12: 15–36. https://doi.org/10.12933/therya-21-1035

Harvey, E.; Holmes, E.C. 2022. Diversity and evolution of the animal virome. Nature Reviews Microbiology 20: 321–334. https://doi.org/10.1038/s41579-021-00665-x

Hoberg, E.P.; Boeger, W.A.; Brooks, D.R.; Trivellone, V.; Agosta, S. 2022. Stepping-stones and mediators of pandemic expansion: a context for humans as ecological super-spreaders. MANTER: Journal of Parasite Biodiversity 18. https://doi.org/10.32873/unl.dc.manter18

Hoberg, E.P.; Boeger, W.A.; Molnár, O.; Földvári, G.; Gardner, S.; Juarrero, A.; et al. 2022. The DAMA protocol, an introduction: finding pathogens before they find us. MANTER: Journal of Parasite Biodiversity 21. https://doi.org/10.32873/unl.dc.manter21

Hoberg, E.P.; Brooks, D.R. 2015. Evolution in action: climate change, biodiversity dynamics and emerging infectious disease. Philosophical Transactions of the Royal Society B 370: 20130553. https://doi.org/10.1098/rstb.2013.0553

Hoberg, E.P.; Cook, J.A.; Agosta, S.J.; Boeger, W.; Galbreath, K.E.; Laaksonen, S.; et al. 2017. Arctic systems in the Quaternary: ecological collision, faunal mosaics and the consequences of a wobbling climate. Journal of Helminthology 91: 409–421. https://doi.org/10.1017/S0022149X17000347

Hoberg, E.P.; Trivellone, V.; Cook, J.A.; Dunnum, J.L.; Boeger, W.A.; Brooks, D.R.; et al. 2022. Knowing the biosphere: documentation, specimens, archives, and names reveal environmental change and emerging pathogens. MANTER: Journal of Parasite Biodiversity 26. https://doi.org/10.32873/unl.dc.manter26

Hoberg, E.P.; Trivellone, V.; Cook, J.A.; Dunnum, J.L.; Boeger, W.A.; Brooks, D.R.; et al. 2023. Document: pathogen diversity—finding them before they find us. In: An Evolutionary Pathway for Coping with Emerging Infectious Disease. S.L. Gardner, D.R. Brooks, W.A. Boeger, E.P. Hoberg (eds.). Zea Books, Lincoln, NE.

Horton, R.; Lo, S. 2015. Planetary health: a new science for exceptional action. Lancet 386: 1921–1922. https://doi.org/10.1016/S0140-6736(15)61038-8

Janik, K.; Panassiti, B.; Kerschbamer, C.; Burmeister, J.; Trivellone, V. 2023. Phylogenetic triage and risk assessment: how to predict emerging phytoplasma diseases. Biology 12: 732. https://doi.org/10.3390/biology12050732

Joly, D.; Kreuder Johnson, C.; Goldstein, T.; Anthony, S.J.; Karesh, W.; Daszak, P.; et al. 2016. The first phase of PREDICT: surveillance for emerging infectious zoonotic diseases of wildlife origin (2009–2014). International Journal of Infectious Diseases 53: 31–32. https://doi.org/10.1016/j.ijid.2016.11.086

Kafle, P.; Peller, P.; Massolo, A.; Hoberg, E.; Leclerc, L.-M.; Tomaselli, M.; Kutz, S. 2020. Range expansion of muskox lungworms track rapid Arctic warming: implications for geographic colonization under climate forcing. Scientific Reports 10: 17323. https://doi.org/10.1038/s41598-020-75358-5

Kelly, T.R.; Machalaba, C.; Karesh, W.B.; Crook, P.Z.; Gilardi, K.; Nziza, J.; et al. 2020. Implementing One Health approaches to confront emerging and reemerging zoonotic disease threats: lessons from PREDICT. One Health Outlook 2: 1. https://doi.org/10.1186/s42522-019-0007-9

Laaksonen, S.; Oksanen, A.; Hoberg E. 2015. A lymphatic dwelling filarioid nematode, *Rumenfilaria andersoni* (Filarioidea; Splendidofilariinae), is an emerging parasite in Finnish cervids. Parasites and Vectors 8: 228. https://doi.org/10.1186/s13071-015-0835-0

Leibler, J.H.; Abdelgadir, A.; Seidel, J.; White, R.F.; Johnson, W.E.; Reynolds, S.J.; et al. 2022. Influenza D virus exposure among US cattle workers: a call for surveillance. Zoonoses and Public Health 70: 166–170. https://doi.org/10.1111/zph.13008

Liu, R.; Sheng, Z.; Huang, C.; Wang, D.; Li, F. 2020. Influenza D virus. Current Opinion in Virology 44: 154–161. https://doi.org/10.1016/j.coviro.2020.08.004

Mayor, P.; El Bizri, H.R.; Morcatty, T.Q.; Moya, K.; Solis, S.; Bodmer, R.E. 2019. Assessing the minimum sampling effort required to reliably monitor wild meat trade in urban markets. Frontiers in Ecology and Evolution 7: 180. https://doi.org/10.3389/fevo.2019.00180

Molnár, O.; Hoberg, E.; Trivellone, V.; Földvári, G.; Brooks, D.R. 2022. The 3P framework: a comprehensive approach to coping with the emerging infectious disease crisis. MANTER: Journal of Parasite Biodiversity 23. https://doi.org/10.32873/unl.dc.manter23

Molnár, O.; Hoberg, E.P.; Trivellone, V.; Földvári, G.; Brooks, D.R. 2023. Prevent-Prepare-Palliate: the 3P framework—integrating the DAMA protocol into global public health systems. In: An Evolutionary Pathway for Coping with Emerging Infectious Disease. S.L. Gardner, D.R. Brooks, W.A. Boeger, E.P. Hoberg (eds.). Zea Books, Lincoln, NE.

Molnár, O.; Knickel, M.; Marizzi, C. 2022. Taking action: turning evolutionary theory into preventive policies. MANTER: Journal of Parasite Biodiversity 28. https://doi.org/10.32873/unl.dc.manter28

Molnár, O.; Knickel, M.; Marizzi, C. 2023. All hands on deck: turning evolutionary theory into preventive policies. In: An Evolutionary Pathway for Coping with Emerging Infectious Disease. S.L. Gardner, D.R. Brooks, W.A. Boeger, E.P. Hoberg (eds.). Zea Books, Lincoln, NE.

Mora, C.; McKenzie, T.; Gaw, I.M.; Dean, J.M.; von Hammerstein, H.; Knudson, T.A.; et al. 2022. Over half of known human pathogenic diseases can be aggravated by climate change. Nature Climate Change 12: 869–875. https://doi.org/10.1038/s41558-022-01426-1

Ortiz, E.; Juarrero, A. 2022. An infectious disease surveillance platform for the 21st century. MANTER: Journal of Parasite Biodiversity 29. https://doi.org/10.32873/unl.dc.manter29

Ortiz, E.; Juarrero, A. 2023. Monitoring: an emerging infectious disease surveillance platform for the 21st century. In: An Evolutionary Pathway for Coping with Emerging Infectious Disease. S.L. Gardner, D.R. Brooks, W.A. Boeger, E.P. Hoberg (eds.). Zea Books, Lincoln, NE.

Parmenter, R.R.; Glass, G.E. 2022. Hantavirus outbreaks in the American Southwest: propagation and retraction of rodent and virus diffusion waves from sky-island refugia. International Journal of Modern Physics B 36: 2140052. https://doi.org/10.1142/S021797922140052X

Plowright, R.K.; Parrish, C.R.; McCallum, H.; Hudson, P.J.; Ko, A.I.; Graham, A.L.; Lloyd-Smith, J.O. 2017. Pathways to zoonotic spillover. Nature Reviews Microbiology 15: 502–510. https://doi.org/10.1038/nrmicro.2017.45

Tinnin, D.S. 2003. Testing the refugia hypothesis: population dynamics of *Peromyscus* and hantavirus seroprevalence across an elevational gradient. Master's thesis, University of New Mexico, Albuquerque.

Trivellone, V. 2022. Let emerging plant disease be predictable. MANTER: Journal of Parasite Biodiversity 30. https://doi.org/10.32873/unl.dc.manter30

Trivellone, V.; Cao, Y.; Dietrich, C.H. 2022. Comparison of traditional and next-generation approaches for uncovering phytoplasma diversity, with discovery of new groups, subgroups and potential vectors. Biology 11: 977. https://doi.org/10.3390/biology11070977

Trivellone, V.; Hoberg, E.P.; Boeger, W.A.; Brooks, D.R. 2022. Food security and emerging infectious disease: risk assessment and risk management. Royal Society Open Science 9: 211687. https://doi.org/10.1098/rsos.211687

USAID [United States Agency for International Development]. 2014. Reducing Pandemic Risk, Promoting Global Health. PREDICT 1 (2009–2014) Final Report.

USAID [United States Agency for International Development]. 2016. Reducing Pandemic Risk, Promoting Global Health, Supporting the Global Health Security Agenda. PREDICT Annual Report. Washington, DC.

Vasbinder, J.W.; Sim, J.Y.H. (eds.). 2022. Buying Time for Climate Action: Exploring Ways around Stumbling Blocks. Exploring Complexity series, vol. 8. World Scientific, Singapore. 222 p.

Vora, N.M.; Hannah, L.; Lieberman, S.; Vale, M.M.; Plowright, R.K.; Bernstein, A.S. 2022. Want to prevent pandemic? Stop spillovers. Nature 605: 419–422. https://doi.org/10.1038/d41586-022-01312-y

Wallace-Wells, D. 2023 February 22. Is the United States ready for back-to-back pandemics? America has failed to do all the things that might have secured lasting normalcy. New York Times.

Whitmee, S.; Haines, A.; Beyrer, C.; Boltz, F.; Capon, A.G.; de Souza Dias, B.F.; et al. 2015. Safeguarding human health in the Anthropocene epoch: report of the Rockefeller Foundation–Lancet Commission on planetary health. Lancet 386: 1973–2028. https://doi.org/10.1016/S0140-6736(15)60901-1

Wille, M.; Geoghegan, J.L.; Holmes, E.C. 2021. How accurately can we assess zoonotic risk? PLOS Biology 19: e3001135. https://doi.org/10.1371/journal.pbio.3001135

7

Document: Pathogen Diversity—Finding Them before They Find Us

Eric P. Hoberg, Valeria Trivellone, Joseph A. Cook, Jonathan L. Dunnum, Walter A. Boeger, Daniel R. Brooks, Salvatore J. Agosta, and Jocelyn P. Colella

Abstract

Cultural and conceptual transformation is essential for recognizing the necessity of placing pathogens in an environmental, evolutionary, and ecological context by incorporating specimens and associated informatics into a foundation for actionable information. Historically, we have been hindered by interdependent, flawed, and deeply held paradigms that have described and determined how emerging infectious diseases (EIDs) have been addressed across the global arena. Conceptually, there is considerable historical baggage posed by assumptions and expectations for coevolution and cospeciation as a dominant force in structuring the biosphere. Operationally, response-based approaches, characterized by "business as usual" (BAU), have continued to codify sequential, often delayed, one-off reactions with each event of emergence or recognition of a new pathogen. One Health programs and trajectories, exemplifying the convergence of these interacting paradigms, are now the apparent standard for exploring the occurrence and distribution of emerging pathogens and disease. By definition, One Health has been characterized as broadly inclusive, collaborative, and transdisciplinary with connectivity across local to global scales, integrating the medical and veterinary community to recognize health outcomes emerging at the environmental nexus for people, animals, plants, and their shared landscapes. One Health has been an incomplete model, conceptually and operationally. A proactive-predictive path is necessary, emanating in part from geographically/taxonomically broad and temporally deep biological collections of pathogen-host assemblages. The DAMA protocol (Document, Assess, Monitor, Act), the operational extension of the Stockholm paradigm (SP), accomplishes this task by emphasizing holistic and strategic biological sampling of pathogens and reservoir host assemblages at environmental interfaces and more extensively through resurveys. The initial or authoritative process of documentation that characterizes biodiversity feeds into development of informatics resources digitally linked to physical specimens held in publicly accessible museum biorepositories. Archives of specimens are the entry point for accumulating interrelated archives of information (the baselines against which change can be identified and tracked), with collections serving as fundamental resources for biodiversity informatics under the conceptual evolutionary and ecological umbrella of the SP. A transformation is essential among the diverse practitioners in the One Health community in abandoning response-based paths (BAU) while looking forward toward proactive transboundary approaches to address EID. SP and DAMA are an opportunity to maximize our conceptual and taxonomic view of diversity across interconnected planetary scales that influence the complexity of pathogen-host interfaces. Evolution, where the past always influences the present and the future, defines our trajectory, as the need for sustained archives that describe the biosphere becomes more acute with each passing day.

Keywords: Stockholm paradigm, DAMA protocol, biorepositories, specimens and archives, pathogens, hosts and emerging disease, One Health

Saving biodiversity and promoting human socioeconomic development is a complex issue that requires networks of both people and research programs. Networks require a common language and discourse, as well as collaborative development of theory and research programs. Modern systematists are the masters of a language powerful enough to facilitate such necessary discourse.

Brooks and Hoberg, 2001

In the absence of taxonomic names there is no information. With the wrong names there is incorrect information. Both situations emphasize consequences for how we identify and understand dynamic change for pathogen-host assemblages under a regime of climate warming.

Brooks and Hoberg, 2007, 2013

A Fundamental Context for Names

Scientific names in biology, encompassing a rich and prolific global cornucopia, serve as the gateway or portal to holistic biodiversity information about history, evolution, ecology, and biogeography (geographic distribution) that is organized and accessible in a phylogenetic context (Brooks and Hoberg, 2000; Wheeler, 2010). Our formalized taxonomies of Latin binomials, emerging nearly 300 years ago, remain incomplete, potentially misleading, and nondimensional without a full appreciation of diversity, phylogenetic connectivity, and natural history. Names link history to the present and can serve as a roadmap to the future. Across the history of humanity, original common and local names for animals and plants were later codified and classified in attempts to systematically characterize life under the umbrella of western science. The contemporary purveyors of names, or nomenclature in biology, are the systematists and taxonomists, who serve to bring a sense of phylogenetic order to the myriad of animals and plants, fungi, protists, eubacteria, archaebacteria, and viruses that have populated the planet across Earth's history in the 6 kingdoms that compose the tree of life (e.g., Woese et al., 1990); despite the pervasive nature of viruses and perhaps prions, their placement within this hierarchy remains incompletely resolved (Harris and Hill, 2021).

The process of naming within a phylogenetic context allows entry to explicit hypotheses about evolution and history—this is curiosity with a purpose (Brooks and McLennan, 2002; Wheeler, 2004). In contrast, curiosity without purpose—the singular act of naming, building taxonomies and nomenclature, that remain without a context of natural history and phylogeny in a world under rapid perturbation—is increasingly a luxury (e.g., Kozloff et al., 2016; Zhu et al., 2022). Taxonomy, anchored in scientific collections, is the international language of biological diversity, establishing the framework necessary for scientific collaboration, cooperation, and the potential for clear communication (Sandall et al., 2022). Biological specimens, with established authoritative identification, are the basis of museum archives, increasingly so in the era of digital biological heritage. Specimens and their associated data facilitate opportunities for global metasynthesis of biodiversity and rigorous investigation of the biosphere. Specimens are the essential cornerstones for taxonomy, as they validate names and validate our theoretical view of the world. Understanding specimens is understanding evolution, where the past always influences the present and the future (e.g., Agosta and Brooks, 2020; Agosta, 2022). That evolutionary perspective, revealing history and structure of global diversity in accelerating change, defines our need for sustained archives, which becomes more acute with each passing day (e.g., Hoberg, Trivellone, et al., 2022, Hoberg et al., 2023).

Although names and phylogeny are the anchors for understanding diversity, a confounding factor is the apparent temporal asynchrony in the process of speciation and formation of species in space and time. Organismal asynchrony is particularly problematic in the realm of viruses and bacteria. Rapid rates of evolutionary change challenge our abilities to recognize the oscillations associated with species formation and subsequent disappearance on fine geographic and temporal scales before adequate study and evaluation is possible (see Souza et al., 2022). Highlighted is the operational challenge and difficulty in creating viral classifications and associated taxonomies that embody a nomenclature also connected to natural history.

As an understanding about the nature of overlapping and synergistic crises for climate, biodiversity, and emerging infectious diseases (EIDs) has become increasingly focused (e.g., Brooks and Hoberg, 2013; Brooks et al., 2019, and references therein), so has explicit recognition of the connectivity between animal, plant, and human disease, facilitated by environmental opportunities. These interactions and outcomes led clinicians, veterinary, and agricultural scientists to propose what is called One Health (Zinsstag, 2011; Cunningham et al., 2017) and more recently Planetary Health (Horton and Lo, 2015). Unfolding in the 1940s, One Health initially brought together a coalition of veterinarians, plant pathologists, physicians, and a broader range of organismal biologists with foundations in the Centers for Disease Control (CDC) and the US Public Health Service. Planetary Health is a complementary and more explicit approach that transcends pathogens to understand the connection between accelerating climate change, unprecedented environmental perturbation, and human health with origins in the Rockefeller Foundation–*Lancet* Commission on Planetary Health in 2014 (Horton and Lo, 2015; Whitmee et al., 2015; Watts et al., 2018). Insights about the complex dynamics of emerging disease are fully dependent on the validity and authority of names for pathogens and hosts, their identities, phylogenies, and natural histories. Named specimens, permanent archives, and associated biodiversity informatics should provide an essential foundation for One Health, although such is currently lacking (Colella et al., 2021; see Cook et al., 2004; Coker et al., 2011; Gebreyes et al., 2014; Astorga et al., 2023, for mainstream definitions of One Health, https://www.cdc.gov/onehealth/index.html). One Health has been characterized as a broadly inclusive, collaborative, and transdisciplinary approach with connectivity across local to global scales, which integrates the medical, veterinary, and agricultural community to recognize health outcomes emerging at the environmental nexus for people, animals, plants, and their shared landscapes. Yet to date, One Health has been an incomplete model, conceptually and operationally, focused on reactive and response-based foundations, to limit the impact of newly recognized pathogens and EIDs, and as such has lacked a powerful proactive capacity (e.g., Giraudoux et al., 2022; Wallace-Wells, 2023).

Species and population-level resolution of diversity across time, from landscape to regional scales, is required to interpret the historical and geographic connectivity for pathogens, host assemblages, and the distribution of disease in the seamless continuum defined as "Audy space" (Brooks et al., 2014, 2019; Hoberg, Boeger, Brooks, et al., 2022; Hoberg, Trivellone, et al., 2022; Hoberg et al., 2023). Authoritative names are too often overlooked in the predominant approaches to zoonoses and broader characterization of pathogen diversity. The potential of a proactive stance for EID is intimately linked to nomenclature and can be realized through the power of specimens and biological collections. Biorepositories connect field biology to permanent archives within a phylogenetic framework that is synthesized and regularly updated in cumulative, digital, and publicly available database resources. A proactive trajectory for addressing EID has been proposed under the evolutionary umbrella of the Stockholm paradigm (SP) and through leveraging the operational components of the DAMA protocol (Document, Assess, Monitor, Act) (Brooks et al., 2014, 2019; Colella et al., 2021; Agosta, 2022; Brooks, Hoberg, et al., 2022; Trivellone, Hoberg, et al., 2022; Brooks, Boeger, et al., 2023; Hoberg et al., 2023; and references therein).

Our current pathways for pathogen detection often disregard authoritative identification and elucidation of the source(s) (e.g., host assemblages and geotemporal context) of essential information, as denoted by inadequacy and limited interoperability of data streams related to assessments of pathogen/host diversity (e.g., Plowright et al., 2019; Ruiz-Aravena et al., 2021). Discordance and disconnects highlight the need for improved communication and shared protocols, not only for optimal sampling across space and time but most critically for development of permanent biological archives and databasing of vast and integrated information streams (Shapiro et al., 2021). Protocols should encompass holistic sampling of pathogens in circulation among diverse assemblages of hosts and parallel expansion of informatics resources directly linked to physical specimens held in fully accessible museum repositories (e.g., Gardner, 1996; Gardner and Jiménez-Ruiz, 2009; Dunnum et al., 2017; Galbreath et al., 2019; Colella et al., 2021). Such an approach is in direct contrast with the current "business as usual" that may target a single pathogen or host and rarely preserves voucher specimens for extended examination. Assumptions about pathogen and host identity are only assumptions in the absence of verifiable voucher specimens.

Ultimately, a primary goal is complete sharing and accessibility of diagnostically relevant data, including sequence-based, genomic, isotopic, behavioral, and morphological information with standardized metadata in public repositories (NASEM, 2020; Sett et al., 2022). As has been noted in the course of the SARS-CoV-2 pandemic:

> *The impact of genome data (or any molecular data) is dependent on their quality, and (how) the reliability and accuracy of such data may influence the global community's ability to track the emergence and spread of variants in a timely manner.*
>
> Chen et al., 2022

Uniform accessibility by the international community to synoptic data sources for diagnostic sequences across the time frame of the current pandemic remains as a bottleneck. A similar circumstance hinders availability of robust and complete sequence data among other viral groups, such as the bunyaviral family Hantaviridae, for which sequence data are incomplete or absent (Kuhn et al., 2023). DNA sequencing—fast genotyping—may indeed represent the only manner to characterize diversity while recognizing evolutionary changes in rapidly evolving lineages of microparasites such as bacteria and viruses (e.g., Simmonds et al., 2023). In a way, sequencing is the nomenclature of the moment. Sequences, like names, connect past and present, allowing recognition of spatial and temporal connectivity for pathogens in the future through phylogeny, but they are not without pitfalls, especially for nonculturable pathogens (Kirdat et al., 2023).

No Specimen, No Name, No Information

Names originate from specimens—the window on the world that is held in museum archives and biorepositories (e.g., Hoberg, 2002; Wheeler, 2010; Dunnum et al., 2017; Cook et al., 2020; Thompson et al., 2021; Naggs, 2022). As scientists we explore the world, collecting specimens, continuing to poke and prod, using increasingly sophisticated toolkits from 3D morphology to isotopic chemistry and genomes in pursuit of insights about the assembly, distribution, interactions, and history of diversity. These integrated pathways formalize the process to **Document**, the initial component of the DAMA protocol, within the operational extension of the SP. Our view of the biosphere begins with specimens.

In current systematic biology, this concept is solidified through permanent archives of whole organisms and associated tissues, genes, metadata, and information derived from specimen-based studies, which are fundamental to an integrated understanding of diversity. The direct observation of a specimen in hand is of pivotal importance and serves as a tool for (re)appraisal; in the absence of physical specimens all information cannot be confirmed or refuted and is immediately suspect (e.g., Hoberg and Soudachanh, 2020, 2021). Specimens validate names and collectively within their evolutionary context validate interpretation of complex data streams. There is an elemental trajectory through specimens to observation and information from which synthesis and narrative emanates, contributing to emerging theories that describe the assembly and nature of the biosphere through space and time. There should always be a pathway from field-based specimens to permanent archives to shared biodiversity informatics and synthesis (Naggs, 2022). Increasingly, the

global community of natural history museums is linking disparate data streams through development of accessible digital formats that integrate physical resources of specimens with deeper insights that bridge natural history, phylogeny, ecology, and biogeography in the realm of *cybertaxonomy* (e.g., Wheeler, 2010; Cook et al., 2020; Colella et al., 2021; references therein).

Specimens and collections resources, despite the dimensions of some prominent international museums, are increasingly and habitually shuttled off to the dark corners of biology in many institutions (Naggs, 2022). Collections do not often achieve or meet their potential as benchmarks and cornerstones in explorations of biodiversity and history, especially in the context of emerging pathogens and disease. Building specimen infrastructure, especially large series that provide extensive and intensive snapshots of the biosphere, has been the focus of a relatively few dedicated systematic and evolutionary biologists. More often, however, such series are never accumulated and preserved because specimens are often an afterthought in the day-to-day practice of ecology, conservation biology, and explorations of planetary biodiversity (Cook et al., 2016). Or, if specimens are accumulated and explored, such may be passively discarded after assumed identification or application to a narrow question, despite the ever-evolving nature of new toolkits to reveal the dimensions of diversity (reviewed in Colella et al., 2020, 2021, references therein). Eventually, often as orphan personal research collections, in the wake of academic and professional turnover, specimen resources are separated from data (or often are not properly digitized and become inaccessible) and critical vouchers are lost in the recesses of slide boxes, bottles, cryovials, and ultracold freezers. Specimen and information loss, either passive or as an active decision, is not new but a pervasive outcome in the continuing trend for inadequate resources and especially limited critical mass for taxonomists and natural history collections encompassing all taxa despite circumstances of increasing global urgency (e.g., Brooks and Hoberg, 2001; Wheeler, 2010; Cook et al., 2020; Bradbury et al., 2022; Naggs, 2022). Specimens seem to fall under the Dangerfield principle—often receiving little respect from a heterogeneous, hubristic community of field biologists, practitioners, disease ecologists, and other scientists whose activities are directly dependent on the availability of permanent archives of specimens, authoritative names, and associated informatics. Specimens validate names, and names enable robust exploration and testing of hypotheses that validate our theories about the assembly and nature of the biosphere through space and time. Ultimately a robust understanding of pathogens and disease can emerge from a rigorous adherence to data collection, feeding into well-supported natural history infrastructure that describes the complex umbrella of global biodiversity (e.g., Brooks and Hoberg, 2013; Drabik and Gardner, 2019).

Documentation of the biosphere through specimens and their metadata deposited and linked in well-connected museums should not be an afterthought in the scientific process but instead a strategic imperative (Hoberg, 2002; McLean et al., 2016; Schindel and Cook, 2018; Miller et al., 2020; Gardner et al., 2021). Although opportunistic activities can provide glimpses of the biosphere, such approaches fail to build a robust, coordinated, and strategic view of spatial and temporal complexity. Limits of host and geographic ranges for pathogens, patterns of diversity, and disease are required to understand distribution and risk space for circulation and emergence (e.g., Audy, 1958; Hoberg et al., 2023). Documentation must be strategic, intentionally accounting for elements of heterogeneity (e.g., mutation rates, environments, etc.) in diversity. Among many viruses and bacteria, diversification is rapid, occurring within weeks, months, or even the stretch of a human lifetime. Consequently, some taxa may necessarily require consideration of the time frames and time series that can maximize exploration of lineages, geographic distribution, evolution, and novel patterns of emergence (e.g., Botero-Cañola et al., 2019; also see Boeger et al. (2022), Hoberg, Boeger, Brooks, et al. (2022), and Holmes (2022) for succinct summaries of the history of SARS-CoV-2). Synoptic data streams that are publicly available can allow us to explore the relationship of pathogen reservoirs and environmental interfaces.

Complex and comprehensive data streams allow for integration across space and time on a planet experiencing accelerating change, as exemplified by the proposal for the **DAMA protocol** (Brooks et al., 2014, 2019; Colella et al., 2021; Brooks, Boeger, et al., 2022; Brooks, Hoberg, et al., 2022; Hoberg, Boeger, Molnár, et al., 2022; Hoberg et al., 2023) (Figure 7.1). Where appropriate, geographically extensive and site-intensive, field-based inventories of communities of organisms, preserved as specimens and digitized information in natural history collections, should be at the core of **Documentation** as the initial phase of DAMA (Hoberg et al., 2023). Incorporated into **Assessment**, specimens with authoritative identification that are placed in a molecular/evolutionary context, are the first step in "phylogenetic triage," which defines the extent of risk space (Hoberg, Boeger, Molnár, et al., 2022; Brooks, Hoberg, et al., 2023; Hoberg et al., 2023; Janik et al., 2023). **Monitoring**

through resampling, resurveys, and reassessment, paired with archival development of infrastructure for pathogens and hosts, contributes to temporal and spatial baselines, providing windows into identified environmental interfaces and the ever-expanding array of host reservoirs under environmental transformation (Botero-Cañola et al., 2019; Hoberg et al., 2023). Collectively, the linked steps of assessment and monitoring provide real-time insights about the distribution of risk and mosaic patterns of pathogen distribution relative to patterns of emergent disease while tracking ecological and evolutionary changes that influence environmental interface(s) and the interaction of opportunity and capacity (e.g., Zhao et al., 2022; Brooks, Hoberg, et al., 2023). Collectively these insights contribute directly into **Act** by using actionable information for dissemination that allows us to anticipate and mitigate enzootic, epidemic, and pandemic emergence (see Molnár, Hoberg, et al., 2022; Molnár, Knickel, et al., 2022; Ortiz and Juarrero, 2022; Brooks, Boeger, et al., 2023; Hoberg et al., 2023). We are left to inquire: If specimens are fundamentally important, why are they so rarely preserved, tabulated in a database, and permanently archived?

A cogent case study of this conundrum is seen in the current spotlight focused on viral species richness among mammals. Seeking to document, predict, and understand the limits of viral diversity and patterns of host and geographic circulation has become a foundation of pandemic biology—that is, efforts to identify direct links between viral pathogens and diseases emergent in humans (Grange et al., 2021, and references therein). Over the past 15 years biological field collections have explored assumed viral hot spots as defined by global geography (e.g., Jones et al., 2008). To be sure, these studies have dramatically broadened our knowledge of viral pathogens and have enabled biologists to refine estimates of total species richness for assemblages of viruses globally in such mammalian groups as bats (chiropterans) and rodents (Young and Olival, 2016; Olival et al., 2017; Mollentze and Streicker, 2020; Zhou et al., 2021; Carlson, Albery, et al., 2022). Critically, however, authoritative identifications of host species have been largely ignored, despite calls for broadening efforts to integrate surveys that link field collection and museum deposition (Colella et al., 2021), especially of bats (Gardner and Jiménez-Ruiz, 2009; Gardner and Whitaker, 2009). Globally, few host or pathogen specimens have actually been permanently archived, and our current and future knowledge of their diversity and distributions effectively has been lost to science and society. For example, nearly 75,000 mammalian specimens were handled or sampled during the duration of

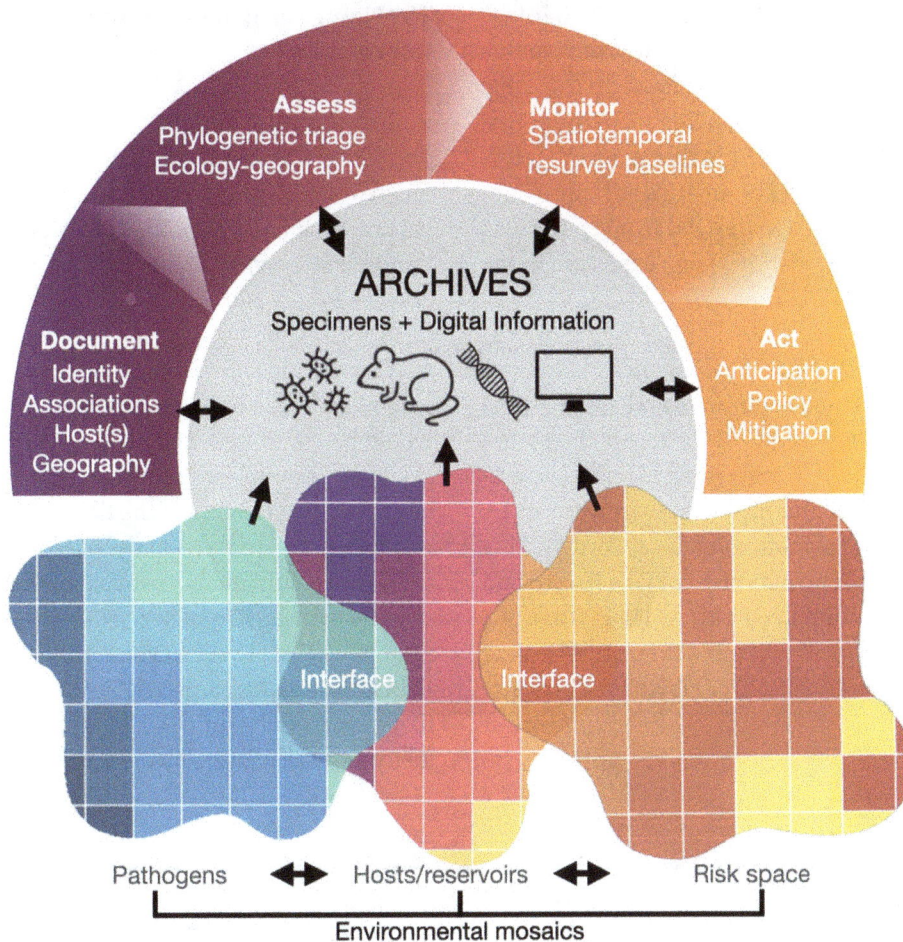

Figure 7.1. Conceptual framework illustrating the centrality of archival research collections to the DAMA protocol: Document, Assess, Monitor, Act. Collections must be representative of the changing mosaic of space, time, and environmental complexity across both pathogens and hosts.

the PREDICT program (e.g., in part summarized by Grange et al., 2021), but relatively few were archived in museum collections, particularly from regions of the world with considerable gaps in knowledge of species richness for rodents, bats, and other groups. Absence of critical archival vouchers for chiropterans and other mammals remains as a common practice in the most recent evaluations of SARS-like viruses circulating among bats in Southeast Asia, the presumed region of origin for the SARS-CoV-2 pandemic (Latinne et al., 2020; Zhou et al., 2021; Temmam et al., 2022). Indeed, over the past 2 years surprisingly few diagnostic sequences for viruses that document the origins and circulation of variants for SARS-CoV-2 have been archived in permanent repositories (Chen et al., 2022); few if any viral specimens or sequences are directly linked to host specimens in museum collections.

We can do better, although such a shift will require a substantial cultural transformation in fieldwork, biological research, public health, and specimen curation that emphasizes expanding transboundary infrastructure, personnel, and informatics resources, driving a new paradigm for integrative collections and archives (Dunnum et al., 2017; Schindel and Cook, 2018; Colella et al., 2020; Thompson et al., 2021; Hoberg et al., 2023) Discussions about the nature of biological (collections) repositories have been in development for more than two decades, with only minimal change in practices and the manner in which most institutions link natural history from fieldwork, to the laboratory where specimens are processed, analyzed, interpreted, and disseminated as vital, useful, accessible, and actionable information (e.g., Brooks and Hoberg, 2000; Hoberg, 2002; Wheeler, 2010; Brooks et al.,

2014; Schindel and Cook, 2018; Cook et al., 2020; Gardner et al., 2021; Chen et al., 2022). Concurrently, during this period of increasing and overlapping global crises, some historically critical repositories, including the Natural History Museum, London, are poised to discard a deep legacy in biological diversity and systematics initially established more than 250 years ago derived from specimens deposited there from the early European explorers in the 17th century (Naggs, 2022).

Urgency for access to synoptic biodiversity information is not simply a question of the dynamics of humans, vertebrates, pathogens, and their links and associations to the animal world. Human well-being across the planet is no less an outcome of food and water security, reflecting the cascading environmental challenges posed by pathogens and diseases of food animals and crop plants in agriculture as a broader component of One Health and Planetary Health (e.g., Whitmee et al., 2015; Watts et al., 2018; Brooks et al., 2019; Benton et al., 2021; Wilcox et al., 2021; Brooks, Hoberg, et al., 2022; Trivellone, Hoberg, et al., 2022) (Figure 7.2).

The Nature of Specimens and the Borders of Information

Philosophically and operationally, typological thinking dominated practices to name and describe species for much of the past 200 years, and adherence to typology in part represented an absence of vision and imagination. Although a static view of biodiversity may have been acceptable a century ago, it is no longer tenable. Among taxonomists, the type was the *one exemplar* and name-holding specimen that served as the representative individual or embodiment of a particular species against which all others would be compared. A type seemed sufficiently emblematic of a species prior to the advent of evolutionary biology and an understanding of the limits for geographical, morphological, and most recently molecular variation. In parasitology, a traditionally host-centric view of parasite diversity and distribution served to further elevate the type. For example, knowing the host could be conceptually equated with knowing the parasites, a core mantra of the past century (e.g., Brooks et al., 2019). An expectation of host specificity and association by descent (cospeciation) provided

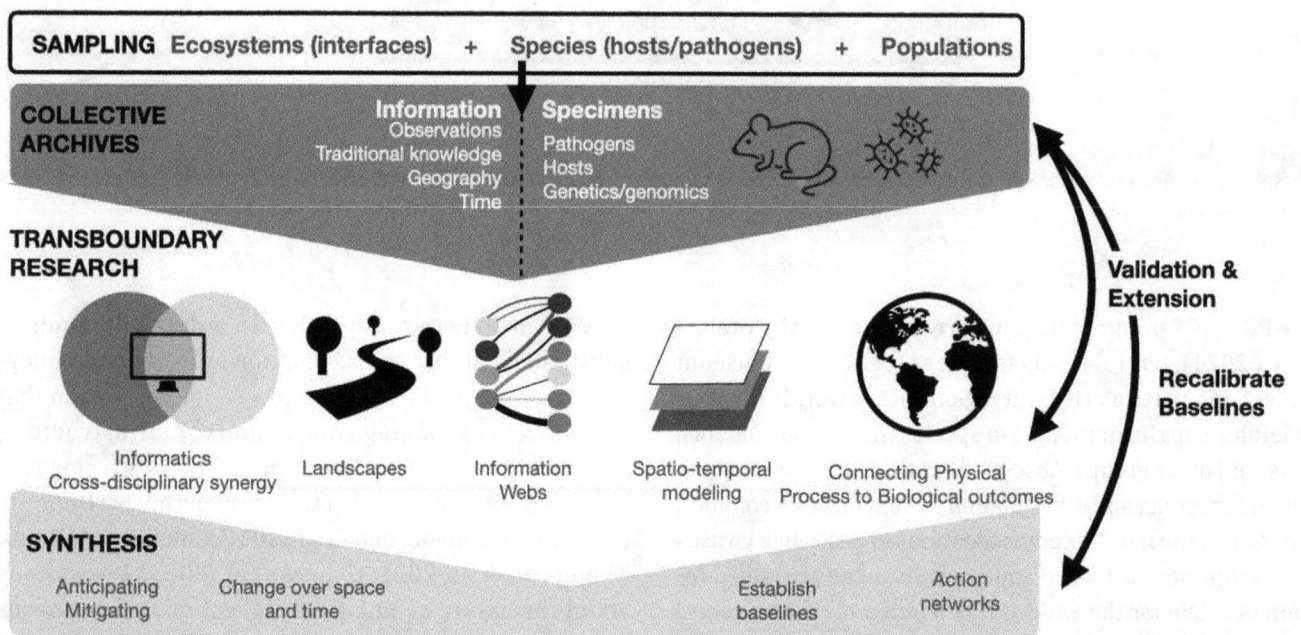

Figure 7.2. Archives of information and physical specimens can be accumulated over time through collective deposition of samples in natural history museum archives. These data fuel transboundary research, including but not limited to cross-disciplinary informatics, landscape-level analyses, information webs, spatiotemporal modeling, and connecting physical processes to biological outcomes. Such research leads to major advances in our synthetic understanding of the biosphere, facilitating anticipation and mitigation of emerging pathogens and change through time, informed establishment and recalibration of baseline conditions, and action in the form of public and health policy and conservation. Importantly, the persistence of research archives is what allows for scientific validation and extension.

a convenient path, a simplified view of diversification and faunal assembly in which to contextualize diversity and limits on community change (e.g., Hoberg and Brooks, 2010; Brooks et al., 2015; Hoberg et al., 2015; Nylin et al., 2018; Trivellone, Cao, et al., 2022).

Fortunately, we are beginning to abandon this simplicity by replacing it with a nuanced picture of complexity as a more inclusive and overarching umbrella for global biodiversity—a story explored elsewhere under the SP (Hoberg and Brooks, 2015; Brooks et al., 2019; Agosta and Brooks, 2020; Agosta, 2022; Brooks, Boeger, et al., 2022; Brooks, Boeger, et al., 2023). We now know that a name by itself, a type, and a mere handful of specimens of either pathogen or host(s) are not sufficient to resolve the critical issues related to zoonotic pathogens and environments that represent risk space for humanity, our domesticated animals, and our crops. Speaking of macroparasites and microparasites, museum collections are more often populated by types and type series (required by the International Code of Zoological Nomenclature; ICZN, 1999) and incomplete series of vouchers that reflect apparently novel host or geographic records. Vouchers are very rarely uniformly represented and archived from geographic or host surveys at any scale (consistent with Colella et al., 2020; Thompson et al., 2021); identification always requires validation and cannot be assumed to be correct, with the possibility of an erroneous view of the biosphere. Published records linked to vouchers are also uneven, seemingly like islands of enlightenment in a wide sea that limits key information and integration. Collections have historically served as gatekeepers, often siloed taxonomically and typically disinterested in archiving materials accumulated from survey and inventory. Further, there are no uniform requirements across a broad spectrum of journals (i.e., another set of gatekeepers) that specify or require deposition of vouchers for either pathogens or hosts (Schilthuizen et al., 2015). Only from the context of specimens archived in biological collections, and through recognition of colonization processes in faunal assembly extending across Earth history can this convoluted historical tapestry be disentangled (Hoberg and Brooks, 2008; Hoberg, 2010; Zhang et al., 2022).

To some extent, how taxonomists and a community of parasitologists in the broadest sense name species continues to represent fundamental assumptions about host associations. These assumptions are consistent with a history of one host (or one host group)–one parasite, ideas codified in the 1920s, and are only reluctantly being discarded among recent practitioners (see Brooks et al., 2019). For some, taxonomy established for a previously unrecognized species linked to a host was the convenience of conferring a name on a pathogen that leaves it easily identifiable. In the end, a name assigned in that context does not reveal insights into history, ecology, or evolution. Notably, such isolated efforts can be further confounded by the point that most species descriptions in parasitology remain the sole report of occurrence, and these isolated records tend to bias ideas (and expectations) about host range (e.g., Carlson et al., 2020). Assumptions are equivocal in the absence of specimens, archives, and verifiable information. Consequently, there is a *danger of convenience* when a worldview is not fully grounded by empirical data that can be verified and extended (Hoberg, Trivellone, et al., 2022). It is hubristic to think that something new cannot be learned from a specimen already examined. Convenience constrains broad-based thinking and misleads when there is reliance on suppositions about the nature and limits of diversity that are in error both conceptually and empirically (Brooks and McLennan, 2002; Hoberg and Brooks, 2015; Hoberg et al., 2015; Brooks et al., 2019).

In parasitology and entomology, and especially relative to macroparasites, comparative morphology must remain a gateway to exploring critical limits of diversity within an integrated framework that is in synergy with molecular data. An increasing shift, however, has been to diminish the essential insights derived from structural comparisons toward reliance on molecular data as the sole or proper source on which to base species definitions and descriptions of faunal diversity; in contrast, molecular-based evaluations provide the only viable pathway in explorations of viral diversity (Simmonds, et al., 2023). Another component of convenience is apparent in that over time cumulative capacities for comparative analyses are neither developed nor transferred to a new generation of scientists (e.g., Brooks and Hoberg, 2001; Bradbury et al., 2022; Poulin and Presswell, 2022). If we continue on this track, our future will then become an artifact of increasingly superficial taxonomy established in the past and extending into the present (e.g., Hoberg and Soudachanh, 2020). A parallel legacy will be information resources that are increasingly derived from noninvasive or nondestructive sampling that will leave us without permanent representation in specimen-based archives as vouchers, effectively limiting any possibility of replication and extension of the science (Colella et al., 2020; Rohwer et al., 2022). Further, apparent trends epitomized by large-scale federally funded ecological programs, such as the National Environmental Observatory Network (NEON) and the Circumpolar Biodiversity Monitoring Program (CBMP), depart from true biological inventories in aspects of specimen collection with articulation of archives. Expanded focus on noninvasive, observational data, including mark-recapture and camera traps, do not effectively provide synoptic, large-scale snapshots that contribute to baselines of the biosphere in change (e.g., Hoberg et al., 2013; Cook et al., 2013, 2016, 2017; Christensen et al., 2021). In the absence of strategic and focused field-based biological collections and specimens, such trajectories will have a detrimental impact on national and international capabilities to identify assemblages of pathogens and hosts in transition.

Archives and Natural History Collections—Truth in the Biosphere?

Operationally, how do we address pathogens and diseases in a world under dynamic transition? How would we track change in the biosphere if not for the insights that specimens can bring to the table with their names that we use as a shortcut to identity? Specimens and archives are the foundations for biological baselines in time and space against which change and transitions can be recognized and measured (Box 1, diversity of *Hantatviridae-Orthohantavirus*). When a valid taxonomy is lacking, the consequences are real. Consider the fact that even now in modern medicine, there is often less concern about the actual identity of an organism that is causing a malady than the emergent disease syndrome (e.g., Robles et al., 2018; Kobayashi et al., 2019; Reuben et al., 2020). Our medical system identifies symptoms and treats them, assuming the presumed pathogen (probably a virus) is unimportant, seldom seen (a rare event), or unlikely to be new (e.g., Molnár, Hoberg, et al., 2022). That a disease will disappear as quickly as it emerged is yet another misconception about the nature of hosts, pathogens, and their intertwined geographic distributions. The notion of disappearing pathogens was effectively dispelled nearly 60 years ago by a British disease ecologist named J.R. Audy who was working in Kuala Lumpur, Malaysia, where he noted the complicated spatial and temporal mosaics that described the shifting occurrences of pathogens and disease (Audy, 1958). We often forge ahead in the absence of a name, which demonstrates the tenuous link between academic taxonomies and clinical taxonomic practices.

Initial emergence of SARS-CoV-2, after all, was reported as a series of unspecified pneumonias among a small number of patients in central China. The short but intense and continuing global history of COVID-19 emphasizes that a common set of symptoms can be produced by more than one pathogen (consider seasonal influenzas and flu, varying sources of pneumonias). Clinically we are challenged to immediately recognize the difference or the etiological agent for a period of time without technological aid. We thus practice deferred diagnostics, and in the absence of definitive information, we make guarded assumptions about disease causation, and thus we remain mired in a response-based approach to EID (e.g., Brooks et al., 2014, 2019; Hoberg and Brooks, 2015; Boeger et al., 2022; Trivellone, Hoberg, et al., 2022).

We pass daily through a world of orphan viruses and among these, how many unknown viral pathogens are potentially emergent (e.g., Carlson, Albery, et al., 2022)? Identifying pathogens and the events associated with circulation, proliferation, and emergence is a considerable challenge, given the interaction of expansion waves and population sizes (of hosts and pathogens) on the margins of rapidly changing distributions. This is the usual dynamic described across pathogen diversity from the viral pandemic of SARS-CoV-2 to epidemics and recurrent outbreaks of Ebola or Zika to obscure lungworms and vector-borne nematodes among Arctic ungulates (e.g., Laaksonen et al., 2015; Kafle et al., 2020; Hoberg, Boeger, Brooks, et al., 2022; Regala-Nava et al., 2022). Because we can't easily observe the initial stages of expansion or emergence, a new term—"silent spillover"—has been proposed (Temmam et al., 2019). Pathogen expansion is "silent" only because for a time it may be hidden. Silent spillover is a misnomer because we are seldom seriously looking (i.e., with rigorous surveys in the context of environmental interfaces), and we apply inexpensive convenient and cursory pathways to documentation and descriptions of biodiversity that are lacking in historical context; thus, the term is completely anthropomorphic (Hoberg et al., 2023). Again, we defer to Audy (1958), who recognized that the distribution of a pathogen is far greater than the distribution of disease caused by that pathogen. The definition of Audy space emphasizes this complexity, described in the wobbling oscillations and seemingly ephemeral mosaics or islands of pathogens and disease, with the understanding that pathogens never really disappear from landscapes and regions (Hoberg et al., 2023).

Names are assumed to be the truth, to be authoritative, and to reflect the reality of situations and circumstances. The foundations and implications of taxonomy apply equally to hosts and to pathogens, and we are continually challenged to get it right because there are distinct consequences, as we have seen, for getting it wrong (Box 2, *Borrelia* diversity). Often, we discover secondarily,

Box 1

Hantaviridae-Orthohantavirus—the Classical Paradigm for Archives

Orthohantaviruses (hantaviruses) are agents of disease among people in Eurasia (hemorrhagic fever with renal syndrome, HFRS) and the Americas (hantavirus pulmonary syndrome, HPS). The history of discovery among rodent hosts, extending over much of the past half century tells a story about evolving perceptions and hidden diversity (Kuhn and Schmaljohn, 2023). Although there were early indications that these viruses may occur in other mammals, the idea that hantas were limited to a narrow spectrum of rodent hosts, and locked in an apparent coevolutionary history with these mammals, appears to have constrained a broader search for diversity until the past few decades (Lee et al., 1978; Yates et al., 2002; Song, Baek, et al., 2007; Song, Kang, et al., 2007)

The revelation that hantas were not solely found among rodents opened the floodgates for an expanding picture of diversity (now including more than 139 distinct strains) that encompass viruses in chiropterans (bats) and eulipotyphlans (shrews and moles), in addition to a growing assemblage of rodents (Jonsson et al., 2010; Yanagihara et al., 2015; Liphardt et al., 2019; Arai and Yanagihara, 2020; Kuhn and Schmaljohn, 2023). Many of these discoveries were based upon mining frozen archives of rodent specimens collected previously for other reasons. Recognition of extensive diversity served to dispel the myth of one rodent host–one virus and histories linked to cospeciation. Increasingly, phylogenetic frameworks demonstrate extensive signatures for bouts of host colonization, initially during early mammalian evolution, extending into shallow evolutionary and ecological time. Among hantaviruses, initial perceptions about host range among rodents and the actual reality across mammalian diversity have served to uncover a potential minefield for emergent pathogens, emphasizing the crossroads opportunity and capacity. Among these viruses fundamental and realized host range, for example among assemblages of rodents, is extensive and influenced directly by ecological contact. As we seek to unravel the global distribution of these potential pathogens not all of the news is good. Again, we run headlong into complications imposed by the absence of archives as well as biodiversity resources that are taxonomically and ecologically narrow. Critical host data for more than two thirds of the currently recognized hantavirus strains are ambiguous and compromised because associated specimens and data are not held in permanent museum repositories, severely limiting our ability to confirm host identity or even extend previous work. Notably, sequence data for bunyavirals, frequently classified solely based on single-segment sequences, are also often incomplete or fragmented and cannot be integrated into robust and comprehensive phylogenetic analyses (Kuhn et al., 2023).

In contrast, the story of the outbreak and discovery of the *Sin Nombre* virus in North America tells us again about the interface for archives and urgent biodiversity discovery, and the integration of biorepositories, names, and critical information. The multiuse nature of biological collections formed the backdrop for an unfolding biological detective story (Yates et al., 2002). Museum collections of rodent specimens and tissues, along with human specimens from undiagnosed disease outbreaks, demonstrated that the virus had been present for decades and likely centuries or millennia. It was literally the stuff of legends, a mysterious disease that had persisted in the oral histories of indigenous people. Archival frozen tissues revealed the history of the virus and were initially the window through which the question of origins could be examined. In the absence of collections, our current knowledge of hanta in the Americas would have been delayed for a considerable period of time. Nor do we always understand the biotic and abiotic factors that catalyze the shift from quiescence to disease. Sometimes it results from changes in ecological conditions, as turned out to be the case with *Sin Nombre*. The El Niño Southern Oscillation, a shift in the ocean and atmosphere that disrupts the Northern Hemisphere and beyond, produces unusually heavy rains that lead to an explosion of vegetative growth, seed production in pinyon and juniper woodlands, and increased arthropod biomass. Many animals, including deer mice, take advantage of this bounty and convert available resources into rodent offspring. Although not directly transferred to young, the increased viral prevalence for *Sin Nombre* came from high densities of rodents in contact and horizontal passage of the pathogen. Expanding rodent populations explore new areas, including the interfaces of human habitations, and contact between humans and hanta is achieved (Yates et al., 2002). Environmental sloshing and host movement under climate forcing are critical in persistence of *Sin Nombre* on landscape scales (Parmenter and Glass, 2022). Dynamics under SP are essential in the broader interactions of compatibility and opportunity and shifts between exploration mode (in expansion) and exploitation mode (in refugial isolation) for viral populations circulating among diverse assemblages of rodents in proximity to people. Archives are the core benchmarks linked to documentation under DAMA for recognizing movement and change in this and other systems.

as archival biological collections are repeatedly and more deeply probed, that the host name attached to a pathogen is not the right one. For example, among the host assemblages for hantaviruses, critical taxonomy among genera and species of rodents has been in flux, reflecting a reassessment of the identities of some pathogen reservoirs following their original descriptions (Thompson et al., 2021). Among pygmy rice rats (genus *Oligoryzomys*), these insights demonstrated the potential for restricted geographic distributions for many reservoir hosts of hantaviruses (e.g., González-Ittig et al., 2014; Weksler et al., 2017; Hurtado and D'Elía, 2019). Vouchered mammal specimens that turned out to be actual host specimens have led to an increasingly refined view of host associations for numerous hantaviruses of significance in public health (Firth et al.,

2012). In all situations, and not simply limited to hantaviruses, taxonomy matters, and apparent errors in identification are a direct impediment to defining host range, potential and realized patterns of circulation, and the elusive but critical definition of risk space. What is the outcome in the absence of biodiversity archives? Our understanding is immediately suspect, we likely will mischaracterize cryptic diversity, and our appreciation of the biosphere suffers in that we are unable to define putative species associations that are often hiding in plain sight.

We further confound descriptions of diversity by the choices we make in naming species, especially the names we propose for pathogens. Perceptions are critical and often become dogma. Thus, as an explanatory identifier, names are often proposed based on a disease syndrome

---

Box 2

A Story of Burgeoning Diversity for *Borrelia* spp.

Lyme disease is among the most significant vector-borne pathogens in the Northern Hemisphere and may have a broader occurrence in South America and Africa than previously understood—no one knows with certainty (Robles et al., 2018). The story of the causative agent of Lyme disease or Lyme borreliosis highlights the significance of accurate and complete taxonomy, which contributes to robust diagnostics (Kobayashi et al., 2019). Over time, our baseline information expands, showing the cumulative process of science in discovery of diversity and in defining pathogen distribution and outcomes of infection. Original concepts established ixodid (hard-bodied) ticks in transmission of a single spirochaete bacterium, *Borrelia burgdorferi*, occurring in an assemblage of mammals and sometimes birds (e.g., Ostfeld and Keesing, 2000). Human infections arose through opportunity at environmental and management interfaces, where ticks, reservoir hosts, and pathogens were in circulation. A single pathogen species, however, seemed incompatible with a growing picture of the variable outcomes for disease depending often on geography in North America, Europe, or South America (e.g., Robles et al., 2018). Diagnostics suitable in the Northern Hemisphere were often confusing or inconclusive beyond this geographic arena. Lyme disease was becoming a mystery. Names are important, underscored by the recent discovery that Lyme disease is actually caused by more than one species of bacteria, or a species complex, often with varying disease presentations (Stone et al., 2017). In excess of 21 genotypic species are now identified in the genus *Borrelia*—or a cluster of pathogens and diseases globally (Stone et al., 2017). Previously unrecognized taxonomic diversity, under a single name, has likely been responsible for considerable confusion attendant upon diagnosing, misdiagnosing, and treating Lyme disease from landscapes to regions and to continents where explanations linked to etiology for a single pathogen are misleading (e.g., Granter et al., 2016; Álvarez-Hernández et al., 2017; Kobayashi et al., 2019). Coinfections with a broader assemblage of tick-borne pathogens may further confuse diagnosis and treatment (e.g., *Anaplasma phagocytophilum*, *Babesia* spp., *Ehrlichia* spp., *Rickettsia* spp.), reflecting variation among ixodid species and geography. Differential diagnosis additionally is challenged by the occurrence of a wider range of pathogens that do not have tick-borne etiology (e.g., dengue, chikungunya, Zika, and leptospirosis). Consequently, misdiagnoses are common—and as Kobayashi et al. (2019) have noted—". . . that regardless of test results, Lyme disease is the diagnosis used to explain the mostly subjective symptoms. Patients and clinicians may be influenced by alternative, non-evidence-based medical practices, or could be confused by non-validated laboratory test results or interpretations." Strategic field-collections and development of archives encompassing pathogens, mammalian reservoir hosts, and arthropod vectors, along with historical clinical information are required to explore the complex dynamics of circulation for multiple pathogens such as *Borrelia* spp. under a DAMA protocol (Brooks et al., 2014; Hoberg et al., 2023). As a generality, accelerating climate change and environmental disruption are anticipated to directly influence distribution of pathogens and disease (e.g., Brooks and Hoberg, 2007; Ostfeld and Brunner, 2015).

in a particular host and become a manifestation of ideas about limited host range detached from natural history; names in the end are a shorthand identifier (Box 3, Phytoplasmas diversity). Names are also often proposed for geography, the idea that the type locality (like the type host) has special biological meaning. Virology has a long tradition of naming viruses that reflect focal localities: consider Ebola (for the Ebola River in 1976), Marburg (for a town in Hesse, Germany, in 1967), and Prospect Hill *Orthohantavirus* (for Prospect Hill, Maryland, in the 1980s), among many others. Unfortunately, the geographic name attached to a first discovery can take on a particular caché that some consider an indicator of limits on (or a limited) spatial distribution, although in all cases the distributions are considerably broader—reflecting the nuances of fundamental and realized fitness space and host range (e.g., Brooks, Hoberg, et al., 2022; Brooks, Boeger, et al., 2023). In conjunction with a geographic source, pathogens may receive names that denote the disease and the regional locality—for example, Bolivian hemorrhagic fever virus (BHFV) (also known as Machupo virus after the river where it was first identified), Middle East respiratory syndrome virus (MERSV), or severe acute respiratory syndrome virus (SARSV)—and especially the disease as it is manifested in the originally recognized host, which is often not the reservoir responsible for maintenance and circulation. Some are denoted to describe special aspects of disease, such as chikungunya virus (CHIKV), found in Tanzania in 1952 and named from a local dialect term for "to become contorted." Lastly, we name parasites based on our assumptions about host association. Much of this taxonomy reflects responses in human hosts for zoonotic viruses for which emergence is often initially recognized. Often it is only later that natural host assemblages, including various arthropods and vertebrate reservoirs for arboviruses, which drive colonization events are documented. Our choices for naming can be misleading in the absence of authoritative information, and there are no authoritative names or informatics for either hosts or pathogens in the absence of archival specimens maintained in well-organized museum collections and databases, as exemplified by the Museum of Southwestern Biology (MSB) and the Harold W. Manter Laboratory of Parasitology (HWML) (Drabik and Gardner, 2019).

Viral taxonomies pose particular challenges and to a degree have remained outside of the bounds of conventions for multicellular diversity. Calisher and Yates (1999) proposed naming hantaviruses with a binomial nomenclature following universal taxonomic norms and based on the species name of their primary host (e.g., *Hantavirus maniculatus* for Sin Nombre virus). A methodology was intended to more closely tie a virus to its reservoir species, recognizing the essential nature of the proper association of a pathogen and its host. In response, and we concur, Bennett et al. (1999) agreed with the need for tying host to pathogen but suggested the issues of misidentified host species and dynamic mammalian taxonomy (as outlined earlier) were hindrances to the success of this naming format. More recently, the International Committee on Taxonomy of Viruses Executive Committee (2020) proposed a new system to establish rank and classification with the following justifications:

> The codified availability of a greater number of ranks in a formal virus classification that emulates a Linnaean framework may also facilitate the comparison, and possibly improve the compatibility of virus taxonomy with the taxonomies of cellular organisms. Although the switching of hosts by viruses may be a complicating factor, . . .

> We expect that the described changes to the hierarchical rank structure will create a new impetus for the exploration of virus macroevolution and a framework for its application to taxonomy. The changes will also stimulate research on the defining characteristics of monophyletic virus lineages and the recognition of historical events that played a decisive role in their origins and evolution.

Do You Know What You Have Named?

Correct names are the lingua franca, and these are derived from deep, fundamental assemblages of specimens and information held in permanent archives. Correct and complete names, when connected to phylogeny, are the linkage to essential context in the biosphere. Phylogeny and a phylogenetic diagnosis (what an organism is in space and time, defined within an array of unique and shared characteristics and relationships to other organisms) are fundamental to establishing the evolutionary, ecological, and biogeographic tapestry of life. From that tapestry we can identify the interactions within the SP (Hoberg and Brooks, 2008, 2015; Araujo et al., 2015; Brooks et al., 2019; Brooks, Boeger, et al., 2023; Hoberg et al., 2023) and the nature of ecological fitting in sloppy fitness space (and limits of phylogenetic conservatism), oscillation, taxon pulses (across hosts and geography), and coevolutionary mosaics that are the essence of a predictive and anticipatory framework for emergent pathogens central to the proposal for DAMA (Janzen, 1985; Brooks et al., 2014; Colella et al., 2021; Agosta, 2022; Boeger et al., 2022; Brooks, Hoberg et al., 2022;

Box 3

Uncovering the Overlooked Extant Diversity of Phytoplasmas

Phytoplasmas are a diverse group of vector-borne obligate intracellular parasitic bacteria (phylum *Mycoplasmatota*, class *Mollicutes*) associated with vascular plants and phloem-feeding hemipteran insects. Ecologically these represent a poorly known but critically important group for food security and environmental integrity. Since their discovery and character-ization (Doi et al., 1967), they have been shown to be associated with severe diseases causing major economic losses in cultivated crops and other plants (Bertaccini et al., 2014; Brooks, Hoberg, et al., 2022). Still phytoplasmas remain among the least known of the *Mollicutes* bacteria. Prior research on phytoplasmas has mainly focused on their role as plant pathogens, aimed toward managing phytoplasma-caused diseases in agro-ecosystems. Previously known strains of phy-toplasma were mostly discovered by screening economically important plants exhibiting disease symptoms. Phytoplasma infections, particularly in noncultivated native plants, are often asymptomatic (Zwolińska et al., 2019) and, therefore, may go undetected by plant pathologists surveying for plant diseases. Consequently, phytoplasmas have been largely named based on the main symptomatic spectrum exhibited by infected plants or based on the host itself. This has reinforced the misconception that specific groups or subgroups of phytoplasmas are restricted to particular host species (host reper-toire) and thus occupy a limited geographic area or habitat (apparent host specialization). As recently summarized in a comprehensive database (Trivellone, 2019), phytoplasma-host data are limited by this shortsighted focus on cultivated plants and agro-ecosystems which introduced bias into preliminary network analyses and highlights shortcomings and gaps in knowledge of pathogen biodiversity (Trivellone and Flores, 2019). As vector-borne pathogens of plants, phyto-plasmas evolved a fine-tuned intimate relationship with hosts and insect vectors on which they rely for survival and dis-persal throughout the environment. Unfortunately, the vectors of most phytoplasmas remain unknown and the known vectors have been studied primarily from the perspective of plant pathology and epidemiology. Specimens of potential vectors are usually analyzed in pooled samples using destructive DNA extraction methods; that is, specimens are homog-enized, and none are retained to serve as vouchers that demonstrate the insects were correctly identified. This method-ology also makes it impossible to track the association between the pathogen strain and the individual insect (particu-larly if more than one strain is present in the pool). Recent research has shown that phytoplasmas have been evolving together with their hosts for more than 300 million years (Cao et al., 2020), manifesting a complex history of diversifica-tion. Notably, associations between phytoplasmas and their native hosts may be older than those documented in agro-ecosystems (Trivellone and Dietrich, 2021), and recent screening of insect specimens from a museum collection/biore-pository has already begun to document new phytoplasma strains and new associations with potential vectors in natural areas worldwide (Trivellone et al., 2021; Wei et al., 2021; Trivellone, Cao, et al., 2022). Moreover, detailed digitized data on collecting events associated with specimens in the biorepository open further opportunities to study temporal series. The phytoplasma-hemipteran system is a critical exemplar of the continuing necessity to build collections infrastructure to establish a global view of dynamic diversity under the umbrella of the Stockholm paradigm.

Hoberg, Boeger, Brooks, et al., 2022; Trivellone, Hoberg, et al., 2022; Hoberg et al., 2023). Field-collected specimens accumulated across extensive geography, tempered in the strategic context of targeted interfaces that interconnect wildland habitats, peri-urban, and urban settings, and ag-ricultural landscapes representing assemblages of recog-nized and potential reservoirs are the core of DAMA (Figure 7.1). Archives of specimens, validated through species-level identification, lead to archives of information, and collec-tions become fundamental resources for biodiversity infor-matics (Hoberg, 2002) under the umbrella of the *holistic specimen* and standardized methodologies (e.g., Frey et al., 1992; Cook et al., 2016, 2020; Galbreath et al., 2019; Phil-lips et al., 2020; D'Andrea et al., 2021).

Proactive approaches to emergent pathogens and dis-eases are derived from an understanding of diversity (phy-logenetic triage, spatial/temporal distribution and model-ing) and capacity on the part of micro- and macroparasites to use historically (evolutionarily) conserved host-based resources (Agosta et al., 2010; Brooks et al., 2019; Brooks, Hoberg, et al., 2023; Hoberg et al., 2023; Janik et al., 2023). These fundamental cornerstones linking empirical data with operational potential provide the ability to anticipate colonization and emergence across changing environmen-tal interfaces and are critically linked to actions (behaviors) that serve to break the links or pathways for opportunity (Molnár, Knickel, et al., 2022). The SP, its operational ex-tension in DAMA, and broadened development of archival resources for specimens and information can transform our understanding and approach to pathogens (Brooks et al., 2019; Colella et al., 2021; Hoberg et al., 2023). Specimen-based trajectories are vital in developing, facilitating, and protecting access to cumulative information representing the baselines for recognizing changing environmental interfaces in the biosphere from natural to managed and agricultural systems and across landscape to regional and global scales (Figure 7.2). A proactive capability unfolds, from natural to managed and agricultural systems, that is fundamental to both animal and human health and to food security (Brooks, Hoberg, et al., 2022; Trivellone, Hoberg, et al., 2022).

Continuing discussions call for program development of national and international scope to address the ecological and anthropogenic causation and increasing frequency of emergent diseases (e.g., Suzán et al. 2015; Brooks et al., 2019; Daszak et al., 2020; Dobson et al., 2020; Gibb et al., 2020; Alimi et al., 2021; Keesing and Ostfeld, 2021; Brooks, Hoberg, et al., 2022; Carlson, Boyce, et al., 2022; Reaser et al., 2022). Yet, there remains a fundamental disconnect relative to those programs that advocate new directions for exploring pathogens and developing informatics resources but which have no apparent interdependence with permanent archival resources (Upham et al., 2021). Concurrently, intensive synthesis about the nature of specimens, archives, and digitized informatics about the biosphere or specifically about pathogens and hosts are seldom incorporated programmatically into disease ecology from landscape to regional scales (e.g., Cook et al., 2020; Colella et al., 2021; Hoberg et al., 2023). Perhaps this disconnect also reflects an outcome of erroneous but prevailing expectations for the rarity and unpredictability of emerging pathogens in the global arena. Again, this situation emphasizes and calls for substantive cultural transformation in a dispersed but digitally connected community.

Conclusions

A cultural and conceptual transformation is essential, one that recognizes the necessity of placing pathogens in an environmental, evolutionary, and ecological context by incorporating specimens and associated informatics into the foundation for actionable information (e.g., Brooks, Boeger, et al., 2022; Hoberg, Boeger, Molnár, et al., 2022; Brooks, Boeger, et al., 2023; Hoberg et al., 2023). Specimens are naturally the nexus of collaborative networks needed to reveal the connectivity and complexity that embodies the biosphere (e.g., Hoberg, 2002; Colella et al., 2021) (Figures 7.1

and 7.2). Collection and archival preservation, as a form of documentation, is consistent with the boundaries of the SP and its operational protocols under DAMA (Brooks et al., 2014; 2019; Boeger et al., 2022; Hoberg et al., 2023). There is an exigency that extends beyond typical research cycles of 3 to 5 years in a world under rapid change and disruption (Colella et al., 2020; Brooks, Hoberg, et al., 2022; Trivellone, Hoberg, et al., 2022). Research groups to some extent have focused on pathogens or free-living assemblages of organisms, including plants and animals, but seldom both and seldom concurrently and almost always in narrow time frames defined by the finite cycles of grants. Business as usual in this arena perpetuates a disconnected and discordant landscape of untestable empirical observations and ideas that cannot be validated nor extended. The possibility of and potential for the holistic specimen, deep synoptic repositories, integrated phylogenetics, and digitized standardized informatics resources linked to natural history are so critically essential yet have rarely been achieved (Brooks and Hoberg, 2000; Wheeler, 2010; Cook et al., 2013, 2017, 2020; Hoberg et al., 2013; Dunnum et al., 2017; Galbreath et al., 2019).

Historically, natural history repositories and archives have served us well in leading characterization of the biosphere and the interrelationships of natural, agricultural, managed, and urban systems. However, few broadly integrated collections drive comprehensive exploration of host-pathogen-disease dynamics. One primary exemplar is the Museum of Southwestern Biology (University of New Mexico, Albuquerque) with decades of sustained development of collections that integrate extensive cryo-preserved specimens and tissues for parasites, pathogens, and potential and actual reservoir hosts (e.g., Yates et al., 2002; Gardner and Jiménez-Ruiz, 2009; Cook et al., 2017). Further, collections held by the Laboratory of Biology and Parasitology of Reservoir Wild Mammals (the Oswaldo Cruz Foundation) centered in Brazil represent a regional resource and counterpart in the Southern Hemisphere and neotropical region (D'Andrea et al., 2021). Even among pathogens and scientists, however, we remain strongly siloed, and the generality of lessons revealed among viruses, other microparasites, and macroparasites are considered in isolation or in an organismal vacuum defined by taxonomy and a lens of assumed specialization (e.g., Astorga et al., 2023). Notably, a substantial conceptual gulf continues to exist between the worlds of animal and plant pathogens and their respective vectors (e.g., Nylin et al., 2018; Trivellone, 2022). Collections are also limited in how they contribute or what they will receive. Programs and curators generally lack a coordinated global vision when a political arena determines infrastructure associated with national strategic plans and especially continuity and predictability of funds, personnel, and essential physical facilities (e.g., Naggs, 2022). As a consequence, curators too often may discourage deposition of large series of specimens from geographically extensive and site-intensive surveys that are central to understanding pathogen-host dynamics.

Historically, natural history collections were not designed to substantially contribute to a broader understanding of pathogen-host and disease dynamics; such was not their original purpose. In the absence of robust and additional support, current freestanding natural history collections may be to a degree ill-suited and poorly positioned to be able to accomplish what is necessary in the arena of EID. Not unlike the blind men and the elephant, our current infrastructure remains strongly partitioned. Although each taxonomic component has the potential for important contributions, each is not sufficient in isolation to constructively contribute to synoptic insights about pathogen distribution and emergence nor to policy actions linked to mitigation or prediction. Critically, this situation requires vision, additional resources, and better integration to effectively contribute to societal needs (NASEM, 2020). Appropriately, existing resources can be co-opted, repurposed, and consolidated to vastly improve descriptions of historical diversity over ecological timeframes and to forecast future conditions. As such these serve in a retrospective and proactive capacity, including pathogen discovery (e.g., Box 1, *Orthohantavirus* diversity, and Box 3, Phytoplasmas diversity), but do not always contribute in substantial and comparative ways to the development of baselines in the context of resurvey to reveal the trajectories of environmental change. Retrospective studies based on a literature in the absence of archives have particular limitations because of the potential for incorrect taxonomy (when voucher specimens are divorced from taxonomy), lack of comparability in methods and the scope of spatial and temporal sampling (e.g., Carlson, Boyce, et al., 2022). Insights are often limited to what we know (an array of known pathogens), not what we need to know nor what is unknown in circumstances of fluid interfaces and changing environments (e.g., Zhao et al., 2022; Hoberg et al., 2023). As a community we are continually surprised when the orientation of our current resources serves to perpetuate a largely response-based stance relative to EID.

A holistic framework for the "extended specimen" and outcomes for comprehensive surveys of pathogen-host associations has been previously articulated (e.g., Gardner and Jiménez-Ruiz, 2009; Cook et al., 2020). Conceptually, such a framework must be expanded to a proposition and then transformation to *next-generation natural history collections,* one that encompasses validated specimens, informatics archives, and digitized resources with broad accessibility and interoperability (Schindel and Cook, 2018). Collections can meet their potential as networked and integrated centers for pathogen-host diversity and biology, phylogenetics, ecology, and biogeography, which coordinate explorations of the nature of EIDs in an arena of accelerating environmental change. Achieving this goal acknowledges considerable caveats requiring careful development of validated information with the recognition that authoritative species level identifications (for pathogens and hosts) are necessary (e.g., Hoberg and Soudachanh, 2020). Original specimen data may be insufficient, and specimen quality, representation, and the geographic and temporal scope of collections may need to be standardized and revisited and may preclude broadbased comparisons and conclusions over space and time (e.g., Wood and Vanhove, 2022; Wood et al., 2023). Beyond these caveats for developing information across space and time is the critical framework and fundamental issue posed by choice of paradigms for exploring the biosphere, a consideration absent from many current discussions (e.g., Brooks et al., 2015, 2019; Agosta, 2022; Brooks, Boeger, et al., 2023).

How and what we name is critical. The truth of a name resides in that essential connection to specimens, evolution, ecology, and natural history; label identifications for specimens held in biorepositories require caution and confirmation. Additionally, an unsubstantiated name is not good enough in our descriptions of diversity—when there is no justification provided from vouchers or specimens in published surveys, with misidentification perpetuated downstream in aspects of ecological, genomic and molecular prospecting, evaluation, and analysis. Names are assumed to be the truth. Misidentification is not only an error but can establish a fallacy that can be further perpetuated by convenience (lack of archival specimens) in a deep global literature about the distribution and history of diversity (e.g., Hoberg et al., 2009; Hoberg and Soudachanh, 2020, 2021; Bush et al., 2021). Thus, how we use taxonomy can be misleading if it is no longer possible to revisit specimens and validate their identification or to explore and probe with new technologies and methods that dissect the world at increasingly fine scales. Misidentification is like a Gordian knot that can never be resolved in the absence of specimens and drives an insidious pattern of error propagation clouding our understanding of the biosphere. Our stories always start with specimens and handwritten field catalogs. Data obtained in the field, with associated host and parasite data, then enter a convoluted path through various museums and archives, settling questions of science, arriving at names and a nomenclature that should be connected to phylogeny and evolution. Documentation under the DAMA protocol is thus the initial foundation for explorations of diversity under the Stockholm paradigm (Brooks, Boeger, et al., 2022, 2023), and in this we would strongly suggest that paradigms matter in our views of the biosphere. How else can we come to agreement about how the biosphere is structured through space and time and especially related to the obscure and hidden minefield that is global pathogen circulation?

Understanding pathogens is essentially one of revealing the complexity and interconnections of planetary biodiversity. Holistic specimen collections would serve as validated clearinghouses for acquisition, interpretation, translation, and dissemination of informatics resources. Not simply an issue of wildland-managed interfaces, conservation, and species extinctions, the broader challenges to food security, domestic food resources, and agriculture must be accommodated within a comprehensive operational landscape for circulation of EID (e.g., Trivellone and Dietrich, 2021; Brooks, Hoberg, et al., 2022; Trivellone, Cao, et al., 2022; Trivellone, Hoberg, et al., 2022). Cultural transformation is conceptual and operational, serving to build and expand a permanent/sustainable infrastructure around biorepositories rather than perpetuating business as usual in cycles of partitioned research networks (Figures 7.1 and 7.2).

A new model and long-term commitments are required for natural history collections to effectively serve informatics resources (inextricably tied to the specimens in their archives) and actionable information. Collections codify a cumulative view across the interconnected arena of pathogens, hosts, reservoirs, vectors, and interfaces through targeted, site-intensive, and geographically extensive explorations of diversity examined in the context of phylogenetic triage, which will help us establish the limits diversity and risk space (Brooks, Hoberg, et al., 2023; Hoberg et al., 2023). A new model encompasses building or extending infrastructure for personnel, facilities, and capacities, especially in biodiverse regions of our planet. Pathogens are a component of broader global diversity; thus to be effective, new trajectories should be in parallel and revitalize (unrealized) expansions of the systematics/phylogenetics/biodiversity community proposed more than two decades ago (e.g., in part summarized in Brooks and Hoberg, 2001). New capacities and policies should also reflect recent calls

for broadened surveillance (but, ironically, biorepositories have not been central to these proposals) in response to emergent public health crises (Alimi et al., 2021; Holmes, 2022). Some components are best achieved on large scales, including global archives and information storage with broad accessibility, supported and coordinated at national and international levels. Appropriately, such commitments require direct governmental infrastructure, cooperation, and new official capacities that recognize and codify EID as a priority across national and international boundaries (Brooks et al., 2019; Colella et al., 2021).

In an absence of a unified culture that values specimens, archives, and collections, we will be challenged to understand the limits of diversity, evolution, and biogeography and will be circumscribed in our abilities to develop integrated, temporal, and spatial snapshots of the world that become critical baselines. As a global and globalized community, it is essential to abandon business as usual while looking forward toward increasingly transboundary approaches that maximize a proactive stance to EID, extending our conceptual and taxonomic view of diversity across interconnected planetary scales that influence the complexity of pathogen-host interfaces (Hoberg and Brooks, 2008; Hoberg et al., 2015; Brooks et al., 2019). The dimensions of the SP with an evolutionary/historical perspective and the DAMA protocol with collections and specimens as an extended empirical foundation reveal a broad inclusive vision and necessity.

Literature Cited

Agosta, S.J. 2022. The Stockholm paradigm explains the dynamics of Darwin's entangled bank, including emerging infectious disease. MANTER: Journal of Parasite Biodiversity 27. https://doi.org/10.32873/unl.dc.manter27

Agosta, S.J.; Brooks, D.R. 2020. The Major Metaphors of Evolution: Darwinism Then and Now. Springer, Cham, Switzerland. 273 p. https://doi.org/10.1007/978-3-030-52086-1

Agosta, S.J.; Janz, N.; Brooks, D.R. 2010. How specialists can be generalists: resolving the "parasite paradox" and implications for emerging infectious disease. Zoologia (Curitiba) 27: 151–162. https://doi.org/10.1590/S1984-46702010000200001

Alimi, Y.; Bernstein, A.; Epstein, J.; Espinal, M.; Kakkar, M.; Jochevar, D.; Werneck, G. 2021. Report of the Scientific Task Force on Preventing Pandemics. Harvard Global Health Institute and the Center for Climate, Health, and the Global Environment at Harvard T.H. Chan School of Public Health. 36 p.

Álvarez-Hernández, G.; Roldán, J.F.G.; Milan, N.S.H.; Lash, R.R.; Behravesh, C.B.; Paddock, C.D. 2017. Rocky Mountain spotted fever in Mexico: past, present, and future. Lancet Infectious Diseases 17: e189–e196. https://doi.org/10.1016/S1473-3099(17)30173-1

Arai, S.; Yanagihara, R. 2020. Genetic diversity and geographic distribution of bat-borne hantaviruses. Current Issues in Molecular Biology 39: 1–28. https://doi.org/10.21775/cimb.039.001

Araujo, S.B.L.; Braga, M.P.; Brooks, D.R.; Agosta, S.J.; Hoberg, E.P.; von Hartenthal, F.W.; Boeger, W.A. 2015. Understanding host-switching by ecological fitting. PLOS ONE 10: e0139225. https://doi.org/10.1371/journal.pone.0139225

Astorga, F.; Groom, Q.; Shimabukuro, P.H.F.; Manguin, S.; Noesgaard, D.; Orrell, T.; et al. 2023. Biodiversity data supports research on human infectious diseases: global trends, challenges, and opportunities. One Health 16: 100484. https://doi.org/10.1016/j.onehlt.2023.100484

Audy, J.R. 1958. The localization of disease with special reference to the zoonoses. Transactions of the Royal Society of Tropical Medicine and Hygiene 52: 308–334. https://doi.org/10.1016/0035-9203(58)90045-2

Bennett, S.G.; Webb, J.P.; Madon, M.; Childs, J.; Ksiazek, T.; Torrez-Martinez, N.; Hjelle, B. 1999. Hantavirus (Bunyaviridae) infections in rodents from Orange and San Diego Counties, California. American Journal of Tropical Medicine and Hygiene 60: 75–84. https://doi.org/10.4269ajtmh.1999.60.75

Benton, T.G.; Bieg, C.; Harwatt, H.; Pudasaini, R.; Wellesley, L. 2021. Food System Impacts on Biodiversity Loss: Three Levers for Food System Transformation in Support of Nature. UN Environment Programme and Compassion in World Farming. Chatham House, London. 75 p. https://www.chathamhouse.org/2021/02/food-system-impacts-biodiversity-loss

Bertaccini, A.; Duduk, B.; Paltrinieri, S.; Contaldo, N. 2014. Phytoplasmas and phytoplasma diseases: a severe threat to agriculture. American Journal of Plant Sciences 5: 1763–1788. https://doi.org/10.4236/ajps.2014.512191

Boeger, W.A.; Brooks, D.R.; Trivellone, V.; Agosta, S.J.; Hoberg, E.P. 2022. Ecological super-spreaders drive host-range oscillations: Omicron and risk space for emerging infectious disease. Transboundary and Emerging Diseases 69: e1280–e1288. https://doi.org/10.1111/tbed.14557

Botero-Cañola, S.; Dursahinhan, A.T.; Racz, S.E.; Lowe, P.V.; Ubelaker, J.E.; Gardner, S.L. 2019. The ecological niche of *Echinococcus multilocularis* in North America: understanding biotic and abiotic determinants of parasite distribution with new records in New Mexico and Maryland, United States. Therya 10: 91–102. https://doi.org/10.12933/therya-19-749

Bradbury, R.S.; Sapp, S.G.H.; Potters, I.; Mathison, B.A.; Frean, J.; Mewara, A.; et al. 2022. Where have all the diagnostic morphological parasitologists gone? Journal of Clinical Microbiology 60: 1–12. https://doi.org/10.1128/jcm.00986-22

Brooks, D.R.; Boeger, W.A.; Hoberg, E.P. 2022. The Stockholm paradigm: lessons for the emerging infectious disease crisis. MANTER: Journal of Parasite Biodiversity 22. https://doi.org/10.32873/unl.dc.manter22

Brooks, D.R.; Boeger, W.A.; Hoberg, E.P. 2023. The Stockholm paradigm: the conceptual platform for coping with the emerging infectious disease crisis. In: An Evolutionary Pathway for Coping with Emerging Infectious Disease. S.L. Gardner, D.R. Brooks, W.A. Boeger, E.P. Hoberg (eds.). Zea Books, Lincoln, NE.

Brooks, D.R.; Hoberg, E.P. 2000. Triage for the biosphere: the need and rationale for taxonomic inventories and phylogenetic studies of parasites. Comparative Parasitology 67: 1–25.

Brooks, D.R.; Hoberg, E.P. 2001. Parasite systematics in the 21st century: opportunities and obstacles. Trends in Parasitology 17: 273–275. https://doi.org/10.1016/s1471-4922(01)01894-3

Brooks, D.R.; Hoberg, E.P. 2007. How will global climate change affect parasite-host assemblages? Trends in Parasitology 23: 571–574. https://doi.org/10.1016/j.pt.2007.08.016

Brooks, D.R.; Hoberg, E.P. 2013. The emerging infectious diseases crisis and pathogen pollution: A question of ecology and evolution. In: The Balance of Nature and Human Impact. K. Rhode (ed.). Cambridge University Press: Cambridge. pp. 215–230. https://doi.org/10.1017/CBO9781139095075.022

Brooks, D.R.; Hoberg, E.P.; Boeger, W.A. 2015. In the eye of the cyclops: the classic case of cospeciation and why paradigms are important. Comparative Parasitology 82: 1–8. https://doi.org/10.1654/4724C.1

Brooks, D.R.; Hoberg, E.P.; Boeger, W.A. 2019. The Stockholm Paradigm: Climate Change and Emerging Disease. University of Chicago Press, Chicago. 423 p. https://doi.org/10.7208/chicago/9780226632582.001.0001

Brooks, D.R.; Hoberg, E.P.; Boeger, W.A.; Trivellone, V. 2022. Emerging infectious disease: an underappreciated area of strategic concern for food security. Transboundary and Emerging Diseases 69: 254–267. https://doi.org/10.1111/tbed.14009

Brooks, D.R.; Hoberg, E.P.; Boeger, W.A.; Trivellone, V. 2023. Assess: using evolution to save time and resources. In: An Evolutionary Pathway for Coping with Emerging Infectious Disease. S.L. Gardner, D.R. Brooks, W.A. Boeger, E.P. Hoberg (eds.). Zea Books, Lincoln, NE.

Brooks, D.R.; Hoberg, E.P.; Gardner, S.L.; Boeger, W.; Galbreath, K.E.; et al. 2014. Finding them before they find us: informatics, parasites, and environments in accelerating climate change. Comparative Parasitology 81: 155–164. https://doi.org/10.1654/4724b.1

Brooks, D.R.; McLennan, D.A. 2002. The Nature of Diversity: An Evolutionary Voyage of Discovery. University of Chicago Press, Chicago. 676 pp.

Bush, S.E.; Gustafsson, D.R.; Tkach, V.V.; Clayton, D.H. 2021. A misidentification crisis plagues specimen-based research: a case for guidelines with a recent example (Ali et al., 2020). Journal of Parasitology 107: 262–266. https://doi.org/10.1645/21-4

Calisher, C.H.; Yates, T.L. 1999. Hantavirus (Bunyaviridae) infections in rodents from Orange and San Diego Counties, California [letter to the editor]. American Journal of Tropical Medicine and Hygiene 61: 863–864. https://doi.org/10.4269/ajtmh.1999.61.863

Cao, Y.; Trivellone, V.; Dietrich, C.H. 2020. A timetree for phytoplasmas (Mollicutes) with new insights on patterns of evolution and diversification. Molecular Phylogenetics and Evolution 149: 106826. https://doi.org/10.1016/j.ympev.2020.106826

Carlson, C.J.; Albery, G.F.; Merow, C.; Trisos, C.H.; Zipfel, C.M.; Eskew, E.A.; et al. 2022. Climate change increases cross-species viral transmission risk. Nature 607: 555–562. https://doi.org/10.1038/s41586-022-04788-w

Carlson, C.J.; Boyce, M.R.; Dunne, M.; Graden, E.; Lin, J.; Abdellatif, Y.O.; et al. 2022. The World Health Organization's disease outbreak news: a retrospective database. medRxiv preprint. https://doi.org/10.1101/2022.03.22.22272790

Carlson, C.J.; Dallas, T.A.; Alexander, L.W.; Phelan, A.L.; Phillips, A.J. 2020. What would it take to describe the global diversity of parasites? Proceedings of the Royal Society B—Biological Sciences 287: 20201841. http://dx.doi.org/10.1098/rspb.2020.1841

Chen, Z.; Azman, A.S.; Chen, X.; Zou, J.; Tian, Y.; Sun, R.; et al. 2022. Global landscape of SARS-CoV-2 genomic surveillance and data sharing. Nature Genetics 54: 499–507. https://doi.org/10.1038/s41588-022-01033-y

Christensen, T.; Coon, C.; Fletcher, S.; Barry, T.; Lárusson, K.F. 2021. Circumpolar Biodiversity Monitoring Program Strategic Plan: 2021–2025. Conservation of Arctic Flora and Fauna International Secretariat, Akureyri, Iceland.

Coker, R.; Rushton, J.; Mounier-Jack, S.; Karimuribo, E.; Lutumba, P.; Kambarage, D.; et al. 2011. Towards a conceptual framework to support one-health research for policy on emerging zoonoses. Lancet Infectious Diseases 11: 326–331. https://doi.org/10.1016/S1473-3099(10)70312-1

Colella, J.P.; Bates, J.; Burneo, S.F.; Camacho, M.A.; Bonilla, C.C.; Constable, I.; et al. 2021. Leveraging natural history biorepositories as a global, decentralized, pathogen surveillance network. PLOS Pathogens 17: e1009583. https://doi.org/10.1371/journal.ppat.1009583

Colella, J.P.; Talbot, S.L.; Brochmann, C.; Taylor, E.B.; Hoberg, E.P.; Cook, J.A. 2020. Conservation genomics in a changing Arctic.

Trends in Ecology and Evolution 35: 149–162. https://doi.org/10.1016/j.tree.2019.09.008

Cook, J.; Brochman, C.; Talbot, S.L.; Fedorov, V.B.; Taylor, E.B.; Väinölä, R.; et al. 2013. Genetics [chapter 17]. In: Arctic Biodiversity Assessment: Status and Trends in Arctic Biodiversity. H. Meltofte (ed.). Conservation of Arctic Flora and Fauna and Arctic Council, Akureyi, Iceland. 460–477 p. https://oaarchive.arctic-council.org/handle/11374/223

Cook, J.A.; Arai, S.; Armién, B.; Bates, J.; Carrion Bonilla, C.A.; de Souza Cortez, M.B.; et al. 2020. Integrating biodiversity infrastructure into pathogen discovery and mitigation of epidemic infectious diseases. Bioscience 70: 531–534. https://doi.org/10.1093/biosci/biaa064

Cook, J.A.; Galbreath, K.E.; Bell, K.C.; Campbell, M.L.; Carrière, S.; Colella, J.P.; et al. 2017. The Beringian Coevolution Project: holistic collections of mammals and associated parasites reveal novel perspectives on evolutionary and environmental change in the North. Arctic Science 3: 585–617. https://doi.org/10.1139/as-2016-0042

Cook, J.A.; Greiman, S.E.; Agosta, S.J.; Anderson, R.P.; Arbogast, B.S.; Baker, R.J.; et al. 2016. Transformational principles for NEON sampling of mammalian parasites and pathogens: a response to Springer and colleagues. BioScience 66: 917–919. https://doi.org/10.1093/biosci/biw123

Cook, R.A.; Karesh, W.B.; Osofsky, S.A. 2004. One World, One Health: Building Interdisciplinary Bridges to Health in a Globalized World. Conference Summary. Wildlife Conservation Society. The Rockefeller University, September 29, 2004 [accessed 12 February 2022]. http://www.oneworldonehealth.org/sept2004/owoh_sept04.html

Cunningham, A.A.; Daszak, P.; Wood, J.N. 2017. One Health, emerging infectious diseases and wildlife: two decades of progress? Philosophical Transactions of the Royal Society B: Biological Sciences 372: 20160167. https://doi.org/10.1098/rstb.2016.0167

D'Andrea, P.S.; Teixeira, B.R.; Gonçalves-Oliveira, J.; Dias, D.; do Val Vilela, R.; dos Santos Lucio, C.; et al. 2021. The mastozoological collection of the Laboratory of Biology and Parasitology of Wild Mammals Reservoirs—Fundação Oswaldo Cruz. Brazilian Journal of Mammalogy 2021: e90202119. https://doi.org/10.32673/bjm.vie90.19

Daszak, P.; das Neves, C.; Amuasi, J.; Hayman, D.; Kuiken, T.; Roche, B.; et al. 2020. Workshop Report on Biodiversity and Pandemics of the Intergovernmental Platform on Biodiversity and Ecosystem Services (IPBES). IPBES Secretariat, Bonn, Germany. https://doi.org/10.5281/zenodo.4147317

Dobson, A.P.; Pimm, S.L.; Hannah, L.; Kaufman, L.; Ahumada, J.A.; Ando, A.W.; et al. 2020. Ecology and economics for pandemic prevention. Science 369: 379–381. https://doi.org/10.1126/science.abc3189

Doi, Y.; Teranaka, M.; Yora, K.; Asuyama, H. 1967. Mycoplasma- or PLT group-like microorganisms found in the phloem elements of plants infected with mulberry dwarf, potato witches' broom, aster yellows, or paulownia witches' broom. Japanese Journal of Phytopathology 33: 259–266. https://doi.org/10.3186/jjphytopath.33.259

Drabik, G.O.; Gardner, S.L. 2019. A new species of *Ancylostoma* (Nemata: Strongylida: Ancylostomatidae) from two species of *Ctenomys* in lowland Bolivia. Journal of Parasitology 105: 904–912. https://doi.org/10.1645/19-100

Dunnum, J.L.; Yanagihara, R.; Johnson, K.M.; Armien, B.; Batsaikhan, N.; Morgan, L.; Cook, J.A. 2017. Biospecimen repositories and integrated databases as critical infrastructure for pathogen discovery and pathobiology research. PLOS Neglected Tropical Diseases 11: e0005133. https://doi.org/10.1371/journal.pntd.0005133

Firth, C.; Tokarz, R.; Simith, D.B.; Nunes, M.R,T.; Bhat, M.; Rosa, E.S.T.; et al. 2012. Diversity and distribution of hantaviruses in South America. Journal of Virology 86: 13756–13766. https://doi.org/10.1128/VNI.02341-12

Frey, J.K.; Yates, T.L.; Duszynski, D.W.; Gannon, W.L.; Gardner, S.L. 1992. Designation and curatorial management of type host specimens (symbiotypes) for new parasite species. Journal of Parasitology 78: 930–932.

Galbreath, K.E.; Hoberg, E.P.; Cook, J.A.; Armién, B.; Bell, K.C.; Campbell, M.L.; et al. 2019. Building an integrated infrastructure for exploring biodiversity: field collections and archives of mammals and parasites. Journal of Mammalogy 100: 382–393 [+ supplementary data SD1: Field Methods for Collection and Preservation of Mammalian Parasites, 36 pp.]. https://doi.org/10.1093/jmammal/gyz048

Gardner, S.L. 1996. Essential techniques for collection of parasites during surveys of mammals. In: Measuring and Monitoring Biological Diversity: Standard Methods for Mammals. D.E. Wilson; F.R. Cole; J.D. Nichols; R. Rudran; M.S. Foster (eds.). Smithsonian Institution Press, Washington, DC, 291–298 p.

Gardner, S.L.; Botero-Cañola, S.; Aliaga-Rossel, E.; Dursahinhan, A.T.; Salazar-Bravo, J.A. 2021. Conservation status and natural history of *Ctenomys*, tuco-tucos in Bolivia. Therya 12: 15–36. https://doi.org/10.12933/therya-21-1035

Gardner, S.L.; Jiménez-Ruiz, F.A. 2009. Methods of endoparasite analysis. In: Ecological and Behavioral Methods for the Study of Bats. 2nd ed. T.H. Kunz, S. Parsons (eds.). Johns Hopkins University Press, Baltimore, MD. 795–805 p.

Gardner, S.L.; Whitaker, J. 2009. Endoparasites of bats. In: Bats in Captivity. S.M. Barnard (ed.). Krieger Publishing Company, Malabar, FL.

Gebreyes, W.A.; Dupouy-Camet, J.; Newport, M.J.; Oliveira, C.J.B.; Schlesinger, L.S.; Saif, Y.M.; et al. 2014. The Global One Health paradigm: challenges and opportunities for tackling infectious diseases at the human, animal, and environment interface in low-resource settings. PLOS Neglected Tropical Diseases 8: e3257. https://doi.org/10.1371/journal.pntd.0003257

Gibb, R.; Redding, D.W.; Chin, K.Q.; Donnelly, C.A.; Blackburn, T.M.; Newbold, T.; Jones, K.E. 2020. Zoonotic host diversity increases in human-dominated ecosystems. Nature 584: 398–402. https://doi.org/10.1038/s41586-020-2562-8

Giraudoux, P.; Bescombes, C.; Bompangue, D.; Guégan, J.-F.; Mauny, F.; Morand, S. 2022. One Health or 'One Health washing'? An alternative to overcome now more than ever. CABI One Health 2022: 1–4. https://doi.org/10.1079/cabionehealth.2022.0006

González-Ittig, R.E.; Rivera, P.C.; Levis, S.C.; Calderón, G.E.; Gardenal, C.N. 2014. The molecular phylogenetics of the genus *Oligoryzomys* (Rodentia: Cricetidae) clarifies rodent host–hantavirus associations. Zoological Journal of the Linnean Society 171: 457–474. https://doi.org/10.1111/zoj.12133

Grange, Z.L.; Goldstein, T.; Johnson, C.K.; Anthony, S.; Gilardi, K.; Daszak, P.; et al. 2021. Ranking the risk of animal-to-human spillover for newly discovered viruses. Proceedings of the National Academy of Sciences USA 118: e2002324118. https://doi.org/10.1073/pnas.2002324118

Granter, S.R.; Ostfeld, R.S.; Milner, D.A., Jr. 2016. Where the wild things aren't: loss of biodiversity, emerging infectious diseases, and implications for diagnosticians. American Journal of Clinical Pathology 146: 644–646. https://doi.org/10.1093/AJCP/AQW197

Harris, H.M.B.; Hill, C. 2021. A place for viruses on the tree of life. Frontiers in Microbiology 11: 604048. https://doi.org/10.3389/fmicb.2020.604048

Hoberg, E.P. 2002. Foundations for an integrative parasitology: collections, archives, and biodiversity informatics. Comparative Parasitology 69: 124–131. https://doi.org/10.1654/1525-2647(2002)069[0124:FFAIPC]2.0.CO;2

Hoberg, E.P. 2010. Invasive processes, mosaics and the structure of helminth parasite faunas. Revue Scientifique et Technique, Office International des Épizooties 29: 255–272. https://dx.doi.org/10.20506/rst.29.2.1972

Hoberg, E.P.; Agosta, S.J.; Boeger, W.A.; Brooks, D.R. 2015. An integrated parasitology: revealing the elephant through tradition and invention. Trends in Parasitology 31: 128–133. http://dx.doi.org/10.1016/j.pt.2014.11.005

Hoberg, E.P.; Boeger, W.A.; Brooks, D.R.; Trivellone, V.; Agosta, S.J. 2022. Stepping-stones and mediators of pandemic expansion: a context for humans as ecological super-spreaders. MANTER: Journal of Parasite Biodiversity 18. https://doi.org/10.32873/unl.dc.manter18

Hoberg, E.P.; Boeger, W.A.; Molnár, O.; Földvári, G.; Gardner, S.L.; Juarrero, A.; et al. 2022. The DAMA protocol, an introduction: finding pathogens before they find us. MANTER: Journal of Parasite Biodiversity 21. https://doi.org/10.32873/unl.dc.manter21

Hoberg, E.P.; Boeger, W.A.; Molnár, O.; Földvári, G.; Gardner, S.L.; Juarrero, A.; et al. 2023. The DAMA protocol: anticipating to prevent and mitigate emerging infectious diseases. In: An Evolutionary Pathway for Coping with Emerging Infectious Disease. S.L. Gardner, D.R. Brooks, W.A. Boeger, E.P. Hoberg (eds.). Zea Books, Lincoln, NE.

Hoberg, E.P.; Brooks, D.R. 2008. A macroevolutionary mosaic: episodic host-switching, geographical colonization and diversification in complex host-parasite systems. Journal of Biogeography 35: 1533–1550. https://doi.org/10.1111/j.1365-2699.2008.01951.x

Hoberg, E.P.; Brooks, D.R. 2010. Beyond vicariance: integrating taxon pulses, ecological fitting and oscillation in evolution and historical biogeography. In: The Biogeography of Host-Parasite Interactions. S. Morand, B.R. Krasnow (eds.). Oxford University Press, Oxford. 7–20 p.

Hoberg, E.P.; Brooks, D.R. 2015. Evolution in action: climate change, biodiversity dynamics and emerging infectious disease. Philosophical Transactions of the Royal Society B 370: 20130553. https://doi.org/10.1098/rstb.2013.0553

Hoberg, E.P.; Kutz, S.J.; Cook, J.A.; Galaktionov, K.; Haukisalmi, V.; Henttonen, H.; et al. 2013. Parasites [chapter 15]. In: Arctic Biodiversity Assessment: Status and Trends in Arctic Biodiversity. H. Meltofte (ed.). Conservation of Arctic Flora and Fauna and Arctic Council, Akureyi, Iceland. 529–557 p. https://oaarchive.arctic-council.org/handle/11374/223

Hoberg, E.P.; Pilitt, P.A.; Galbreath, K.E. 2009. Why museums matter: a tale of pinworms (Oxyuroidea: Heteroxynematidae) among pikas (Ochotona princeps and O. collaris) in the American West. Journal of Parasitology 95: 490–501. https://doi.org/10.1645/GE-1823.1

Hoberg, E.P.; Soudachanh, K.M. 2020. Insights about diversity of Tetrabothriidae (Eucestoda) among Holarctic Alcidae (Charadriiformes): what is Tetrabothrius jagerskioeldi? MANTER: Journal of Parasite Biodiversity 11. 42 pp. https://doi.org/10.32873/unl.dc.manter11

Hoberg, E.P.; Soudachanh, K.M. 2021. Diversity of Tetrabothriidae (Eucestoda) among Holarctic Alcidae (Charadriiformes): resolution of the Tetrabothrius jagerskioeldi cryptic species complex: Cestodes of Alcinae—provides insights on the dynamic nature of tapeworm and marine bird faunas under the Stockholm paradigm. MANTER: Journal of Parasite Biodiversity 17. 76 pp. https://doi.org/10.32873/unl.dc.manter17

Hoberg, E.P.; Trivellone, V.; Cook, J.A.; Dunnum, J.L.; Boeger, W.B.; et al. 2022. Knowing the biosphere: documentation, specimens, archives, and names reveal environmental change and emerging pathogens. MANTER: Journal of Parasite Biodiversity 26. https://doi.org/10.32873/unl.dc.manter26

Holmes, E.C. 2022. COVID-19—lessons for zoonotic disease. Science 375: 1114–1115. https://doi.org/10.1126/science.abn2222

Horton, R.; Lo, S. 2015. Planetary health: a new science for exceptional action. Lancet 386: 1921–1922. https://doi.org/10.1016/S0140-6736(15)61038-8

Hurtado, N.; D'Elía, G. 2019. An assessment of species limits of the South American mouse genus Oligoryzomys (Rodentia, Cricetidae) using unilocus delimitation methods. Zoologica Scripta 48: 557–570. https://doi.org/10.1111/zsc.12365

ICZN [International Commission on Zoological Nomenclature]. 1999. International Code of Zoological Nomenclature. 4th ed. W.D.L. Ride (chair) et al. (eds.) International Trust for Zoological Nomenclature, London. 106 p. https://www.iczn.org/the-code/the-code-online/

International Committee on Taxonomy of Viruses Executive Committee. 2020. The new scope of virus taxonomy: partitioning the virosphere into 15 hierarchical ranks. Nature Microbiology 5: 668–674. https://doi.org/10.1038/s41564-020-0709-x

Janik, K.; Panassiti, B.; Kerschbamer, C.; Burmeister, J.; Trivellone, V. 2023. Phylogenetic triage and risk assessment: how to predict emerging phytoplasma diseases. Biology 12: 732–754. https://doi.org/10.3390/biology12050732

Janzen, D.H. 1985. On ecological fitting. Oikos 45: 308–310.

Jones, K.E.; Patel, N.G.; Levy, M.A.; Storeygard, A.; Balk, D.; Gittleman, J.L.; Daszak, P. 2008. Global trends in emerging infectious diseases. Nature 451: 990–993. https://doi.org/10.1038/nature06536

Jonsson, C.B.; Moraes Figueiredo, L.T.; Vapalahti, O. 2010. A global perspective on hantavirus ecology, epidemiology, and disease. Clinical Microbiology Reviews 23: 412–441. https://doi.org/10.1128/CMR.00062-09

Kafle, P.; Peller, P.; Massolo, A.; Hoberg, E.; Leclerc, L.-M.; Tomaselli, M.; Kutz, S. 2020. Range expansion of muskox lungworms track rapid Arctic warming: implications for geographic colonization under climate forcing. Scientific Reports 10: 17323. https://doi.org/10.1038/s41598-020-74358-5

Keesing, F.; Ostfeld, R.S. 2021. Impacts of biodiversity and biodiversity loss on zoonotic diseases. Proceedings of the National Academy of Sciences USA 118: e2023540118. https://doi.org/10.1073/pnas.2023540118

Kirdat, K.; Tiwarekar, B.; Sathe, S.; Yadav, A. 2023. From sequences to species: charting the phytoplasma classification and taxonomy in the era of taxogenomics. Frontiers in Microbiology 14: 1123783. https://doi.org/10.3389/fmicb.023.1123783

Kobayashi, T.; Higgins, Y.; Samuels, R.; Moaven, A.; Sanyal, A.; Yenokyan, G.; et al. 2019. Misdiagnosis of Lyme disease with unnecessary antimicrobial treatment characterizes patients referred to an academic infectious diseases clinic. Open Forum Infectious Diseases 6: ofz299. https://doi.org/10.1093/ofid/ofz299

Kozlov, A.M.; Zhang, J.; Yilmaz, P.; Glöckner, F.O.; Stamatakis, A. 2016. Phylogeny-aware identification and correction of taxonomically mislabeled sequences. Nucleic Acids Research 44: 5022–5033. https://doi.org/10.1093/nar/gkw396

Kuhn, J.H.; Bradfute, S.B.; Calisher, C.H.; Klempa, B.; Klingström, J.; Laenen, L.; et al. 2023. Pending reorganization of Hantaviridae to include only completely sequenced viruses: a call to action. Viruses 15: 660–671. https://doi.org/10.3390/v15030660

Kuhn, J.H.; Schmaljohn, C.S. 2023. A brief history of Bunyaviral family Hantaviridae. Diseases 11: 38–49. https://doi.org/10.3390/diseases110110038

Laaksonen, S.; Oksanen, A.; Hoberg, E. 2015. A lymphatic dwelling filarioid nematode, *Rumenfilaria andersoni* (Filarioidea; Splendidofilariinae), is an emerging parasite in Finnish cervids. Parasites and Vectors 8: 228. https://doi.org/10.1186/s13071-015-0835-0

Latinne, A.; Hu, B.; Olival, K.J.; Zhu, G.; Zhang, L.; Li, H.; et al. 2020. Origin and cross-species transmission of bat coronaviruses in China. Nature Communications 11: 4235. https://doi.org/10.1038/s41467-020-17687-3

Lee, H.W.; Lee, P.W.; Johnson, K.M. 1978. Isolation of the etiologic agent of Korean hemorrhagic fever. Journal of Infectious Diseases 137: 298–308. https://doi.org/10.1093/infdis/137.3.298

Liphardt, S.W.; Kang, H.J.; Dizney, L.J.; Ruedas, L.A.; Cook, J.A.; Yanahihara, R. 2019. Complex history of codiversification and host switching of a newfound soricid-borne orthohantavirus in North America. Viruses 11: 637. https://doi.org/10.3390/v11070637

McLean, B.S.; Bell, K.C.; Dunnum, J.L.; Abrahamson, B.; Colella, J.P.; Deardorff, E. R.; et al. 2016. Natural history collections–based research: progress, promise, and best practices. Journal of Mammalogy 97: 287–297. https://doi.org/10.1093/jmammal/gyv178

Miller, S.E.; Barrow, L.N.; Ehlman, S.M.; Goodheart, J.A.; Greiman, S.E.; Lutz, H.L.; et al. 2020. Building natural history collections for the 21st century and beyond. BioScience 70: 674–687. https://doi.org/10.1093/biosci/biaa069

Mollentze, N.; Streicker, D.G. 2020. Viral zoonotic risk is homogenous among taxonomic orders of mammalian and avian reservoir hosts. Proceedings of the National Academy of Sciences USA 117: 9423–9430. https://doi.org/10.1073/pnas.1919176117

Molnár, O.; Hoberg, E.; Trivellone, V.; Földvári, G.; Brooks, D.R. 2022. The 3P framework: a comprehensive approach to coping with the emerging infectious disease crisis. MANTER: Journal of Parasite Biodiversity 23. https://doi.org/1032873/unl.dc.manter23

Molnár, O.; Knickel, M.; Marizzi, C. 2022. Taking action: turning evolutionary theory into preventive policies. MANTER: Journal of Parasite Biodiversity 28. https://doi.org/10.32873/unl.dc.manter28

Naggs, F. 2022. The tragedy of the Natural History Museum, London. Megataxa 007: 85–112. https://doi.org/10.11646/megataxa.7.1.2

NASEM [The National Academies of Sciences, Engineering, and Medicine]. 2020. Biological Collections: Ensuring Critical Research and Education for the 21st Century. The National Academies Press, Washington, DC. 210 pp. https://doi.org/10.17226/25592

Nylin, S.; Agosta, S.; Bensch, S.; Boeger, W.A.; Braga, M.P.; Brooks, D.R.; et al. 2018. Embracing colonizations: a new paradigm for species association dynamics. Trends in Ecology and Evolution 33: 4–14. https://doi.org/10.1016/j.tree.2017.10.005

Olival, K.J.; Hosseini, P.R.; Zambrana-Torrelio, C.; Ross, N.; Bogich, T.L.; Daszak, P. 2017. Host and viral traits predict zoonotic spillover from mammals. Nature 546: 646–650. https://doi.org/10.1038/nature22975

Ortiz, E.; Juarrero, A. 2022. An emerging infectious disease surveillance platform for the 21st century. MANTER: Journal of Parasite Biodiversity 29. https://doi.org/10.32873/unl.dc.manter29

Ostfeld, R.S.; Brunner, J.L. 2015. Climate change and *Ixodes* tick-borne diseases of humans. Philosophical Transactions of the Royal Society B—Biological Sciences 370: 20140051. https://doi.org/10.1098/rstb.2014.0051

Ostfeld, R.S.; Keesing, F. 2000. Biodiversity and disease risk: the case of Lyme disease. Conservation Biology 14: 722–728. https://doi.org/10.1046/j.1523-1739.2000.99014.x

Parmenter, R.R.; Glass, G.E. 2022. Hantavirus outbreaks in the American Southwest: propagation and retraction of rodent and virus diffusion waves from sky-island refugia. International Journal of Modern Physics B 36: 2140052. https://doi.org/10.1142/S021797922140052X

Phillips, V.C.; Zieman, E.A.; Kim, C.-H.; Stone, C.M.; Tuten, H.C.; Jiménez, F.A. 2020. Documentation of the expansion of the Gulf Coast Tick (*Amblyomma maculatum*) and *Rickettsia parkeri*: first report in Illinois. Journal of Parasitology 106: 9–13. https://doi.org/10.1645/19-118

Plowright, R.K.; Becker, D.J.; McCallum, H.; Manlove, K.R. 2019. Sampling to elucidate the dynamics of infections in reservoir hosts. Philosophical Transactions of the Royal Society B—Biological Sciences 374: 20180336. http://dx.doi.org/10.1098/rstb.2018.0336

Poulin, R.; Presswell, B. 2022. Is parasite taxonomy really in trouble? A quantitative analysis. International Journal for Parasitology 52: 469–474. https://doi.org/10.1016/j.ijpara.2022.03.001

Reaser, J.K.; Hunt, B.E.; Ruiz-Aravena, M.; Tabor, G.M.; Patz, J.A.; Becker, D.J.; et al. 2022. Fostering landscape immunity to protect human health: a science-based rationale for shifting conservation policy paradigms. Conservation Letters 15: e12869. https://doi.org/10.1111/conl.12869

Regala-Nava, J.A.; Wang, Y-T.; Fontes-Garfias, C.R.; Liu, Y.; Syed, T.; Susantono, M.; et al. 2022. A Zika virus mutation enhances transmission potential and confers escape from protective dengue virus immunity. Cell Reports 39: 110655. https://doi.org/10.1016/j.celrep.2022.110655

Reuben, R.C.; Danladi, M.M.A.; Pennap, G.R. 2020. Is the COVID-19 pandemic masking the deadlier Lassa fever epidemic in Nigeria? Journal of Clinical Virology 128: 104434. https://doi.org/10.1016/j.jcv.2020.104434

Robles, A.; Fong, J.; Cervantes, J. 2018. *Borrelia* infection in Latin America. Revista de Investigación Clínica 70: 158–163. https://doi.org/10.24875/RIC.18002509

Rohwer, V.G.; Rohwer, Y.; Dillman, C.B. 2022. Declining growth of natural history collections fails future generations. PLOS Biology 20: e3001613. https://doi.org/10.1371/journal.pbio.3001613

Ruiz-Aravena, M.; Mckee, C.; Gamble, A.; Lunn, T.; Morris, A.; Snedden, C.E.; et al. 2021. Ecology, evolution and spillover of coronaviruses from bats. Nature Reviews Microbiology 20: 299–314. https://doi.org/10.1038/s41579-021-00652-2

Sandall, E.L.; Maureaud, A.A.; Guralnick, R.; McGeoch, M.A.; Sica, Y.V.; Rogan, M.S.; et al. 2022. A globally integrated structure of taxonomy supporting biodiversity science and conservation. EcoEvoRxiv preprint. https://doi.org/10.32942/X2WC74

Schilthuizen, M.; Vairappan, C.S.; Slade, E.M.; Mann, D.J.; Miller, J.A. 2015. Specimens as primary data: museums and "open science." Trends in Ecology and Evolution 30: 237–238. https://doi.org/10.1016/j.tree.2015.03.002

Schindel, D.E.; Cook, J.A. 2018. The next generation of natural history collections. PLOS Biology 16: e2006125. https://doi.org/10.1371/journal.pbio.2006125

Sett, S.; dos Santos Ribeiro, C.; Prat, C.; Haringhuizen, G.; European Virus Archive principal investigators; Scholz, A.H. 2022. Access and benefit sharing by the European Virus Archive in response to COVID-19. Lancet Microbe 3: E316–E323. https://doi.org/10.1016/S2666-5247(21)00211-1

Shapiro, J.T.; Víquez-R, L.; Leopardi, S.; Vicente-Santos, A.; Mendenhall, I.H.; Frick, W.F.; et al. 2021. Setting the terms for zoonotic diseases: effective communication for research, conservation, and public Policy. Viruses 13: 1356. https://doi.org/10.3390/v13071356

Simmonds, P.; Adriaenssens, E.M.; Zerbini, F.M.; Abrescia, N.G.A.; Aiewsakun, P.; Alfenas-Zerbini, P.; et al. 2023. Four principles to establish a universal virus taxonomy. PLOS Biology 21: e3001922. https://doi.org/10.1371/journal.pbio.3001922

Song, J.-W.; Baek, L.J.; Schmaljohn, C.S.; Yanagihara, R. 2007. Thottapalayam virus, a prototype shrewborne hantavirus. Emerging Infectious Diseases 13: 980–985. https://doi.org/10.3201/eid1307.070031

Song, J.-W.; Kang, H.J.; Song, K.-J.; Truong, T.T.; Bennett, S.N.; Arai, S.; et al. 2007. Newfound hantavirus in Chinese mole shrew, Vietnam. Emerging Infectious Diseases 13: 1784–1787. https://doi.org/10.3201/eid1311.070492

Souza, A.T.C.; Araujo, S.B.L.; Boeger, W.A. 2022. The evolutionary dynamics of diseases on an unstable planet: insights from modeling the Stockholm paradigm. MANTER: Journal of Parasite Biodiversity 25. https://doi.org/10.32873/unl.dc.manter25

Stone, B.L.; Tourand, Y.; Brissette, C.A. 2017. Brave new worlds: the expanding universe of Lyme disease. Vector-Borne andn Zoonotic Diseases 17: 619–629. https://doi.org/10.1089/vbz.2017.2127

Suzán, G.; García-Peña, G.E.; Castro-Arellano, I.; Rico, O.; Rubio, A.V.; Tolsá, M.J.; et al. 2015. Metacommunity and phylogenetic structure determine wildlife and zoonotic infectious disease patterns in time and space. Ecology and Evolution 5: 865–873. https://doi.org/10.1002/ece3.1404

Temmam, S.; Chrétien, D.; Bigot, T.; Dufour, E.; Petres, S.; Desquesnes, M.; et al. 2019. Monitoring silent spillovers before emergence: a pilot study at the tick/human interface in Thailand. Frontiers in Microbiology 10: 2315. https://doi.org/10.3389/fmicb.2019.02315

Temmam, S.; Vongphayloth, K.; Baquero, E.; Munier, S.; Bonomi, M.; Regnault, B.; et al. 2022. Bat coronaviruses related to SARS-CoV-2 and infectious for human cells. Nature 604: 330–336. https://doi.org/10.1038/s41586-022-04532-4

Thompson, C.W.; Phelps, K.L.; Allard, M.W.; Cook, J.A.; Dunnum, J.L.; Ferguson, A.W.; et al. 2021. Preserve a voucher specimen! The critical need for integrating natural history collections in infectious disease studies. mBio 12: e02698-20. https://doi.org/10.1128/mBio.02698-20

Trivellone, V. 2019. An online global database of Hemiptera-phytoplasma-plant biological interactions. Biodiversity Data Journal 7: e32910. https://doi.org/10.3897/BDJ.7.e32910

Trivellone, V. 2022. Let emerging plant diseases be predictable. MANTER: Journal of Parasite Biodiversity 30. https://doi.org/10.32873/unl.dc.manter30

Trivellone, V.; Cao, Y.; Dietrich, C.H. 2022. Comparison of traditional and next-generation approaches for uncovering phytoplasma diversity, with discovery of new groups, subgroups and potential vectors. Biology 11: 977. https://doi.org/10.3390/biology11070977

Trivellone, V.; Dietrich, C.H. 2021. Evolutionary diversification in insect vector–phytoplasma–plant associations. Annals of the Entomological Society of America 114: 137–150. https://doi.org/10.1093/aesa/saaa048

Trivellone, V.; Flores, G.C.O. 2019. Network analyses of a global Hemiptera-phytoplasma-plant biological interactions database. Phytopathogenic Mollicutes 9: 35–36. https://doi.org/10.5958/2249-4677.2019.00018.5

Trivellone, V.; Hoberg, E.P.; Boeger, W.A.; Brooks, D.R. 2022. Food security and emerging infectious disease: risk assessment and risk management. Royal Society Open Science 9: 211687. https://doi.org/10.1098/rsos.211687

Trivellone, V.; Wei, W.; Filippin, L.; Dietrich, C.H. 2021. Screening potential insect vectors in a museum biorepository reveals undiscovered diversity of plant pathogens in natural areas. Ecology and Evolution 11: 6493–6503. https://doi.org/10.1002/ece3.7502

Upham, N.S.; Poelen, J.H.; Paul, D.; Groom, Q.J.; Simmons, N.B.; Vanhove, M.P.M.; et al. 2021. Liberating host-virus knowledge from biological dark data. Lancet Planetary Health 5: e746–e750. https://doi.org/10.1016/S2542-5196(21)00196-0

Wallace-Wells, D. 2023 February 22. Is the United States ready for back-to-back pandemics? America has failed to do all the things that might have secured lasting normalcy. New York Times.

Watts, N.; Amann, M.; Ayeb-Karlsson, S.; Belesova, K.; Bouley, T.; Boykoff, M.; et al. 2018. The Lancet Countdown on health and climate change: from 25 years of inaction to a global transformation for public health. Lancet 391: 581–630. https://doi.org/10.1016/S0140-6736(17)32464-9

Wei, W.; Trivellone, V.; Dietrich, C.H.; Zhao, Y.; Bottner-Parker, K.D.; Ivanauskas, A. 2021. Identification of phytoplasmas representing multiple new genetic lineages from phloem-feeding leafhoppers highlights the diversity of phytoplasmas and their potential vectors. Pathogens 10: 352. https://doi.org/10.3390/pathogens10030352

Weksler, M.; Lemos, E.M.S.; D'Andrea, P.S.; Bonvicino, C.R. 2017. The taxonomic status of *Oligoryzomys mattogrossae* (Allen 1916) (Rodentia: Cricetidae: Sigmodontinae), reservoir of Anajatuba hantavirus. American Museum Novitates 2017: 1–32. https://doi.org/10.1206/3880.1

Wheeler, Q. 2010. What would NASA do? Mission-critical infrastructure for species exploration. Systematics and Biodiversity 8: 11–15. https://doi.org/10.1080/14772001003628075

Wheeler, Q.D. 2004. Taxonomic triage and the poverty of phylogeny. Philosophical Transactions of the Royal Society B 359: 571–583. https://doi.org/10.1098/rstb.2003.1452

Whitmee, S.; Haines, A.; Beyrer, C.; Boltz, F.; Capon, A.G.; de Souza Dias, B.F.; et al. 2015. Safeguarding human health in the Anthropocene epoch: report of the Rockefeller Foundation–Lancet Commission on planetary health. Lancet 386: 1973–2028. https://doi.org/10.1016/S0140-6736(15)60901-1

Wilcox, J.J.S.; Lopez-Cotto, J.J.; Hollocher, H. 2021. Historical contingency, geography and anthropogenic patterns of exposure drive the evolution of host switching in the *Blastocystis* species-complex. Parasitology 148: 985–993. https://doi.org/10.1017/S003118202100055X

Woese, C.R.; Kandler, O.; Wheelis, M.L. 1990. Towards a natural system of organisms: proposal for the domains Archaea, Bacteria, and Eucarya. Proceedings of the National Academy of Sciences USA 87: 4576–4579. https://doi.org/10.1073/pnas.87.12.4576

Wood, C.L.; Leslie, K.L.; Claar, D.; Mastick, N.; Preisser, W.; Vanhove, M.P.M.; Welicky, R. 2023. How to use natural history collections to resurrect information on historical parasite abundances. Journal of Helminthology 97: e6. https://doi.org/10.1017/S0022149X2200075X

Wood, C.L.; Vanhove, M.P.M. 2022. Is the world wormier than it used to be? We'll never know without natural history collections. Journal of Animal Ecology 92: 260–262. https://doi.org/10.1111/1365-2656.13794

Yanagihara, R.; Gu, S.H.; Song, J.-W. 2015. Expanded host diversity and global distribution of hantaviruses: implications for identifying and investigating previously unrecognized hantaviral diseases. In: Global Virology I—Identifying and Investigating Viral Diseases. P. Shapshak; J.T. Sinnott; C. Somboonwit; J.H. Kuhn (eds.). Springer, New York. 161–198 p. https://doi.org/10.1007/978-1-4939-2410-3_9

Yates, T.L.; Mills, J.N.; Parmenter, C.A.; Ksiazek, T.G.; Parmenter, P.R.; Vande Castle, J.R.; et al. 2002. The ecology and evolutionary history of an emergent disease: hantavirus pulmonary syndrome: evidence from two El Niño episodes in the American Southwest suggest that El Niño–driven precipitation, the initial catalyst of a trophic cascade that results in a delayed density-dependent rodent response is sufficient to predict heightened risk for human contraction of hantavirus pulmonary syndrome. BioScience 52: 989–998. https://doi.org/10.1641/0006-3568(2002)052[0989:TEAEHO]2.0.CO;2

Young, C.C.W.; Olival, K.J. 2016. Optimizing viral discovery in bats. PLOS ONE 11: e0149237. https://doi.org/10.1371/journal.pone.0149237

Zhang, L.; Rohr, J.; Cui, R.; Xin, Y.; Han, L.; Yang, X.; et al. 2022. Biological invasions facilitate zoonotic disease emergences. Nature Communications 13: 1762. https://doi.org/10.1038/s41467-022-29378-2

Zhao, M.; Yue, C.; Yang, Z.; Li, Y.; Zhang, D.; et al. 2022. Viral metagenomics unveiled extensive communications of viruses within giant pandas and their associated organisms in the same ecosystem. Science of the Total Environment 820: 153317. http://dx.doi.org/10.1016/j.scitotenv.2022.153317

Zhou, H.; Ji, J.; Chen, X.; Bi, Y.; Li, J.; Wang, Q.; et al. 2021. Identification of novel bat coronaviruses sheds light on the evolutionary origins of SARS-CoV-2 and related viruses. Cell 184: 4380–4391.e14. https://doi.org/10.1016/j.cell.2021.06.008

Zhu, Q.; Huang, S.; Gonzalez, A.; McGrath, I.; McDonald, D.; Haiminen, N.; et al. 2022. Phylogeny-aware analysis of metagenome community ecology based on matched reference genomes while bypassing taxonomy. mSystems 7: e00167-22. https://doi.org/10.1128/msystems.00167-22

Zinsstag, J.; Schelling, E.; Waltner-Toews, D.; Tanner, M. 2011. From "one medicine" to "one health" and systemic approaches to health and well-being. Preventive Veterinary Medicine 101: 148–156. https://doi.org/10.1016/j. prevetmed.2010.07.003

Zwolińska, A.; Krawczyk, K.; Borodynko-Filas, N.; Pospieszny, H. 2019. Non-crop sources of Rapeseed Phyllody phytoplasma ('*Candidatus* Phytoplasma asteris': 16SrI-B and 16SrI-(B/L) L), and closely related strains. Crop Protection 119: 59–68. https://doi.org/10.1016/j.cropro.2018.11.015

8

Assess: Using Evolution to Save Time and Resources

Daniel R. Brooks, Eric P. Hoberg, Walter A. Boeger, and Valeria Trivellone

Abstract

Assessing the risk a pathogen poses before there is a disease outbreak is the lynchpin of the DAMA (Document, Assess, Monitor, Act) protocol. By taking advantage of the predictive power of evolutionary conservatism and of the Stockholm Paradigm (SP) modeling platform, we can narrow our focus to particular pathogens in particular places and times, particular habitat interfaces, and particular assemblages of hosts, saving time and resources. *Assess* is a three-part process. Phylogenetic triage allows us to determine if a newly discovered potential pathogen is (1) known to cause disease, (2) closely related to species known to cause disease, or (3) closely related to species not known to cause disease. Phylogenetic assessment of the second category of pathogens uses historical ecological techniques to infer potential host range, transmission dynamics, and microhabitat preferences based on what is known about disease-causing close relatives. Dynamic assessment of the current ecology and population genetics of target pathogens helps determine reservoir hosts and habitat interfaces where disease outbreaks are likely to occur. Risk assessment is summarized in an evolutionary-based risk heat map to pinpoint foci and environmental interfaces for monitoring activities, helping establish the potential impact, prevent outbreaks, and in mitigating adverse outcomes when outbreaks occur.

Keywords: Stockholm Paradigm, DAMA protocol, assess, monitor, phylogenetic triage, phylogenetic assessment, dynamic assessment, population genetics, reservoirs, habitat interfaces, heat maps, priorities for monitoring, systematic biology, archival collections, historical ecology, modeling

Introduction

Simply documenting pathogen species across the globe will not help us cope with the emerging infectious disease (EID) crisis. It is critical to assess the nature and extent of any risk represented by any given pathogen encountered. We must make effective decisions in a timely fashion, and that calls not only for rapid intensive and extensive documentation but also for rapid assessments of the risk posed by pathogens we encounter. The EID crisis looms large and is immediate, and we do not have the time or resources to painstakingly study each species as it comes to us to determine its functions in the biosphere. The Assess portion of the DAMA (Document, Assess, Monitor, Act) protocol aims to save time and resources in coping with the EID crisis by identifying pathogens that pose an elevated risk for disease emergence before outbreaks occur (Brooks et al., 2014; Brooks et al., 2019; Hoberg, Boeger, Molnár, et al., 2022; Hoberg, Boeger, et al., 2023).

Based on Audy's (1958) insight that the geographic range of a pathogen always exceeds the geographic range of disease caused by the pathogen, we expect that there are always hosts that are infected but not exhibiting symptoms occurring in places where disease outbreaks have not yet occurred (e.g., Leibler et al., 2022; Salazar-Hamm et al., 2022; Sun et al., 2023). Assessment therefore involves understanding the host and geographic associations, along with the basic biology associated with transmission and disease signs and symptoms of high-risk pathogens documented before they cause disease outbreaks. Uncovering that information requires knowledge of deep evolutionary history as well as contemporary population ecological and genetic dynamics.

Phylogenetic Triage

As anticipated by prominent advocates of large-scale biodiversity inventories in the late 20th century (e.g., Wilson, 1988; Ehrlich and Wilson, 1991; Janzen and Hallwachs, 1994; SA2000, 1994), ongoing inventory efforts have documented an enormous number of potential pathogen species not previously known to science as well as previously known species occurring in novel geographic areas and in association with new hosts (e.g., summaries of a diverse literature in Brooks et al., 2019, 2022; Hoberg, Boeger, Brooks, et al., 2022; Hoberg, Trivellone, et al., 2023). We are being overwhelmed by basic documentation.

The phylogenetics revolution in the late 20th century allowed systematic biology to develop and refine tools that allow us to focus our research beams, illuminating pathogens of varying levels of risk. Twenty-first century systematics is a multifaceted science capable of providing a wide range of essential biological information in an explicitly evolutionary framework (see Brooks and McLennan, 2002; Brooks et al., 2019). This begins with answering three questions (hence the term "triage") under an umbrella established by interactions for diversity in the biosphere. Is the newly documented species: (1) a known pathogen; (2) closely related to a known pathogen, in which case it needs to be investigated further; or (3) closely related to species that are not known to cause disease, in which we case can provisionally ignore it, although we must archive its existence for future reference. This triage is accomplished by locating any documented species within a phylogenetic framework, identifying its closest relatives (Figure 8.1), and then, using our knowledge of those closest relatives, quickly categorizing each species according to the three questions.

Well-entrenched methods of molecular taxonomy, in conjunction with ever-growing and readily accessible phylogenetic databases, allow the first question—*Is the species a known pathogen?*—to be answered readily, by matching data collected during the inventory with archived data from known pathogens. The power of a comparative context is linked to validated and archived specimens and data for pathogens and hosts with authoritative identification held in globally accessible biorepositories (Dunnum et al., 2017; Colella et al., 2021; Hoberg, Trivellone, et al., 2023). The hope is to find pathogens known to cause disease in an area where they were not previously known to occur before they produce a disease outbreak,

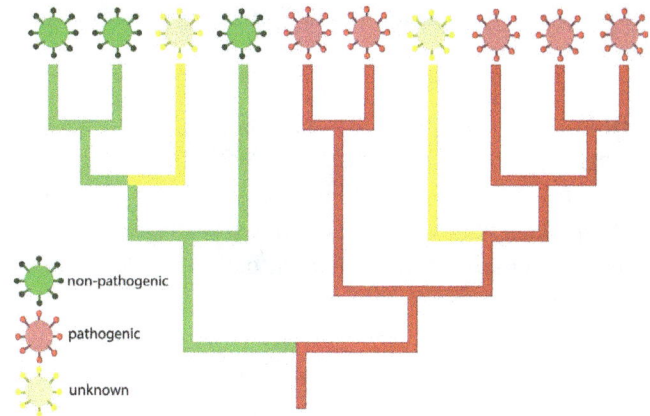

Figure 8.1. Diagrammatic representation of how phylogenetic relationships can help assess the threat of an uncharacterized, newly discovered pathogen. In this hypothetical case, a group of ciliated microbes of some sort includes two that are "unknowns." The one on the left belongs to a group of three other species, none of which cause disease. This species is assessed as low risk, or "archive, but ignore for now." The one on the right belongs to a group of three other species, all of them are disease-causing pathogens. This species is assessed as high risk, or "obtain more information." (Redrawn and modified from Brooks et al., 2019.)

rather than waiting for them to announce themselves by making us or our livestock and crops sick. Appropriate public, agricultural, or wildlife health agencies can then be alerted to activate prevention and mitigation activities. Answering the third question—*Is the species closely related to species that are not known to cause disease?*—allows us to avoid wasting time investigating pathogens that pose no socioeconomic risk to humans and the species on which they depend. Recognizing that evolution is an ongoing process, however, we cannot afford to forget such species entirely. Basic information about such species—who they are, what are their hosts, and where to find them—must be archived for possible future reference.

Answering the second question—*Is the newly documented species closely related to a known pathogen?*—is the primary focus of assessment in the DAMA protocol. It is more time-consuming than answering the other two questions and requires particular skills on the part of those charged with providing answers. By eliminating pathogens corresponding to questions (1) and (3), the number of species requiring such close scrutiny should be greatly reduced. Phylogenetic triage is only the first stage in reducing costs and speeding up the delivery of essential services for effectively coping with the EID crisis.

Phylogenetic Assessment—Establishing Fundamental Insights

Historical ecology (sensu Brooks, 1985; for extended treatments, see Brooks and McLennan, 1991, 1993, 2002) is a powerful tool for fine-tuning risk assessment of known pathogens or close relatives of known pathogens that are present but have not caused disease outbreaks. Phylogenies are essential for assessing how evolutionarily conservative traits will influence the ways in which pathogens behave in novel hosts and novel geographic locations, sometimes called *evolutionary lagload* (Folk et al., 2019; Lande and Shannon, 1996).

Inferring the host(s) in which a pathogen originated as well as hosts that have been added to the pathogen's repertoire range provides insights into the potential for host range expansion in novel environmental contexts.

Cophylogeny analysis is the historical ecological tool both for inferring fundamental host range and in assessing realized host range, the outcomes of capacity, opportunity, and compatibility (Brooks et al., 2019; see also Trivellone and Panassiti, 2023, this volume; Trivellone et al., 2023).

The evolutionary history of host associations for species of *Haemonchus* nematodes (summarized in a tanglegram shown in Figure 8.2) depicts the expected many cases of host range oscillation, with marked departures from cospeciation (Hoberg et al., 2004; Hoberg and Zarlenga, 2016). If we remove *H. contortus*, the species that has been moved around the globe with domestic livestock most extensively, the total host range for the remaining members of *Haemonchus* still encompasses the known host range for *H. contortus*. Assuming that cophylogeny studies tell us something about fundamental fitness space, which, among other things, means potential host range, then the broad

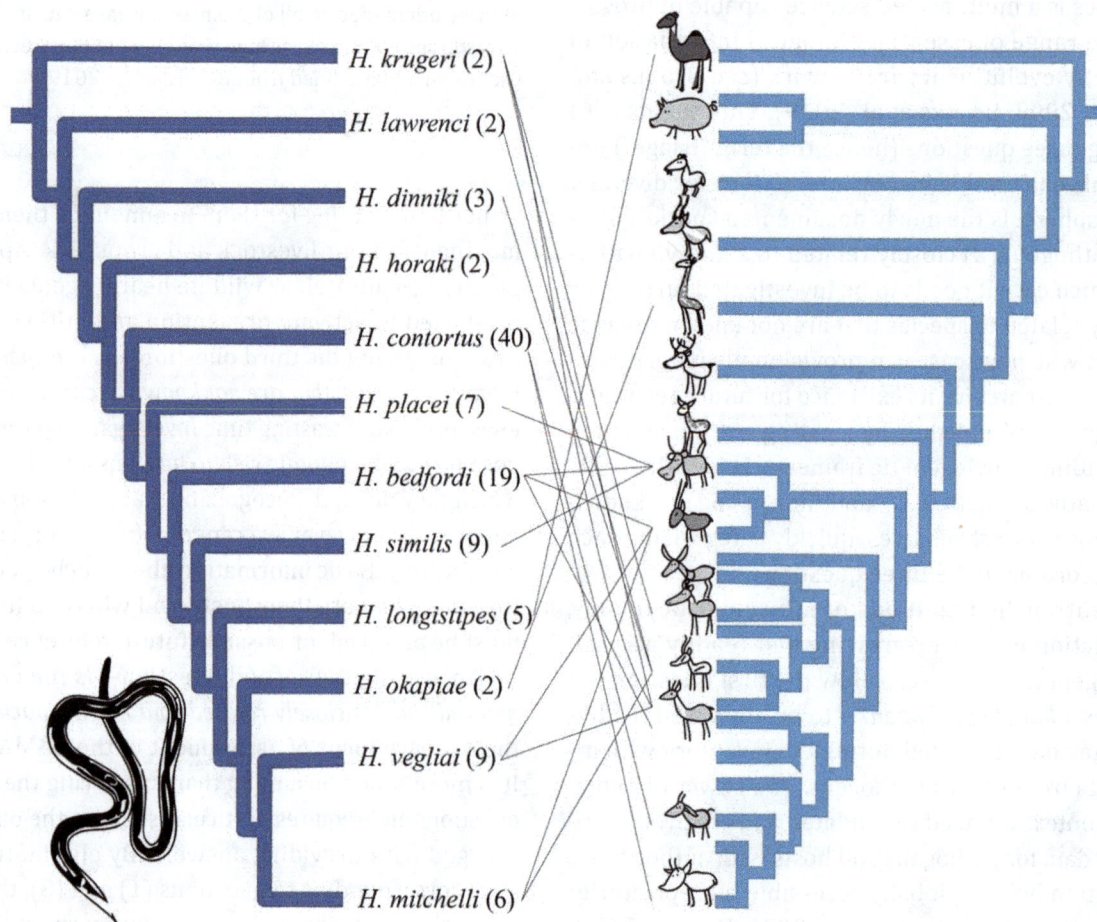

Figure 8.2. Cophylogeny tanglegram comparing the phylogeny of the barber-pole nematodes *Haemonchus* spp. (*left*) and a simplified phylogeny of their hosts. Numbers in parentheses denote numbers of host genera inhabited by each species. (Data from Hoberg et al., 2004; Hoberg and Zarlenga, 2016; redrawn and modified from Brooks et al., 2019.)

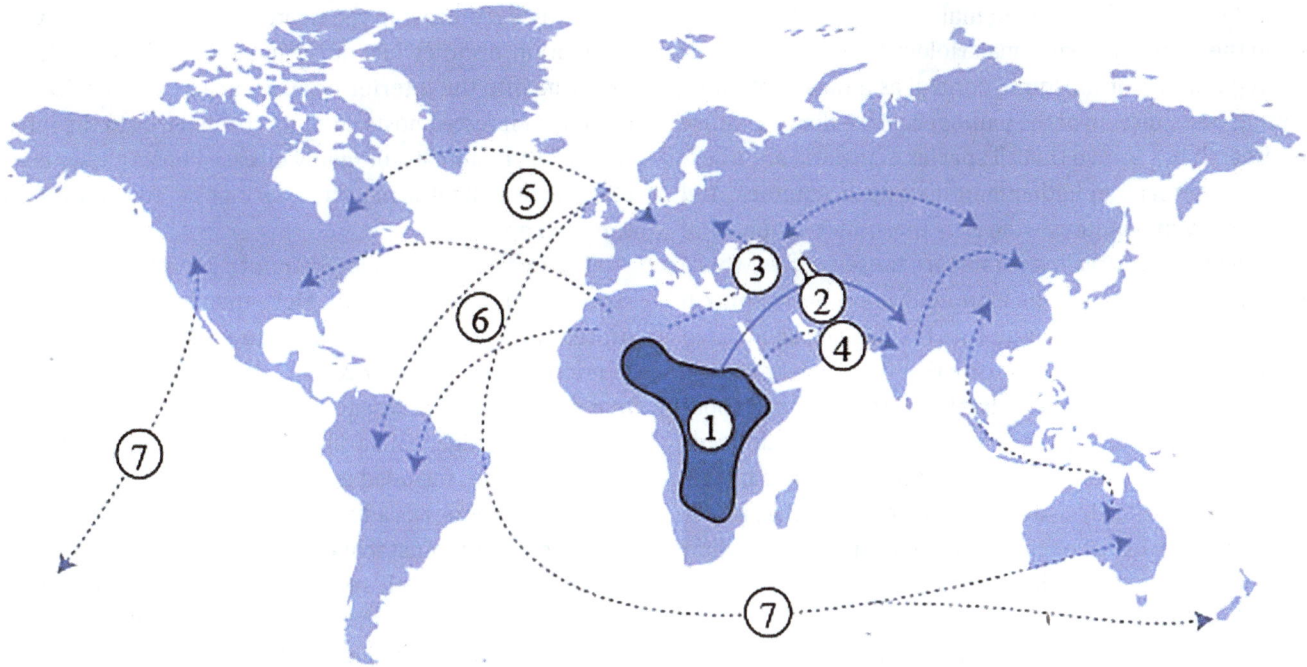

Figure 8.3. Geographic history for the species of *Haemonchus*. Solid lines are historical events that may not be directly linked to human involvement; dashed lines denote anthropogenic expansion and introductions connecting regions and continents; bidirectional arrows indicate recurring exchange. The group originated during the Miocene in Africa, where it diversified among antelopes and other bovids and remained until recently (1). Four lineages moved out of Africa into Eurasia, associated with host range expansion into domesticated cattle, sheep, goats, and camels. *Haemonchus longistipes* expanded from Africa across southern Eurasia into India coincidental with host colonization and domestication of the dromedary (2). Neolithic domestication of cattle, sheep, and goats led to subsequent expansion into Asia for *H. contortus*, *H. placei*, and *H. similis* (3). Later trade routes such as the Silk Road linking Africa, Europe, and India further disseminated those species (4). European contact and exploration introduced *H. contortus*, *H. placei*, and *H. similis* to South America and North America, with recurring or overlapping introductions late into the 20th century (5, 6). *Haemonchus* was introduced to Australia and New Zealand from European and other sources in the 19th century (7).

host range of *H. contortus* should be no surprise. As well, we should anticipate that other species in the genus, such as *H. bedfordi*, have a high potential for host range expansion, given the opportunity. Application of the methods of historical ecology and ecological niche modeling is a critical method for inferring where a pathogen originated, what the most likely initial host range was, and, by extension, patterns of geographic occurrence. These combined methods will enable researchers to determine whether the pathogen and, potentially, its vector (if it has one) is expanding its host and geographic range or is remaining relatively static (Figure 8.3).

In a complementary manner, our evidence indicates that critical traits associated with transmission dynamics and microhabitat preferences (which affect signs and symptoms of disease) are both specific and simultaneously evolutionarily conservative. Many of the properties of an unknown species are likely to be shared with its closest relatives. Faced with an unknown species whose closest relatives are

pathogens—*making it a high-level threat*—we can use what we already know about the natural history of those closest relatives. Identity and phylogeny allow informed inferences about ways in which the newly recognized species might be transmitted, where it might live in its host(s), and even a number of the signs and symptoms of the disease it might provoke that can serve as early warnings of emergent outbreaks. Achieving this potential for resolution of pathogen dynamics requires a fundamental shift in how the broader community views emerging disease (Brooks et al., 2019, 2022; Trivellone et al., 2022; Hoberg, Boeger, et al., 2023).

Asking specialists in the allied health professions to refocus their work to encompass an evolutionary foundation may seem to represent an unwarranted intrusion into another's professional territory. We believe this is not so. Epidemiologists have unconsciously made use of the most fundamental principle of Darwinism, which is that *the more closely related species are to each other, the more inherited traits they will share*. Informed of an outbreak of malaria, regardless of which species of malaria it is or where in the world the outbreak occurs, malariologists will immediately begin planning for mosquito control as a means of interrupting transmission of the pathogen. More than a century of research has shown that all species of malaria are transmitted to humans and other mammals by mosquitoes. And not just any mosquitoes—the only mosquitoes capable of helping malaria parasites persist are those species whose females must feed twice on mammals to properly nourish their eggs. When those mosquitoes first feed on an infected mammal, they pick up the parasite that then undergoes sexual recombination and multiplies in the insect, and in about 10 to 18 days is infective when the female mosquito feeds again on a mammal. The essential mosquito feeding behavior is phylogenetically widespread and conservative, just as the alternating use of mammals and mosquitoes as habitat is for the pathogens. Such principles are universal and not limited to arthropods and protozoans.

Haematoloechus floedae, for example, is a digenetic trematode species described originally in bullfrogs from Georgia, USA. In its native locale, it uses local pulmonate snails and dragonfly naiads (larval aquatic forms of the dragonfly) as intermediate hosts. The parasite species also occurs in northwestern Costa Rica (Brooks, McLennan, et al., 2006), where aquatic frogs, pulmonate snails, and dragonflies occur, none of which are the same species as those found in the southeastern USA. All these assemblages share the same ecological and behavioral traits as a result of inheritance from ancient ancestors. A high degree of evolutionary lagload in both host and pathogen behavior creates enormous amounts of potential pathogen fitness space. This accounts for substantial commonality in the ecological structure of flatworm parasite communities in frogs from temperate deciduous forests in Bohemia, Czechia, and North Carolina, USA; the temperate grasslands of Nebraska, USA; and the dry and wet forests in Mexico and Costa Rica (Brooks, McLennan, et al., 2006). And while it suggests the risk space for pathogens is enormous, it also provides a powerful predictive tool for efforts to identify and stifle outbreaks in a timely fashion.

Pathogens exhibit specific and phylogenetically conservative microhabitat preferences; this is commonly known as *site specificity* or *tissue tropism*. All cestodes (tapeworms) live as adults in the intestines of vertebrates, with subgroups living in particular parts of the intestines of particular groups of hosts. Coronaviruses, including SARS-CoV-2, are fundamentally intestinal pathogens; most coronaviruses familiar to veterinarians are manifested as diarrhea with dehydration and weight loss. Severe disease,

sometimes resulting in the death of the host and its viral population, occurs when the viruses penetrate outside the intestine into the interior organs, such as the respiratory system, of infected hosts. Finding coronavirus in the intestines of asymptomatic or mildly diseased hosts allows more effective treatment than waiting for signs of respiratory distress to appear.

The analysis of the platyhelminth parasite community structure of frogs by Brooks, McLennan, et al. (2006) elaborates the manner in which phylogenetic conservatism in microhabitat preference can allow multiple pathogens that originated in different hosts to coexist in novel hosts. Among adult digeneans (digenetic trematodes), individuals of all species included in the genus *Haematoloechus* live in amphibian lungs as adults; members of *Halipegus* live in the buccal cavity, Eustachian tubes, and esophagus; while members of the closely related *Megalodiscus*, *Diplodiscus*, and *Catadiscus* live in the rectum; and members of various closely related plagiorchioid genera live in the small intestine. In the latter case, when more than one species co-occurs in a frog's small intestine (e.g., *Cephalogonimus americanus* and *Glypthelmins quieta* in the northern leopard frog, *Rana pipiens*), members of each species tend to clump together, demonstrating substantial potential for ecological fitting even at the level of a single host individual. As well, few individual frogs are infected simultaneously with all species of parasites, an indication that host fitness space at the microhabitat level is substantially sloppy, even when microhabitat preferences are highly specific (Brooks, León-Règagnon, et al., 2006).

Dynamic Risk Assessment

Once we have identified pathogens of special interest and have assembled fundamental evolutionary information about original and contemporary host range as well as potential for adding hosts and the geographic distribution, transmission dynamics, and microhabitat preferences, we can reveal the scope of their immediate threat. We can focus our attention toward identifying the range of genetic variation and the basic natural history of transmission for targeted pathogens and the intersections with various hosts and habitat interfaces where those hosts come into contact with each other.

Assessing genetic variation
Unveiling the total genetic diversity or limits of genetic variation within a species is not an easy task to accomplish. Sampling has been the zeitgeist in biology for a long time. Using

traditional sampling and analytical methods, researchers have not been able to describe the totality of the genetic diversity of a species, and we continue to lack critical insights about the limits of variation and pathogen fitness. The Stockholm Paradigm (SP) modeling platform (Souza et al., 2023, this volume and references therein) gives us three important population genetic insights into assessing the risk of disease outbreak by a targeted pathogen. The first and most fundamental insight is that even if different hosts in completely different localities contain the same overall set of genetic variants of a pathogen, the fitness of these pathogens will be variable among hosts and among areas. This is another way of saying that there is no special "parasite evolution," just evolution that applies universally to all species, *hosts and parasites alike* (Brooks and McLennan, 1993).

The second insight is that while the most fit variant in a given host and location may seem to be the one with the greatest chance of transmitting itself to other hosts, this is not necessarily true—some variants with low fitness in one host may achieve much higher fitness in another. Empirical data and simulation strongly suggest that low-frequency variants (e.g., marginal variants) of potential pathogens may represent an important element for colonization and the emergence of new disease syndromes (Araujo et al., 2015). Upon opportunity, the establishment of new antagonistic associations (i.e., EID) resulting from these marginal and often cryptic variants are clearly unexpected and likely influences the resulting population bottleneck observed in simulations of emergence (Araujo et al., 2015) and reported empirically in many studies (e.g., Li and Roossinck, 2004; Ali et al., 2006; Gutierrez et al., 2012). Rare variants are difficult to sample in the original source or donor population, and very often their existence is not recognized. These dynamics contribute to the prevalent (and incorrect) conclusion that emergence may be the result of the "right mutation" allowing expansion into a new host in a particular spatial and temporal arena. See Brooks et al. (2019) for a more detailed summary and analysis of why this former paradigm is wrong. We expect that this process of colonization is likely less frequent because of the reduced chance of encounter between the compatible host and the marginal variant of the pathogen; such host range expansion may succeed but is expectedly rare. Subsequently, the expected bottleneck in the variability of the new pathogen population does not apparently preclude the demographic growth in the new host. Paradoxically, the "rare" variant is now the prevalent variant in this new host, and this increases its ability to explore the same or even new compatible hosts in the area. This is indeed a game changer, as propagule size

(and pressure focused at interfaces), which is now expectedly greater, is considered an important feature that determines the success of new host expansions (colonization and exploitation) (Feronato et al., 2022).

The final major insight from the modeling platform is support for the feasibility of the stepping-stone dynamic (Araujo et al., 2015), in which a low-fitness (rare) variant of a pathogen in host A gets into host B and becomes amplified into a high-fitness (common) variant, from which it may produce novel low-fitness variants that may be capable of infecting and becoming common in host C. Braga et al. (2015) first empirically suggested the expansion of host repertoires by stepping-stones. Population assessment using the modeling platform can determine the genetic status of the pathogen in the host(s) discovered and assess the potential for expansion through a stepping-stone dynamic. This is important because whenever pathogens colonize distantly related novel hosts, the potential for disease is greater because the new host will always initially represent marginal fitness space for the pathogen. The rare variants in the tails of overall genetic diversity profiles for pathogens are potentially of greater concern than the most common variants.

These insights remind us that we must document the genetic variation of targeted pathogens both *intensively* (to find the rare variants) and *extensively* (to find a more accurate assessment of host and geographic range of variants).

Finding the focus of potential outbreaks

What do pathogens do when they are not causing a disease outbreak? They do not disappear from the face of the earth, only to reappear later and cause new outbreaks. The SP allows us to assert with confidence that whenever "Audy space" includes a host species that is diseased, there is at least one other, usually asymptomatic, host—often called the reservoir (among animal pathogens) or the source of inoculum (among plant pathogens)—living in habitats adjacent to the one in which the diseased hosts reside (Hoberg, Boeger, et al., 2023). For that reason, we cannot overemphasize the importance of determining the extent of these reservoir hosts and in what habitats they reside. For example, a concentration of many virus families, with known zoonotic members, occurs in a relatively restricted number of chiropteran (bat) families that live in habitat interfaces with humans and livestock (e.g., Evans et al., 2023; see also references in Brooks et al., 2019). There is encouraging evidence of a recent focus on finding the reservoirs and the habitat interfaces where pathogens persist (e.g., Ortiz-Baez et al., 2023; Latinne et al., 2023).

Audy (1958) recognized that the geographic extent of a pathogen is greater than the geographic extent of disease, and this insight continues to have substantial implications. Essentially, this relationship represents invisible (or latent) spread in asymptomatic hosts, or in reservoirs for which there is limited data or which do not receive our attention. The errors of misdiagnoses or convenience of superficial diagnoses by clinicians not capable of recognizing something new further contribute to a clouded picture of distribution. Reservoir assemblages in landscapes adjacent to those in which the hosts of interest occur are significant as sources of reemergence, indicating the need for broadened documentation and monitoring. Once inside a new landscape, a pathogen will cope with the new conditions according to its capacity for ecological fitting and thus may persist in the new surroundings in ways that were not predictable based on the assumption of tightly coevolved association with the original host. Fundamental and realized fitness space for pathogens ("Audy space" of Hoberg, Boeger, et al., 2023, this volume) includes portions where disease is occurring and portions where disease is not occurring. Diseased hosts represent marginal fitness space for a pathogen, and while it is essential to treat and contain high-risk pathogens, to focus only on diseased hosts is insufficient for preventing as many outbreaks as possible and in mitigating their impact when

outbreaks do occur. Reservoir hosts that are widespread among habitat interfaces are prime candidates for becoming ecological super-spreaders. This means we need to focus on habitat interfaces and the host assemblages in each that are critical for pathogen persistence; there is growing evidence that the interfaces are more significant than overall landscape features in assessing EID risk (Klain et al., 2023), supporting the predictions of the SP. How are host assemblages, including reservoirs, focused at habitat interfaces creating an active nexus with animals and plants of immediate economic interest to us? How do we effectively incorporate an examination of critical habitat interfaces involving wildlands and agroscapes, agroscapes and urban and peri-urban landscapes, and urban and the relatively new urban green spaces (Zeller et al., 2019; Trivellone et al., 2022; Xu et al., 2022; Evans et al., 2023; Janik et al., 2023; Ortiz-Baez et al., 2023)?

From Assess to Monitor: Generating Heat Maps

As a foundation of tool development, there is a need to summarize assessment data in a form that is readily understandable and transferable to those who allocate funds for monitoring (also called surveillance) and Action within DAMA (Hoberg, Boeger, et al., 2023; Hoberg, Trivellone, et al., 2023). As an essential process, all the data from the

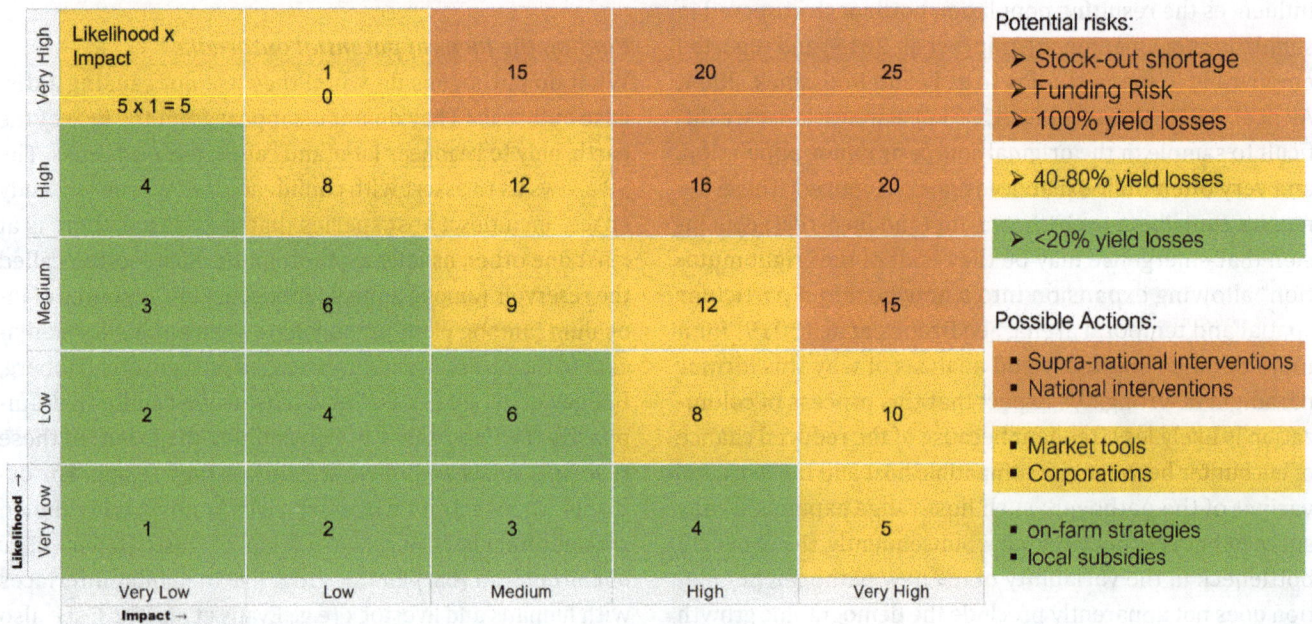

Figure 8.4. Example of a classical risk heat map for risk assessment and management. In this case, impact and likelihood of a potential disease outbreak are color coded in a risk quantitative scale from 1 to 25. The colors depict the qualitative scale of the interventions translated in specific actions. On the right are some examples for a potential plant disease outbreak. The scheme may be generalized to any kind of disease or association.

Document stage along with phylogenetic and dynamic assessment (Assess stage) are summarized into risk heat maps built upon diversity, phylogenies, and historical ecologies of the interacting species (pathogen and the repertoire of hosts for transmission). The classical risk heat map is established by evaluating (qualitatively and/or quantitatively) the probability and the impact of a usually unwanted event to occur (McKay, 2016). For emerging diseases, the heat map defines probability and impact for host-pathogen associations to occur in a specific area and the associated risk of outbreaks. The heat map is color coded and defines the space and the magnitude of the risk (Figure 8.4). The likelihood or probability of an event or new association is determined in a multidimensional context; defined based on transmission dynamics, microhabitat preferences, and phylogenetic relatedness; and connected with the specific biologic system under study. The impact may be ranked qualitatively, or it may depict a precise financial burden in the event of an outbreak.

All together, the risk for an event of disease outbreak, related to a specific biological association, will be estimated case by case and will generate potential actions for risk management to be undertaken (see Figure 8.4 for some hypothetical examples). Considered, for example, within the potential of a tri-trophic interaction space (the generality posed by vector-borne plant pathogens, Figure 8.5), we can use a gridded quadrat whose sides fit a phylogenetic tree resulting from triage. The heat map is then readdressed to fit an evolutionary perspective, which we can describe as an evolutionary-based risk heat map. Finally, each cell encompasses specific phylogenetic groups for the pathogens and their hosts, and it is colored based on the level of risk assigned for a particular area, time, and assemblage (Figure 8.5). Recently, a risk heat map of this design has been applied to a vector-borne plant pathogen (phytoplasma) model to evaluate the risk of outbreaks from a region in South Germany (Janik et al., 2023). This biological model for phytoplasma is idiosyncratic, representing a pathogen with primarily horizontal transmission and a remarkable trans-kingdom interaction (plants and insects) expressed in each life cycle. The generality with applications tailored for variation in fitness space and diversity including pathogens with direct or indirect (complex life history) patterns of circulation is apparent.

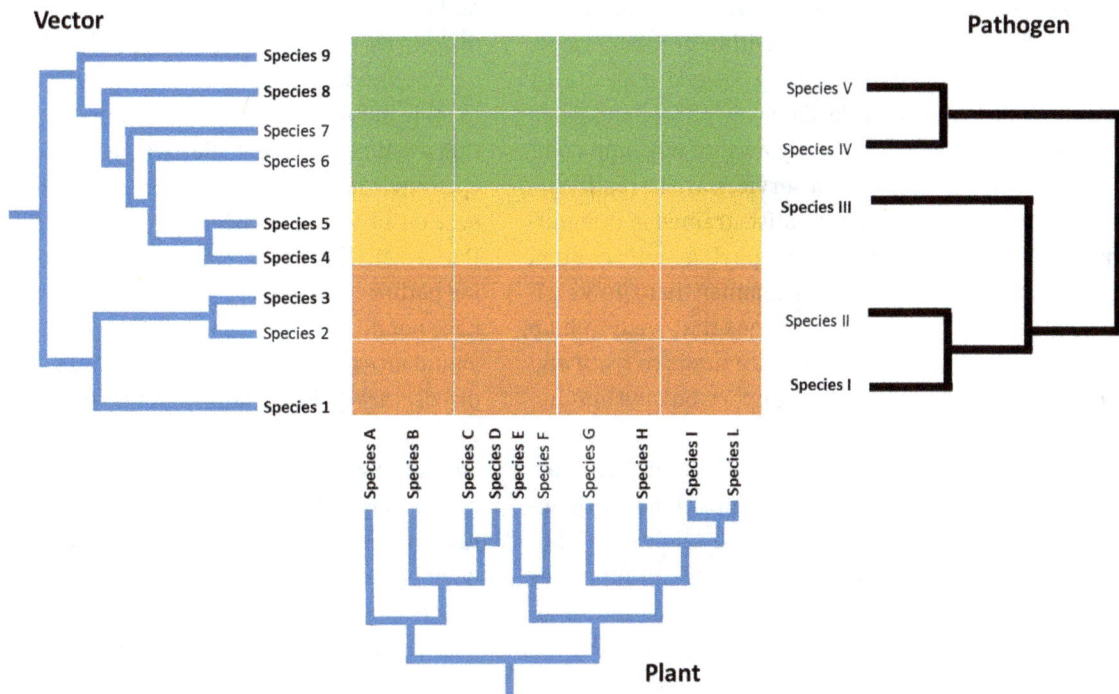

Figure 8.5. Evolutionary-based risk heat map derived from the results of phylogenetic triage. Here the example of a vector-borne pathogen (phylogeny on the right in black) and its hosts (vector and plant phylogenies on the left and bottom). Three levels of risk (color codes) are represented for different outcomes derived from the tripartite associations potentially established in the area under study. Taxa highlighted in bold represent species that co-occur in an area and, in the case of the plant, with an associated high economic importance. The risk associated with each association is derived by evaluating tanglegrams and historical ecology of the associations. The final risk level is assigned in each cell by depicting the outcomes of the tripartite association derived from the three-way entry.

Assessment Urgent Needs

Implementation of the Assess portion of DAMA makes use of many of the same personnel and infrastructure resources used for intially documenting pathogen diversity, which are already in short supply, underfunded, underappreciated, and underutilized. Effective assessment requires input and analysis from an array of specialists, from scientists who maintain archival collections to those who perform molecular and bioinformatics analyses, to those who place documented information in both real-time (e.g., specialists working on population dynamics of hosts and pathogens) and deep-history (phylogenetic) contexts, and finally those who not only maintain these complex information databases but are capable of updating them rapidly and providing as close to real-time access as possible (summarized in Dunnum et al., 2017; Colella et al., 2021; Hoberg, Boeger, Brooks, et al., 2022; Hoberg, Trivellone, et al., 2023).

There is a real and burgeoning need for training and employing increasing numbers of taxonomists as well as improving the infrastructure for archival collections where their findings reside. We must develop and effectively use valuable, forward-looking, and innovative archival collections of hosts and pathogens, along with informatics resources, to describe pathogen risk within broader trajectories for change in the biosphere (e.g., Colella et al., 2021). We must emphasize how we develop views of coping with a world in change rather than simply a list of wildland commodities for our use. Ecosystem services are in reality archival collections services. Systematists trained in integrating natural history information in a phylogenetic context can present these discoveries in a manner that allows triage decisions to be made. All countries that wish to help mitigate disease threats must allocate funds to train and employ more taxonomists and to support, bolster, and improve museum infrastructure.

Those essential people require sophisticated infrastructure in addition to economically viable positions for themselves and support personnel: molecular taxonomy laboratories, phylogenetics laboratories, and natural history repositories for specimens, tissues, sequences, and digitized natural history information. Networks must be developed that link field-based collections to deposition and maintenance of specimens, tissues, and nucleotide products in archival repositories (Dunnum et al., 2017). The latter are the baselines for documenting pathogen and host distribution and provide the capacity to recognize (often retrospectively) the occurrence of novel pathogens or new host and geographic distributions (Hoberg, Trivellone, et al., 2022; Hoberg, Boeger, et al., 2023; Hoberg, Trivellone, et al., 2023). As well, much of the critical infrastructure necessary for documenting pathogen diversity also supports activities associated with producing information necessary to make informed risk assessments, recommending pathogens, habitat interfaces, and likely reservoir hosts for intensive monitoring.

Summary

The need for comprehensive risk assessment before outbreaks occur is great because it sets the stage for monitoring high-risk pathogens before they cause disease outbreaks and mitigating the impact of outbreaks before they become pandemics. The more you can anticipate and generalize, the more cost-effective prevention becomes—evolution is more general than ecology because evolutionary conservatism makes ecology predictable and persistent across a range of conditions; ecology makes no sense except in the light of evolution. In evolution, the past is always with us and makes it possible to predict the future to some extent. The SP gives us reason for optimism when it comes to predicting disease outbreaks, preventing some of them, and mitigating the impacts of those that do occur, that is, making certain that outbreaks do not become pandemics.

The fundamental role of Assess within the DAMA protocol is in focusing essential monitoring activities in ways that are time- and cost-effective, that is, on hosts and places (the environmental interfaces) where the risk of new disease outbreaks is greatest (e.g., Brooks et al., 2019; Hoberg, Boeger, et al., 2023). Of paramount importance is identifying pathogens that could cause disease outbreaks but as yet have not done so. The places and hosts of greatest pathogen abundance and of marginal existence, of disease and of nondisease, are all important—supporting the notion that populations in marginal habitats can be highly resilient (Plunkett and Swindles, 2022). These relationships establish a monitoring trajectory to determine the current genetic status of variants in each host and location: what is common, what is rare but capable of infecting other hosts; what are potential super-spreaders in a broad ecological context and which are stepping-stones for expansion of host range (e.g., Araujo et al., 2015; Boeger et al., 2022; Hoberg, Boeger, Brooks, et al., 2022; Latinne et al., 2023).

Having done that, we can begin to ask questions, such as "Why have they not yet triggered an outbreak and what could trigger one?" "What host species might serve as stepping-stones or as ecological super-spreaders?" (Araujo et al., 2015; Boeger et al., 2022; Hoberg, Boeger, Brooks, et al.,

2022). "What behaviors will make newly arrived pathogens more or less capable of establishing themselves in the new location?" "What behaviors will make newly arrived humans and their associated plants and animals more or less susceptible to colonization, expanding host range, and disease outbreaks?"

At least some elements of Assess are being applied in many particular cases, but truly effective assessment can be done in a timely and economical manner only if there is an increased commitment to comprehensive and global efforts. And that requires augmented content relating to evolutionary biology in medical veterinary and agricultural science curricula, training and employing taxonomic and systematic specialists and support staff, and increased funding to expand infrastructure for analysis and for archiving and making relevant information widely available. Effective assessment is a foundation for DAMA, providing critical insights about pathogen distribution, host range, and behavior within the broader arena of the biosphere and Audy space.

Literature Cited

Ali, A.; Li, H.; Schneider, W.L.; Sherman, D.J.; Gray, S.; Smith, D.; Roossinck, M.J. 2006. Analysis of genetic bottlenecks during horizontal transmission of cucumber mosaic virus. Journal of Virology 80: 8345–8350. https://doi.org/10.1128/jvi.00568-06

Araujo, S.B.L.; Braga, M.P., Brooks, D.R., Agosta, S.J.; Hoberg, E.P.; von Hartenthal, F.W.; Boeger, W.A. 2015. Understanding host-switching by ecological fitting. PLOS ONE 10: e0139225. https://doi.org/10.1371/journal.pone.0139225

Audy, J.R. 1958. The localization of disease with special reference to the zoonoses. Transactions of the Royal Society of Tropical Medicine and Hygiene 52: 308–328.

Boeger, W.A.; Brooks, D.R.; Trivellone, V.; Agosta, S.J.; Hoberg, E.P. 2022. Ecological super-spreaders drive host-range oscillations: Omicron and risk space for emerging infectious disease. Transboundary and Emerging Diseases 69: e1280–e1288. https://doi.org/10.1111/tbed.14557

Braga, M.P.; Razzolini, E.; Boeger, W.A. 2015. Drivers of parasite sharing among Neotropical freshwater fishes. Journal of Animal Ecology 84: 487–497. https://doi.org/10.1111/1365-2656.12298

Brooks, D.R. 1985. Historical ecology: a new approach to studying the evolution of ecological associations. Annals of the Missouri Botanical Garden 72: 660–680. https://doi.org/10.2307/2399219

Brooks, D.R.; Hoberg, E.P.; Boeger, W.A. 2019. The Stockholm Paradigm: Climate Change and Emerging Disease. University of Chicago Press, Chicago.

Brooks, D.R.; Hoberg, E.P.; Boeger, W.A.; Gardner, S.L.; Galbreath, K.E.; Herczeg, D.; et al. 2014. Finding them before they find us: informatics, parasites, and environments in accelerating climate change. Comparative Parasitology 81: 155–164. https://doi.org/10.1654/4724b.1

Brooks, D.R.; Hoberg, E.P.; Boeger, W.A.; Trivellone, V. 2022. Emerging infectious disease: an underappreciated area of strategic concern for food security. Transboundary and Emerging Diseases 69: 254–267. https://doi.org/10.1111/tbed.14009

Brooks, D.R.; León-Règagnon, V.; McLennan, D.A.; Zelmer, D. 2006. Ecological fitting as a determinant of the community structure of platyhelminth parasites of anurans. Ecology 87 (supplement): S76–S85. https://doi.org/10.1890/0012-9658(2006)87[76:efaado]2.0.co;2

Brooks, D.R.; McLennan, D.A. 1991. Phylogeny, Ecology, and Behavior: A Research Program in Comparative Biology. University of Chicago Press, Chicago.

Brooks, D.R.; McLennan, D.A. 1993. Parascript: Parasites and the Language of Evolution. Smithsonian Institution Press, Washington DC.

Brooks, D.R.; McLennan, D.A. 2002. The Nature of Diversity: An Evolutionary Voyage of Discovery. University of Chicago Press, Chicago.

Brooks, D.R.; McLennan, D.A.; León-Règagnon, V.; Hoberg, E. 2006. Phylogeny, ecological fitting and lung flukes: helping solve the problem of emerging infectious diseases. Revista Mexicana de Biodiversidad 77: 225–234.

Colella, J.P.; Bates, J.; Burneo, S.F.; Camacho, M.A., Carrion Bonilla, C.; Constable, I.; et al. 2021. Leveraging natural history biorepositories as a global, decentralized, pathogen surveillance network. PLOS Pathogens 17: e1009583. https://doi.org/10.1371/journal.ppat.1009583

Dunnum, J.L.; Yanagihara, R.; Johnson, K.M.; Armien, B.; Batsaikhan, N.; Morgan, L.; Cook, J.A. 2017. Biospecimen repositories and integrated databases as critical infrastructure for pathogen discovery and pathobiology research. PLOS Neglected Tropical Diseases 11: e0005133. https://doi.org/10.1371/journal.pntd.0005133

Ehrlich, P.R.; Wilson, E.O. 1991. Biodiversity studies: science and policy. Science 253: 758–762. https://doi.org/10.1126/science.253.5021.758

Evans, T.S.; Tan, C.W.; Aung, O.; Phyu, S.; Lin, H.; Coffey, L.L.; et al. 2023. Exposure to diverse sarbecoviruses indicates frequent zoonotic spillover in human communities interacting with wildlife. International Journal of Infectious Diseases 131: 57–64. https://doi.org/10.1016/j.ijid.2023.02.015

Feronato, S.G.; Araujo, S.; Boeger, W.A. 2022. 'Accidents waiting to happen'—insights from a simple model on the emergence of infectious agents in new hosts. Transboundary and Emerging Diseases 69: 1727–1738. https://doi.org/10.1111/tbed.14146

Folk, R.A.; Stubbs, R.L.; Mort, M.E.; Cellinese, N.; Allen, J.M.; et al. 2019. Rates of niche and phenotype evolution lag behind diversification in a temperate radiation. Proceedings of the National Academy of Sciences USA 116: 10,874–10,882. https://doi.org/10.1073/pnas.1817999116

Gutiérrez, S.; Michalakis, Y.; Blanc, S. 2012. Virus population bottlenecks during within-host progression and host-to-host transmission. Current Opinion in Virology (Virus evolution/ Antivirals and resistance special issue) 2: 546–555. https://doi.org/10.1016/j.coviro.2012.08.001

Hoberg, E.P.; Boeger, W.A.; Brooks, D.R.; Trivellone, V.; Agosta, S.J. 2022. Stepping-stones and mediators of pandemic expansion—a context for humans as ecological super-spreaders. MANTER: Journal of Parasite Diversity 18. https://doi.10.32873/unl.dc.manter18

Hoberg, E.P.; Boeger, W.A.; Molnár,O.; Földvári, G.; Gardner, S.; Juarrero, A.; et al. 2022. The DAMA protocol, an introduction: finding pathogens before they find us. MANTER: Journal of Parasite Diversity 21. https://doi.org/10.32873/unl.dc.manter21

Hoberg, E.P.; Boeger, W.A.; Molnár, O.; Földvári, G.; Gardner, S.L.; Juarrero, A.; et al. 2023. The DAMA protocol: anticipating to prevent and mitigate emerging infectious diseases. In: An Evolutionary Pathway for Coping with Emerging Infectious Disease. S.L. Gardner, D.R. Brooks, W.A. Boeger, E.P. Hoberg (eds.). Zea Books, Lincoln, NE.

Hoberg, E.P.; Lichtenfels, J.R.; Gibbons, L. 2004. Phylogeny for species of *Haemonchus* (Nematoda: Trichostrongyloidea): considerations of their evolutionary history and global biogeography among Camelidae and Pecora (Artiodactyla). Journal of Parasitology 90: 1085–1102. https://doi.org/10.1645/GE-3309

Hoberg, E.P.; Trivellone, V.; Cook, J.A.; Dunnum, J.L.; Boeger, W.A.; Brooks, D.R.; et al. 2022. Knowing the biosphere: documentation, specimens, archives, and names reveal environmental change and emerging pathogens. MANTER: Journal of Parasite Diversity 26. https://doi.org/10.32873/unl.dc.manter26

Hoberg, E P.; Trivellone, V.; Cook, J.A.; Dunnum, J.L.; Boeger, W.A.; Brooks, D.R.; et al. 2023. Document: pathogen diversity—finding them before they find us. In: An Evolutionary Pathway for Coping with Emerging Infectious Disease. S.L. Gardner, D.R. Brooks, W.A. Boeger, E.P. Hoberg (eds.). Zea Books, Lincoln, NE.

Hoberg, E.P.; Zarlenga, D.S. 2016. Evolution and biogeography of *Haemonchus contortus*: linking faunal dynamics in space and time. Advances in Parasitology 93: 1–30. https://doi.org/10.1016/bs.apar.2016.02.021

Janik, K.; Panassiti, B.; Kerschbamer, C.; Burmeister, J.; Trivellone, V. 2023. Phylogenetic triage and risk assessment: how to predict emerging phytoplasma diseases. Biology 12: 732. https://doi.org/10/3390/biology12050732

Janzen, D.H.; Hallwachs, W. 1994. All-Taxa Biodiversity Inventory (ATBI) of Terrestrial Ecosystems: A Generic Protocol for Preparing Wildland Biodiversity for Non-damaging Use. Report of a National Science Foundation Workshop, 16–18 April 1993, Philadelphia, PA. 132 pp.

Klain, V.; Mentz, M.B.; Bustamante-Manrique, S.; Bicca-Marues, J.C. 2023. Landscape structure has a weak influence on the parasite richness of an arboreal folivorous-frugivorous primate in anthropogenic landscapes. Landscape Ecology 38: 1237–1247. https://doi.org/10.1007/s10980-023-01603-3

Lande, R.; Shannon, S. 1996. The role of genetic variation in adaptation and population persistence in a changing environment. Evolution 50: 434–437. https://doi.org/10.1111/j.1558-5646.1996.tb04504.x

Latinne, A.; Nga, N.T.T.; Long, N.V.; Ngoc, P.T.B.; Thuy, H.B.; PREDICT Consortium; et al. 2023. One Health surveillance highlights circulation of viruses with zoonotic potential in bats, pigs, and humans in Viet Nam. Viruses 15: 790. https://doi.org/10.3390/v15030790

Leibler, J.H.; Abdelgadir, A.; Seidel, J.; White, R.F.; Johnson, W.E.; Reynolds, S.J.; et al. 2022. Influenza D virus exposure among US cattle workers: a call for surveillance. Zoonoses and Public Health 70: 166–170. https://doi.org/10.1111/zph.13008

Li, H.; Roossinck, M.J. 2004. Genetic bottlenecks reduce population variation in an experimental RNA virus population. Journal of Virology 78: 10582–10587. https://doi.org/10.1128/JVI.78.19.10582-10587.2004

McKay, S. 2016. CGMA [Chartered Global Management Accountant] tools: how to communicate risks using a heat map. Journal of Accountancy (June): 35–40. 7 p. See http://maaw.info/ArticleSummaries/ArtSumMcKay2016.htm

Ortiz-Baez, A.S.; Jaenson, T.G.T.; Holmes, E.C.; Pettersson, J.H.-O.; Wilhelmsson, P. 2023. Substantial viral and bacterial diversity at the bat-tick interface. Microbial Genomics 9: 000942. https://doi.org/10.1099/mgen.0.000942

Plunkett, G.; Swindles, G.T. 2022. Bucking the trend: population resilience in a marginal environment. PLOS ONE 17: e0266680. https://doi.org/10.1371/journal.pone.0266680

SA2000. 1994. Systematics Agenda 2000: Charting the Biosphere. Technical Report. American Museum of Natural History, New York. 34 pp.

Salazar-Hamm, P.S.; Montoya, K.N.; Montoya, L.; Cook, K.; Liphardt, S.; Taylor, J.W.; et al. 2022. Breathing can be dangerous: opportunistic fungal pathogens and the diverse community of the small mammal lung mycobiome. Frontiers in Fungal Biology 3: 996574. https://doi.org/10.3389/ffunb.2022.996574

Souza, A.T.C.; Araujo, S.B.L.; Boeger, W.A. 2023. Modeling the Stockholm Paradigm: insights for the nature and dynamics of emerging infectious diseases. In: An Evolutionary Pathway for Coping with Emerging Infectious Disease. S.L. Gardner,

D.R. Brooks, W.A. Boeger, E.P. Hoberg (eds.). Zea Books, Lincoln, NE.

Sun, W.; Shi, Z.; Wang, P.; Zhao, B.; Li, J.; Wei, X.; et al. 2023. Metavirome analysis reveals a high prevalence of porcine hemagglutination encephalomyelitis virus in clinically healthy pigs in China. Pathogens 12: 510. https://doi.org/10.3390/pathogens12040510

Trivellone, V.; Araujo, S.B.L.; Panassiti, B. 2023. HostSwitch: an R package to simulate the extent of host-switching by a consumer. The R Journal 14: 179–194. https://doi.org/10.32614/RJ-2023-005

Trivellone, V.; Hoberg, E.P.; Boeger, W.A.; Brooks, D.R. 2022. Food security and emerging infectious disease: risk assessment and risk management. Royal Society Open Science 9: 211687. https://doi.org/10.1098/rsos.211687

Trivellone, V.; Panassiti, B. 2023. Pathogen-host phylogenetic congruence varies with paradigmatic assumptions, analytical method, and type of association. In: An Evolutionary Pathway for Coping with Emerging Infectious Disease. S.L. Gardner, D.R. Brooks, W.A. Boeger, E.P. Hoberg (eds.). Zea Books, Lincoln, NE.

Wilson, E.O. 1988. The current state of biological diversity. In: Biodiversity. E.O. Wilson (ed.). 1–8 p. National Academy Press, Washington DC.

Xu, F.; Yan, J.; Heremans, S.; Somers, B. 2022. Pan-European urban green space dynamics: a view from space between 1990 and 2015. Landscape and Urban Planning 226: 104477. https://doi.org/10.1016/j.landurbplan.2022.104477

Zeller, U.; Perry, G.; Göttert, T. (eds.). 2019. Biodiversity and the Urban-Rural Interface: Conflicts vs. Opportunities. Proceedings of an International Workshop in Linde, Germany, 24–27 September 2018. https://doi.org/10.18452/19995

9

Monitoring: An Emerging Infectious Disease Surveillance Platform for the 21st Century

Eloy Ortíz and Alicia Juarrero

Abstract

Current vector surveillance programs are insufficient for coping with emerging infectious disease crises. In particular, current practices do not deploy appropriate IT tools to generate insights that can inform interventions and contain and mitigate the spread of vector-borne diseases. Vector-Analytica has developed a highly configurable and adaptable IT platform that imports, harmonizes, and integrates a range of data sources. The state-of-the-art back-end development allows users to fully appreciate and understand the complexity of infectious disease dynamics without having to resort to additional statistical and GIS software packages. The platform yields actionable insights into outbreak patterns of vector-borne diseases that could not have been obtained without integrating pertinent contextual information at a granular spatiotemporal scale. Doing so empowers local scientists to participate actively in an ongoing manner in protecting their communities. The platform can also empower regional and national scientists and other scholars and agencies by providing them with a shared platform through which to access continuously updated real time data with which to formulate, validate, and update new dynamic forecasting and simulating models that support timely and appropriate decision-making and interventions on a wide range of scales. VectorAnalytica's effectiveness is described in three case studies for which integrating and visualizing unexpected but pertinent variables (in addition to those traditionally studied for vector disease monitoring and management) yielded critical information about significant correlations between weather variables and disease incidence, and between complaints to local call centers, weather conditions, and emergent vector hotspots.

Keywords: Stockholm Paradigm, DAMA protocol, dengue, Zika, mosquito-borne disease, vector-borne disease, rodent control, 311 hotlines, trap surveillance, data entry, mobile app, weather, census tracts, vector surveillance, infectious diseases surveillance, vector monitoring, infectious diseases monitoring

Introduction

Each documented pathogen species that is assessed as warranting special attention must be monitored (after it is so identified) to more fully document any changes in genetic variation associated with changes in host and geographic range. When pathogens are given an invitation to explore, they are relentless. The easiest way for them to explore, and coincidentally, create the most havoc, is biotic expansion triggered by climate change. Trophic changes within existing biological communities bring previously unexposed animals or plants (potential hosts) into contact with resident pathogens. Geographic spread of pathogens without host movements may occur along the margins of two ecosystems, where the parasite might expand geographically because there is a related host in the adjacent area. The exploratory ability of pathogens explains why the distribution of a pathogen is broader than the distribution of disease attributable to it (Audy, 1958). Environmental perturbations that create new connections in fitness space are associated with amplification of pathogen populations, host range expansion, and disease emergence (Brooks et al., 2019; Agosta, 2022, 2023; Brooks et al., 2022, 2023, and references therein). Special attention for monitoring activities will be given to new pathogen-host associations emerging in the interfaces among wildlands and managed landscapes, agricultural, peri-urban, and urban areas (Hoberg et al., 2022, 2023, and references herein). When such information is incorporated into the SP modeling platform (Souza et al., 2022, 2023, and references herein) as well as models of anthropogenic climate forcing, we hope to identify indications that can help prevent outbreaks or to mitigate their impact—that is, preventing outbreaks from becoming pandemics (Hoberg et al., 2022, 2023, and references herein).

Outbreaks of vector-borne diseases catch us unprepared time and again. Response by the public health system is usually ad hoc, and business as usual resumes as soon as the immediate crisis wanes. Although the problem extends most egregiously to climate change, here we turn to vector-borne diseases to outline and emphasize outbreak cases. As outbreaks of dengue, malaria, chikungunya, and Zika have demonstrated, these infections are exemplars of complex adaptive dynamics. Complex systems are combinations of nonliving, living, and social patterns of organization that form far from equilibrium when multiple variables and integrated ecological relationships intertwine into complex patterns both at the individual and the population or collective levels in response to a variety of constraints (Juarrero, 1999, 2023). Once formed, organisms and ecosystems, individuals and cultures adapt to and evolve in light of past experiences and current conditions. Critically, the mutually dependent relations that constitute such emergent patterns of organization and behavior display novel properties (such as herd immunity) that can be more significant than the individual events and processes that compose them. These overarching patterns and correlations cannot be clearly discerned, much less understood, and managed without taking into account the pertinent contextual constraints that generate and govern them. In order to reveal the emergent properties and behaviors of the complex dynamics produced and perpetuated by vector-borne diseases, the framework of the IT Surveillance Platform developed at VectorAnalytica (and deployed in the use cases described below; see also Ortiz and Juarrero, 2022) is informed by complexity theory.

Infectious diseases can be better contained and mitigated if they are anticipated (Brooks et al., 2014, 2019, 2023). Doing so, however, requires a proper monitoring and tracking system. Historically, this meant expensive state-of-the-art computers managed by experts in both statistics and software. Alas, despite the explosive expansion in computational power and reduction in processing costs, IT monitoring of vector-borne diseases, even in the developed world, still has not lived up to its potential. Public health agencies, especially those at the municipal and county level, inevitably rely on outdated hardware and software, and data management is highly siloed in labs, clinics, and political departments. As a case in point, systematic and digitized data collection is rare. Integrating field data obtained by vector-control departments with laboratory, clinical, and public health data is even rarer.

Meager resources of public health departments mean that locally gathered data is returned after processing with a lag of several weeks. Despite mosquito lifespans of approximately 2 weeks, laboratory analyses of larvae and blood samples are many times weeks old by the time the results arrive back at the local vector-control or public health departments. Consequently, as was vividly exposed by both the Zika and COVID outbreaks, local public health officials depend on post hoc analyses of incidence and prevalence trends. The problem of lag time is compounded by the fact that the information provided by statewide or national organizations is aggregated and coarse grained; granular correlation between the prevalence of mosquitoes, weather conditions, and case incidence in a particular municipality and local cases with which to support decision-making is unavailable. This point can be generalized to all complex systems. Because they are exquisitely context-dependent, aggregate analysis misses subtle and significant nuances of local and idiosyncratic outbreak patterns. Maps and other stats reports generated from that information is therefore not "fit for purpose," if that phrase is understood as supporting actionable, effective recommendations that local agencies can rely on to implement interventions and contain and mitigate the infection's spread.

To illustrate the process, we begin with data collection. Even in major US cities, data about cases and vector specimens commonly consists of a handwritten entry on a paper form. Handwriting is error-fraught: answers are incomplete, illegible, misidentified, and so forth. Furthermore, paper data entry is not only inaccurate and incomplete, it has the following drawbacks:

- Paper entry usually does not include precise and automated geolocation or time stamp of data site. As just mentioned, vector control and lab data end up being aggregated, submitted, and processed with such significant time delay as to be useless in establishing medium and long-term trends, much less acting preemptively on them.

- In those rare cases in which data are entered digitally, software packages used by vector control departments, local laboratories, and various hospitals and physicians are not compatible. Comprehensive insight of current conditions becomes impossible.

- Additional relevant information about weather and climate data, population density, vegetation indices, and neighborhood socioeconomic demographics, all of which are readily available from third-party providers such as satellite feeds and census tracts, are not taken into account except by research facilities.

- Available and archived data in a range of formats (from APIs through Excel files to satellite feeds) are not incorporated, much less harmonized systematically with vector-control, laboratory, and clinical data. For example, troves of archived data in Excel format languish in desk drawers for lack of integration and processing with current data collection efforts.

- Potentially valuable geolocated and time-stamped citizen scientist reports are typically not even on the radar. If they do exist, they are often siloed in the records of the organization that sponsored that project.

- Seemingly tangential but potentially valuable and public information available from other databases such as those compiled by nationwide 311 call centers or United Way agencies are not integrated for processing with vector control and clinical and lab data. We include visualizations from case studies using 311 data and census tracts in the following illustrations.

The analytic stage of existing surveillance systems is riddled with analogous deficiencies. This stage commonly involves the following steps:

- Some local communities send paper data records collected and entered manually to the nearest university or to a separate office to be "cleaned up" and then entered manually on a computer. Because of paper data entry, digitally mapping the data must then also be performed manually. Even these data, however, are rarely available at a sufficient granular resolution (read municipal, much less zip code) to support significant and effectual interventions. Even though zip code scale analysis is easily obtainable, local decision-makers commonly have no access to real-time simulations and forecasts of local risk factors for that granularity.

- Academic studies that rely on data supplied by public health agencies are published with a one-year time lag and, moreover, are rarely fed back to decision-makers in support of actual interventions. Correspondingly, technology transfer is woefully inadequate.

- Relevant processing requires purchasing additional software packages for statistics and mapping (Matlab, SPSS, SAS, ArcGis, Tableau, etc.). But this typically means local health agencies must hire additional personnel with the advanced math and stats skills to integrate, process, and analyze data collected for analysis by these packages. As always, pitifully meager local budgets for public health surveillance and the lack of skilled personnel in rural communities compound the problems.

- Failure to build real-time data collection and analytics into existing surveillance systems exacerbates local bottlenecks and supply chain problems. Consequence: no state-of-the-art actionable project management recommendations based on machine learning or even simple simulations is available to local decision makers. These failures range from anticipating and monitoring inventory stock of vaccines, bed nets, and repellents, fumigation chemicals, and more. Significantly, there is no way to track the effectiveness of interventions (such as comparing one form of intervention with another, much less understanding any synergies that might arise from the combination of different types of intervention).

- Current surveillance analytics are often limited to passive point maps of dated reports; these cannot be incorporated into digital processing to inform policy and decision-making at the local level, much less to anticipate outbreak trajectory and severity and inform interventions.

The Bottom Line

There are structural and systematic reasons why significant delays are common between discovery of suspected clinical or entomological events and implementation of interventions such as vaccination campaigns or fumigation schedules. This lag time in anticipation delays and obstructs containment that is necessary to prevent transmission. Negative public health outcomes are real consequences of the long intervals between handwritten, paper-based data capture, processing, and returning analytic output to local decision makers. These slow inputs to the necessary data stream result in unavoidable delays and consequently ineffective interventions.

Failure to implement an accurate and efficient system with all these functionalities inevitably results in increased morbidity and mortality. Because complex systems are exquisitely context and path dependent, and because their dynamics have properties and consequences that the individual components that make them up do not, the effects of these deficiencies are often worse than anticipated.

Granular and timely insight, in real time, about current local conditions and their likely trajectory must be communicated to decision makers who can undertake appropriate and timely interventions. Real time alert "push notifications" to the general public are therefore impossible.

The consequences of communication failures can be worse than no information at all. Stale and out-of-date information can be worse that no communication at all. It misleads the community about the reality of the situation, leaving the playing field open to misinformation spread by unreliable and inaccurate sources. As noted, communication failures make accurate local risk assessment and actions impossible for city and county public health managers. Consequently, medium to long-term "planning" by local authorities and the public alike becomes a series of conjectures inspired at best by historical and inaccurate anecdotes rather than data-based analysis.

Understanding the challenges to transitioning to fully digital monitoring and tracking systems is not difficult, nor need it be time-consuming if set up properly from the beginning. It requires:

- Enhancing (not replacing) existing governmental and corporate tools

- Converting existing historical paper-based data into digital format

- Transitioning new data entry to digital format

- Integrating data from multiple sources into one comprehensive system while protecting privacy and ensuring data security at every scale

- Developing the IT backend that incorporates state-of-the-art mapping, statistical processing, simulation, and forecast modeling with machine-learning algorithms to eliminate the need for additional software or additional skilled personnel

- Generating forecasts and risk assessments down to the neighborhood block scale

- Visualizing analytic output of all integrated data with one click

- Empowering local authorities by placing full control of the software in their hands

- Empowering academic research to contribute actively to real-time decision-making support and interventions

The goal, of course, is to establish a shared information framework through which health-care agencies, their suppliers, and the public can more effectively monitor and track, anticipate, manage, and contain outbreaks of infectious diseases in general and of vector-borne diseases in particular. The platform must be useful at scales ranging from local public health decision makers to national organizations; it must also provide input to county and state research and support organizations and receive feedback from them to inform the models and simulations. Because, to repeat, complex systems are context-dependent, granular information about local conditions is necessary. Data aggregated at the national data level cannot provide insight and understanding (much less recommendations for action) about locally unique conditions.

Solutions

VectorAnalytica's management platform for vector-borne diseases was developed to support decision-making by health care agencies, their suppliers, and the communities they serve. On that platform, users

- enter official primary (vector, clinical, lab, etc.) data through mobile devices on mobile app query forms that users create on easy drag-and-drop interfaces and which they build to their specifications. To dramatically improve the quality, timeliness, and thoroughness of data collection, accurate VectorAnalytica mobile apps make data entry possible even without internet connectivity.

- integrate and analyze archived data, such as third-party demographic information, satellite feeds of vegetation and weather variables, and so forth while respecting the owner's privacy and proprietary requirements. Because VectorAnalytica software is meant to complement and not replace existing governmental or corporate surveillance efforts, its IT platform does not store third-party data.

- automate data entry of APIs as well as other tangential databases such as those of 311 or other nationwide crisis response networks.

- empower citizen scientist projects in order to generate community buy-in for improved public health surveillance.

- process and map all data without additional software packages or GIS skills. Users must also be able to disaggregate analytic output by date range, variable, correlation, and especially by local geographic area.

- produce (in real time) visualizations of statistical, mapping, and machine learning analytics, at scales ranging from neighborhood block to global levels, with the capability of disaggregation, as just mentioned. Visualization significantly increases situational assessment and insight into local, current, and anticipated conditions.

- generate scientifically validated risk assessments and recommendations for local communities in real time.

- quickly disseminate information and recommendations to other public health agencies and the community at large through SMS push notifications or email.

- establish a trusted and transparent interactive communications network between authorities and the public, and between health care agencies and their suppliers.

Three Case Studies

The following are three examples of insights revealed by VectorAnalytica's platform integrative and analytic capabilities.

Case Study 1. Zika outbreak in Miami-Dade County, 2016

At the start of the Zika outbreak in 2016, Miami-Dade County's vector-control department had fewer than 30 mosquito traps placed around its 2,400 square miles. Given the impossibility of establishing credible prior probabilities from data from such few traps in such a large geographical expanse, VectorAnalytica turned to the existing rich trove of data of "mosquito nuisance complaints" recorded on the county's 311 hotline.

Because of the low number of permanent residents in certain well-trafficked areas such as Miami International Airport, relying on absolute numbers of 311 calls alone would not have provided actionable insight. To remedy the inability to formulate traditional indices such as the Breteau index, household indices, and others due to the paucity of trap data, VectorAnalytica's chief scientist created an indirect transmission risk index by integrating variables from multiple sources (see Figure 9.1.a). The fact that Zika transmission had already begun in the county supported the assumption that a certain percentage of those phone complaints reported the presence of *Aedes aegypti* mosquitoes. To obtain that index, VectorAnalytica used a quotient obtained by dividing the number of 311 calls per block by the population density of that block (per 2010 census tract). The resulting thematic or choropleth map (see Figure 9.1.b) uncovered differences in transmission risk in different areas of the county.

Integrating census information of residential density with location and frequency of 311 "mosquito nuisance" complaints was revelatory. As the area of the map colored in darkest blue shows, a region adjacent to the airport unmistakably emerged as the county's zone with highest transmission risk. This was initially puzzling, for it is not a densely populated residential area. However, it is the location of the Miami Intermodal Center, where the airport's major parking garages and rental car center are housed, along with the region's Amtrak and Tri-Rail hubs linking Ft. Lauderdale airport with the Miami airport. Major county metrorail, metrobus, and automated people-mover stations are also located there. In addition, the area has a number of small hotels and motels that host overnighting airline crew members. To complete the picture, airport authorities maintain extensive and attractive landscaping of tropical foliage, nestled into several large containment ponds. It is also an area prone to flooding and standing water after summer downpours.

The VectorAnalytica index thus unexpectedly revealed a realistic and significant central convergence point with respect to mosquito-borne transmission risk. Although it has few permanent residents (therefore a low denominator for the index's quotient), the area identified by the thematic map experiences a large daily volume of international and national travelers arriving at the Miami airport or transferring to or from the Ft. Lauderdale airport, from which they disperse by bus, people movers, and trains to other areas of the county, or travel beyond South Florida. A high quotient for transmission risk for these few blocks was almost obvious in retrospect. The software's visualization made it apparent.

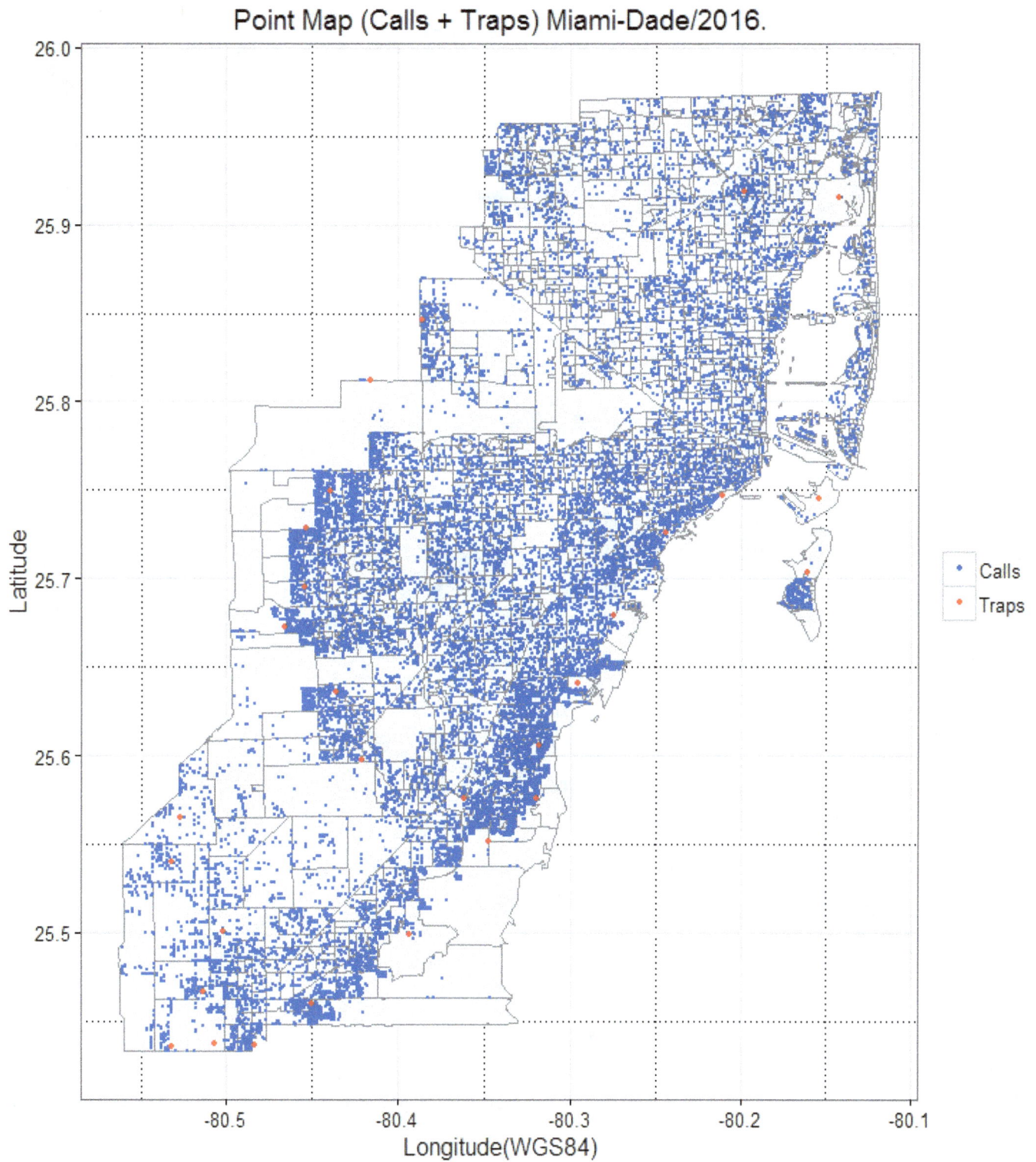

Figure 9.1.a. Point map of Miami-Dade County, Florida, 2016. *Blue dots:* "Mosquitoes nuisance" complaints to the 311 hotline. *Red dots:* Mosquito traps.

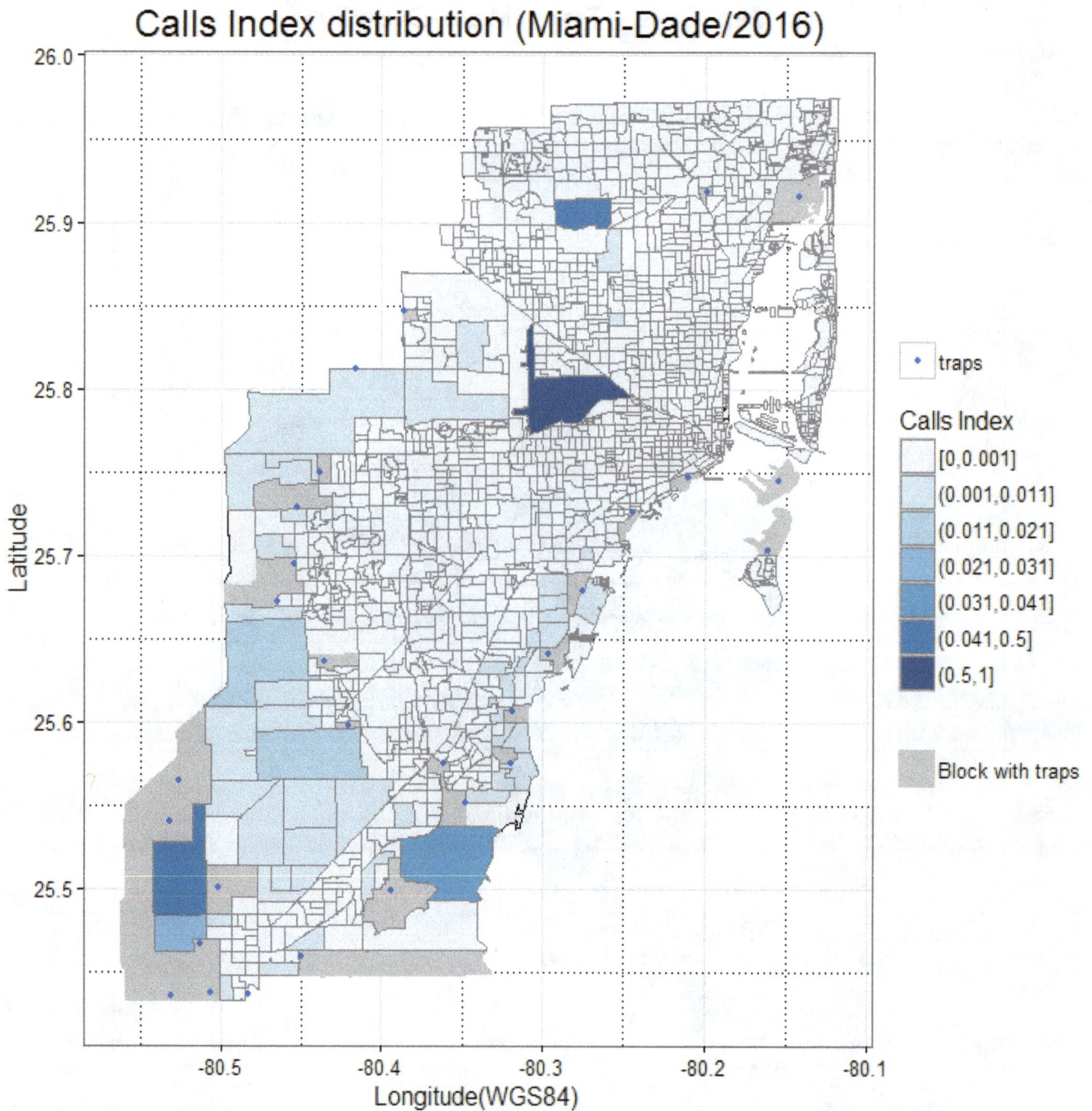

Figure 9.1.b. Choropleth or thematic map. Call index in various shades of blue. Trap locations in gray. Miami-Dade County, Florida, 2016.

Subsequent iterations of these maps revealed the following interesting property indirectly suggestive of the direction in which transmission would spread: Wynwood, the site of the first major Zika outbreak in the county is a nearby neighborhood accessed on the Airport Expressway (Route 112) adjacent to the Miami Rental Car Center.

Census information states that the neighborhood is historically identified as Miami's Little Puerto Rico, with many longtime residents continuing to travel often to the island (which by early 2016 was already a serious focus of Zika infections). Add to that information the fact that, although still composed for the most part of rundown and abandoned warehouses and older residences, Wynwood had experienced new openings of trendy restaurants, designer shops, artists' studios, and nightspots in the last decade. These attractions had been drawing increasing customers from Miami Beach, a short causeway drive away. It had also become a place where travelers to and from Miami International Airport would stop, shop, and grab a bite to eat on their way to or from the beaches.

It is therefore reasonable to conjecture that those traveling from the Miami Intermodal Center to Miami Beach (or vice versa) might have become exposed to the virus during their Wynwood stop and shop. This conjecture is reinforced by a corollary, that it is likely that the Airport Expressway served as a transmission channel for contagion and spread not only to Wynwood but also to North Miami Beach, which became the second Zika hotspot in the county. It is curious to note as an aside that Route 112's terminus on the barrier island is located in an area some call Little Brazil. As a reminder, Brazil was already experiencing a huge Zika outbreak when this study was done.

Despite the absence of other more direct and pertinent data (such as species of mosquito specimens collected from numerous traps and confirmed in a lab as infected, or as gravid females—only female mosquitoes bite, etc.), and the fact that the index utilized is not a validated index, under such conditions of uncertainty, integrating and visualizing disparate sets of data (GPS-identified calls of mosquito biting combined with census information) revealed insightful information that narrowed the search for a common source of dramatically increased transmission risk to a very circumscribed neighborhood. The thematic map generated from the call index also suggested a possible channel of transmission.

Having this information earlier would have supported a far more targeted fumigation campaign (and probably as effective but less ecologically harmful) than the blanket aerial spraying the county chose to embark on instead.

Case Study 2. Mosquito-borne diseases in Costa Rica

It is common for studies of historical outbreaks of dengue and other mosquito borne diseases to note the correlation between prevalence of clinical cases and ambient temperature (typically increasing in summer months, e.g.). When vector-control data of mosquito-borne infections is entered by hand, however, this correlation appears to be anecdotal at best.

The ability to import from OpenWeather APIs of records of 23 local weather measurements in Costa Rica (temperature, humidity, dewpoint, atmospheric pressure, etc.)

allowed VectorAnalytica's integrative IT platform to combine official Costa Rica public health department reports of clinical cases of mosquito-borne diseases with weather data for that country during 2016–17.

Time series visualizations of these multiple variables clearly indicate that, contrary to traditional reports in the literature, the one factor with which clinical case numbers is not systematically and universally correlated in Costa Rica is ambient temperature, reported in Sri Lanka by Goto et al. (2013). Also in contradiction to the literature (Polwiang, 2020), which rarely mentions dewpoint in connection with these infections, the correlation from these analytics does show a systematic—if time-lagged—correlation for Costa Rica between humidity and dewpoint on the one hand and clinical case increase on the other.

Comparing Zika and dengue with weather conditions revealed that the different pathogens appear to be correlated differently with different weather conditions in different localities. This suggests a complex spatio-temporal context dependence indexed to a particular virus in a particular microclimate and/or geographic location. It is likely, therefore, that the coevolutionary dynamics of vectors, pathogens, and hosts intertwine in radically complex manner with local weather trends and conditions, sui generis vegetation cover, and so forth. This dynamic makes local vector surveillance and monitoring the key to containing and mitigating spread.

The following graphs illustrate the value of integrating disparate variables as proposed by complexity theory. As noted, the literature reports a correlation between transmission of vector-borne diseases such as dengue and Zika and annual temperature peaks. Although studies pertaining to the role of dewpoint in increases and decreases of clinical cases are rare and recent (Nygren et al., 2014), these graphs generated through VectorAnalytica's platform show maximum correlation between disease incidence and relative humidity and dewpoint, especially the latter. In contradiction to the literature, on the other hand, the visualized output of integrating weather variables as shown in Figures 9.2.a–d shows that peak spikes for dengue incidence do not correspond directly with temperature spikes. But in a remarkable and novel result that could not have been predicted or derived otherwise, analytic output clearly shows that for two years in a row, and for both Zika and dengue, there is a systematic and consistent lag between temperature and case incidence: temperature peaks around weeks 10–12, while dengue cases do not peak until approximately 20 weeks later.

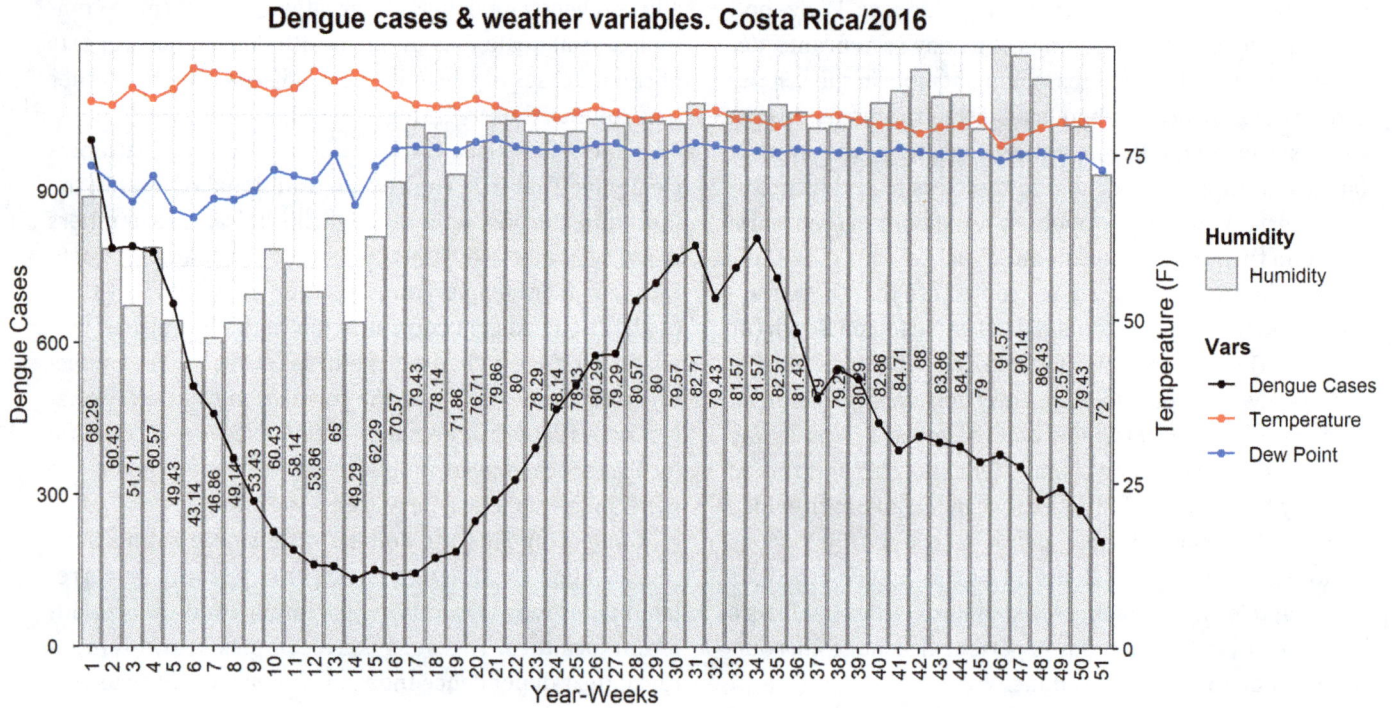

Figure 9.2.a. Time series of dengue and weather variables for Costa Rica, 2016.

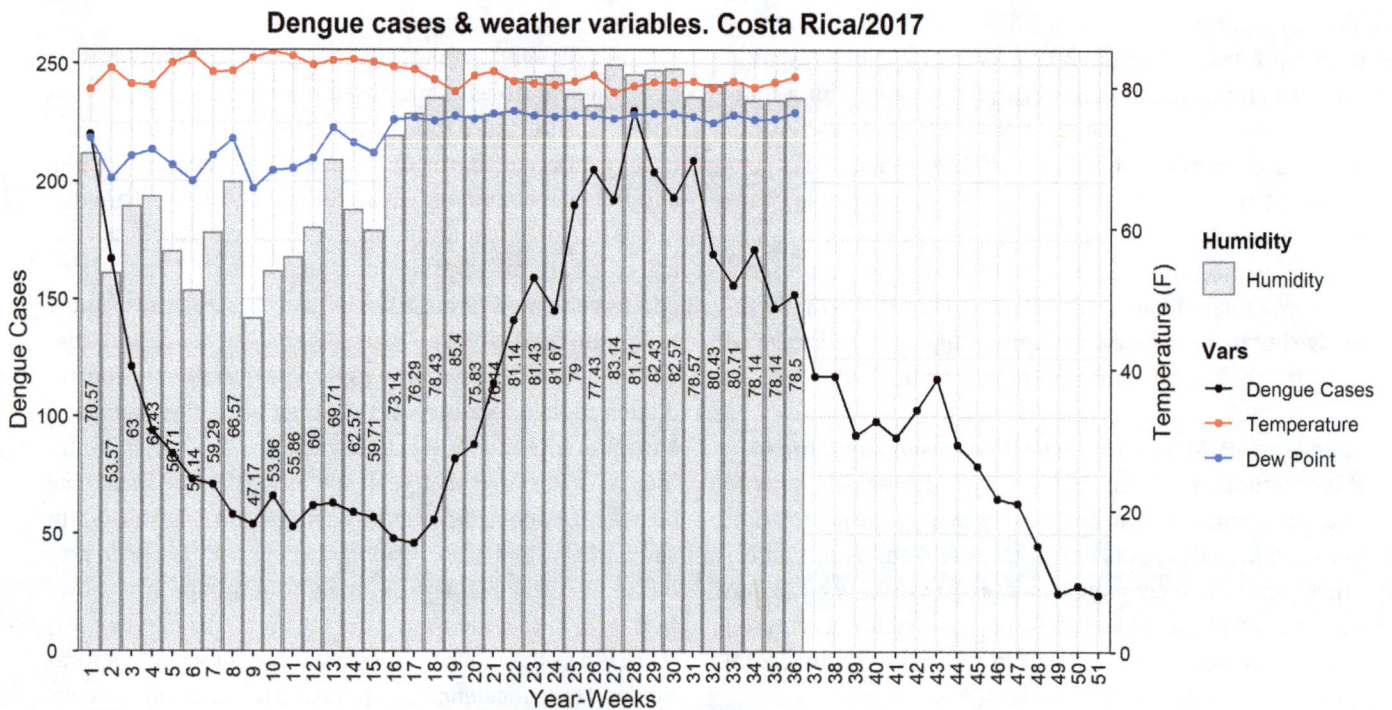

Figure 9.2.b. T-series correlating official weekly reports of dengue cases in Costa Rica with weather variables, 2017.

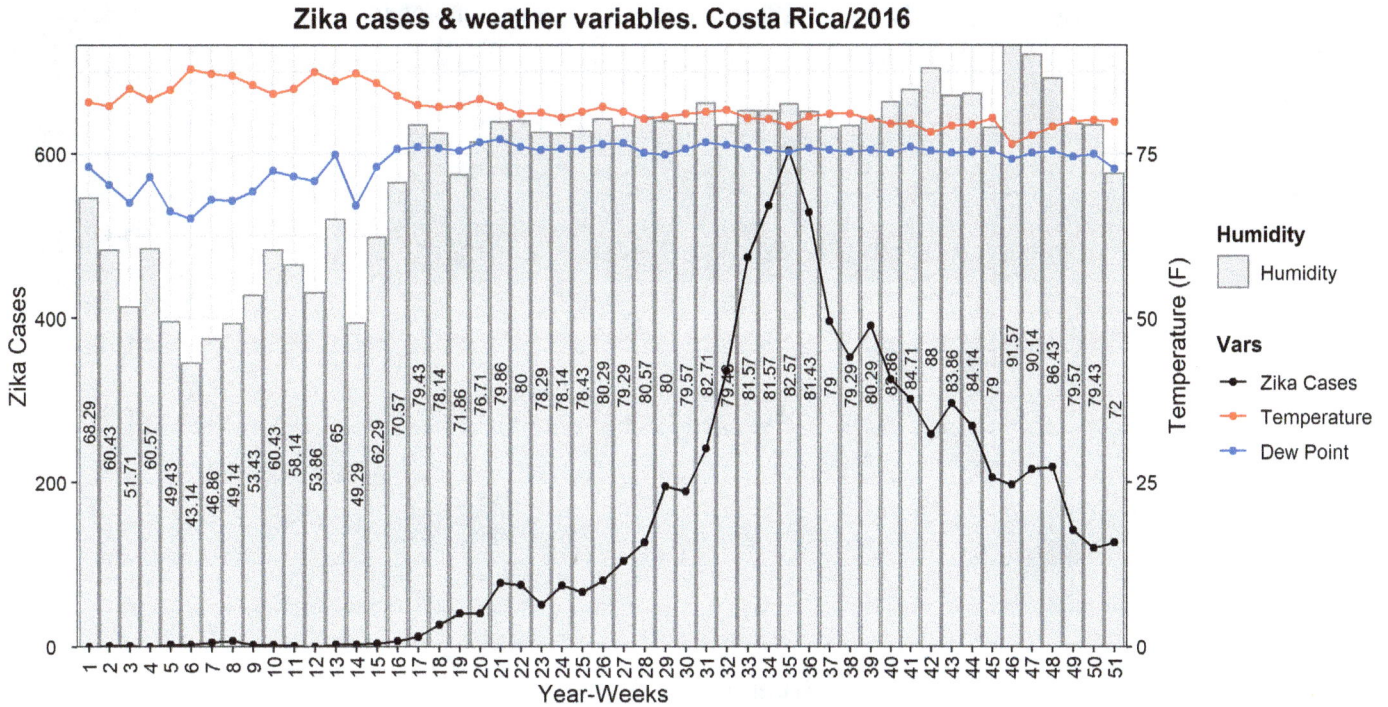

Figure 9.2.c. Time series of Zika and weather variables for Costa Rica 2016.

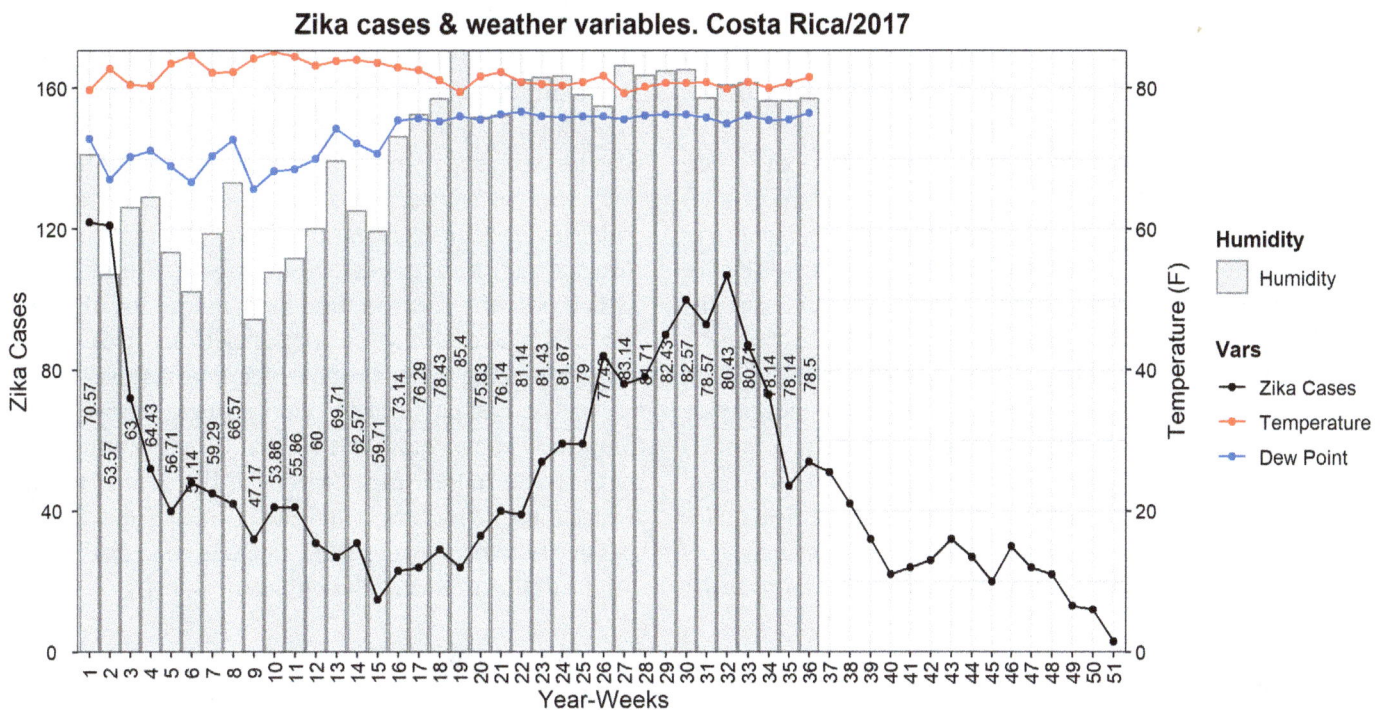

Figure 9.2.d. T-series correlating official weekly reports of Zika cases in Costa Rica with weather variables, 2017.

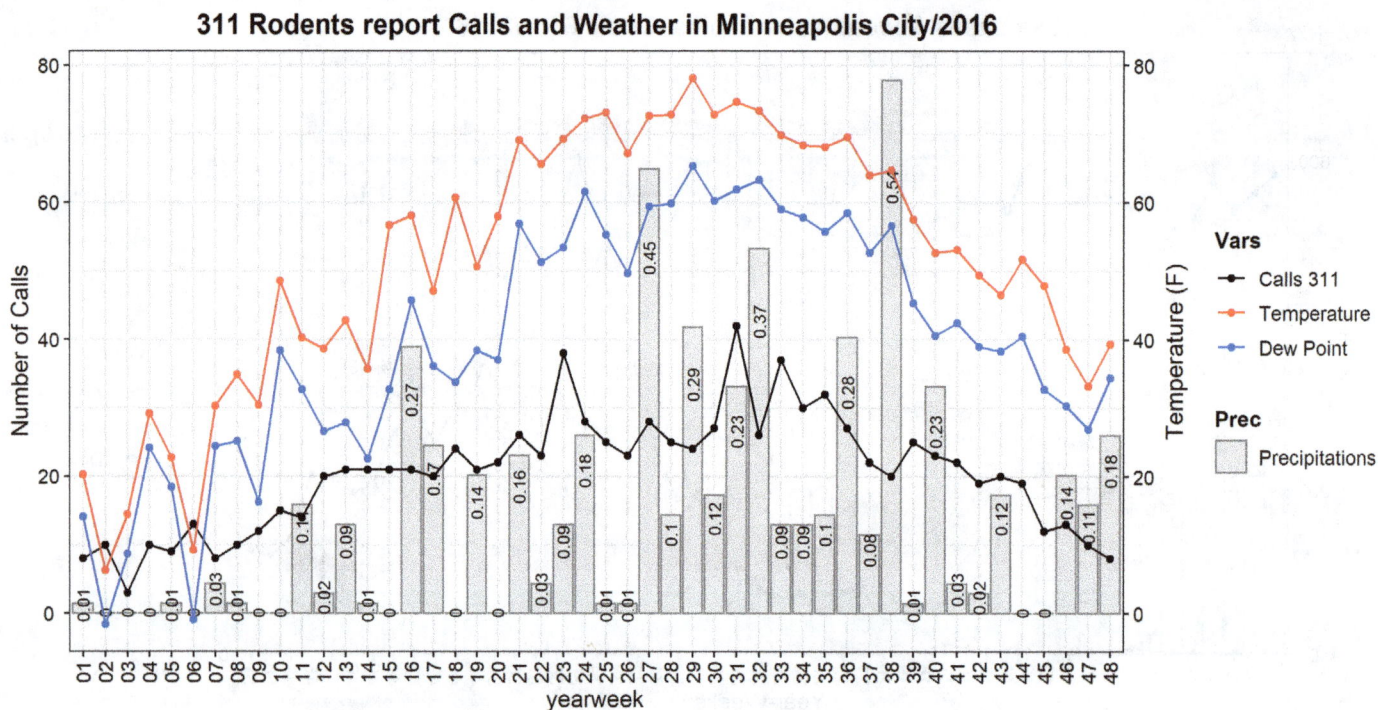

Figure 9.3. Time series graph of relations between rodent-related complaints to the Minneapolis, Minnesota, 311 hotline (*black line*) and measures of precipitation, dew point, and temperature in the Minneapolis metropolitan area (see legend).

Case Study 3. Complaints to Minneapolis 311 hotline about rodent presence

For our final use case we present a study of rodent infestation in Minneapolis, Minnesota, USA. Visualization of statistical processing showed a dramatic correlation between trendlines of rodent complaints to 311 hotline and those of dewpoint and temperature (not so much with precipitation, possibly due to the much wider range of ambient temperatures in Minneapolis compared to Costa Rica) (Figure 9.3). Remarkably and in contrast to the mosquito-borne disease cases in Costa Rica noted earlier, in the case of rodent-related complaints in Minneapolis there is no time lag at all between the number of complaints and either temperature or dewpoint. In these particular conditions, for this particular vector, dewpoint and temperature can therefore serve as predictive variables; they faithfully anticipate increases in rodent complaints (as well as the presumed corresponding increase in transmission risk of rodent-related zoonotic disease).

Conclusions

The three case studies clearly show that integrated information supports decision-making and is actionable. In the case of rodent presence in Minneapolis, private firms with this insight obtained from integrating information about dewpoint and temperature with frequency of pest complaints to 311 can systematically anticipate trends in demand for their services throughout the city. In the case of Costa Rica, it suggests the need for further study of the relation between dewpoint and humidity and dengue and Zika incidence. Such local, timely, and nuanced insight can provide public health agencies and private pest control service firms alike with data-based anticipatory awareness with which private firms can time their marketing and sales campaigns, and with which public health agencies can "anticipate to mitigate" and contain zoonotic disease spread.

An IT-based public health monitoring and tracking system today is the equivalent of a nation's highway system or the National Hurricane Center. It is the channel through which information and action can be managed to advance public health nationwide; it is a common good that must be publicly supported and disseminated.

Literature Cited

Agosta, S.J. 2022. The Stockholm paradigm explains the dynamics of Darwin's entangled bank, including emerging infectious disease. MANTER: Journal of Parasite Diversity 27. https://doi.org/10.32873/unl.dc.manter27

Agosta, S.J. 2023. The Stockholm Paradigm explains the eco-evolutionary dynamics of the biosphere in a changing world, including emerging infectious disease. In: An Evolutionary Pathway for Coping with Emerging Infectious Disease. S.L. Gardner, D.R. Brooks, W.A. Boeger, E.P Hoberg (eds.). Zea Books, Lincoln, NE.

Audy, J.R. 1958. The localization of disease with special reference to the zoonoses. Transactions of the Royal Society of Tropical Medicine and Hygiene 52: 308–334. https://doi.org/10.1016/0035-9203(58)90045-2

Brooks, D.R.; Boeger, W.A.; Hoberg, E.P. 2022. The Stockholm paradigm: lessons for the emerging infectious disease crisis. MANTER: Journal of Parasite Diversity 22. https://doi/10.32873/unl.dc.manter22

Brooks, D.R.; Boeger, W.A.; Hoberg, E.P. 2023. The Stockholm Paradigm: the conceptual platform for coping with the emerging infectious disease crisis. In: An Evolutionary Pathway for Coping with Emerging Infectious Disease. S.L. Gardner, D.R. Brooks, W.A. Boeger, E.P. Hoberg (eds.). Zea Books, Lincoln, NE.

Brooks, D.R.; Hoberg, E.P.; Boeger, W.A. 2019. The Stockholm Paradigm: Climate Change and Emerging Disease. University of Chicago Press, Chicago.

Brooks, D.R.; Hoberg, E.P.; Boeger, W.A.; Gardner, S.L.; Galbreath, K.E.; Herczeg, D.; et al. 2014. Finding them before they find us: informatics, parasites, and environments in accelerating climate change. Comparative Parasitology 81: 155–164. https://doi.org/10.1654/4724b.1

Goto, K.; Kumarendran, B.; Mettananda, S.; Gunasekara, D.; Fujii, Y.; Kaneko, S. 2013. Analysis of effects of meteorological factors on dengue incidence in Sri Lanka using time series data. PLOS ONE 8: e63717. https://doi.org/10.1371/journal.pone.0063717

Hoberg, E.P.; Boeger, W.A.; Molnár, O.; Földvári, G.; Gardner, S.L.; Juarrero, A.; et al. 2022. The DAMA protocols, an introduction: finding pathogens before they find us. MANTER: Journal of Parasite Diversity 21. https://doi.org/10.32873/unl.dc.manter21

Hoberg, E.P.; Boeger, W.A.; Molnár, O.; Földvári, G.; Gardner, S.L.; Juarrero, A.; et al. 2023. The DAMA protocol: anticipating to prevent and mitigate emerging infectious diseases. In: An Evolutionary Pathway for Coping with Emerging Infectious Disese. S.L. Gardner, D.R. Brooks, W.A. Boeger, E.P. Hoberg (eds.). Zea Books, Lincoln, NE.

Juarrero, A. 1999. Dynamics in Action: Intentional Behavior as a Complex System. MIT Press, Boston.

Juarrero, A. 2023. Context Changes Everything: How Constraints Create Coherence. MIT Press, Boston.

Nygren, D.; Stoyanov, C.; Lewold, C.; Månsson, F.; Miller, J.; Kamanga, A.; Shiff, C.J. 2014. Remotely-sensed, nocturnal, dew point correlates with malaria transmission in Southern Province, Zambia: a time-series study. Malar Journal 13: 231. https://doi.org/10.1186/1475-2875-13-231

Ortiz, E.; Juarrero, A. 2022. An emerging infectious disease surveillance platform for the 21st century. MANTER: Journal of Parasite Biodiversity 29. https://doi.org/10.32873/unl.dc.manter29

Polwiang, S. 2020. The time series seasonal patterns of dengue fever and associated weather variables in Bangkok (2003–2017). BMC Infectious Diseases 20: 208. https://doi.org/10.1186/s12879-020-4902-6

Souza, A.T.C.; Araujo, S.B.L.; Boeger, W.A. 2022. The evolutionary dynamics of diseases on an unstable planet: insights from modeling the Stockholm paradigm. MANTER: Journal of Parasite Diversity 25. https://doi.org/10.32873/unl.dc.manter25

Souza, A.T.C.; Araujo, S.B.L.; Boeger, W.A. 2023. Modeling the Stockholm Paradigm: insights for the nature and dynamics of emerging infectious diseases. In: An Evolutionary Pathway for Coping with Emerging Infectious Disease. S.L. Gardner, D.R. Brooks, W.A. Boeger, E.P. Hoberg (eds.). Zea Books, Lincoln, NE.

10

All Hands on Deck: Turning Evolutionary Theory into Preventive Policies

Orsolya Molnár, Marina Knickel, and Christine Marizzi

Abstract

The emerging infectious disease (EID) crisis has been challenging global health security for decades, dealing substantial damage to all socioeconomic landscapes. Control measures have failed to prevent or even mitigate damages of an accelerating wave of EIDs, leading to the emergence and devastation caused by the COVID-19 pandemic. In the wake of the pandemic, we must critically review our public health policies and approaches. Current health security measures are based on the evolutionary theorem of host-parasite coevolution, which falsely deems EIDs as rare and unpredictable. The DAMA protocol (Document, Assess, Monitor, Act) is nested in a novel evolutionary framework that describes how emergence can be prevented before the onset of an outbreak. In this paper, we discuss the importance of establishing efficient communication channels between various stakeholders affected by EIDs. We describe implementation strategies for preventive interventions on global, regional, and local scales and provide guidelines for using such strategies in relevant policy environments of human, livestock, and crop diseases.

Keywords: infectious disease, emerging infectious disease (EID) prevention, DAMA protocol, policy implementation, Living Labs, citizen science

The Crisis of Emerging Infectious Diseases

The past decades have seen a striking rise in the number of emerging infectious diseases (EIDs) across the globe, including known diseases appearing in previously unknown areas (e.g., West Nile virus, diphtheria, measles), new variants becoming resistant to treatment (e.g., malaria, MRSA), and completely novel pathogens infecting novel hosts (e.g., SARS, African swine fever, phytoplasma). With the ever increasing rate of globalization and international trade and travel, EIDs have spread faster than ever before in human history, resulting in a staggering US$1 trillion per year for containment costs and production losses before 2020 (Brooks et al., 2019). This figure was further elevated by the recent COVID-19 pandemic, which resulted in one of the largest economic recessions since the mid-1900s (Blake and Wadhwa, 2020; COVID-19 to Plunge, 2020). A more significant reason for concern is that the damages to both economic production and human life were highest in the United States and United Kingdom, whose health-care systems were announced to have been best prepared for such an event (Cameron et al., 2019; Singh et al., 2020). These controversial patterns highlight the dissonance between how we try to control EIDs and what we should be focusing on instead.

As is the case with most epidemics, important conclusions are drawn after the fact that should lead to better preparedness for the next such event. However, our investments into disease control and surveillance methods have not been able to slow down, let alone stop, the acceleration of the EID crisis thus far, nor prevent the COVID-19 pandemic. To understand why our efforts have not been efficient or successful, we must understand the requirements and limitations of current disease management strategies and identify the gaps that allow for new diseases to emerge.

(Wrong) lessons learned

The main way we approach an outbreak today is to gather the maximum amount of data available to suggest pathways of containment. Although effects of an epidemic are felt throughout all sectors from tourism to education all the way to the job market, proactive initiatives are typically assigned to only two key fields: research and public health (DeSalvo et al., 2021; WHO, 2020b). Consistent throughout the majority of the reports and studies is the general aim of increasing preparedness measures (Quaglio et al., 2016; WHO, 2020b), which translates into two major suggested outputs:

Rapid response: The main direction for developments in public health is improving preparedness. Better preparedness involves decreasing the time required to identify and respond to an emerging disease. Identification of a public health event of international concern (PHEIC) requires investment into outbreak surveillance, health-care data management, reaction protocols, and real-time communication channels between local health authorities and regional or global organizations (Lakoff, 2017; WHO, 2020b). Response, on the other hand, warrants sufficient capacity in health-care infrastructure to treat incoming waves of patients and therefore relies on stockpiling equipment and medication, and increasing the number of trained health-care professionals (Cheng et al., 2013; DeSalvo et al., 2021).

Focused research: The main direction for developments in science is increasing our knowledge of the emerging disease. A typical reaction to an epidemiological emergency is the reallocation of research funds toward studies that target the emergent pathogen, which shifts the focus of novel and existing labs. Unfortunately, analyses show that this heightened attention and support wanes with the decreasing sense of emergency and proves to be inefficient in the long term; thus, fund reallocation is often referred to as "boom-and-bust funding" (Funding boom or bust?, 2009; Kading et al., 2020).

Both outputs have had significant results in managing reemergence of known diseases and epidemics that are considered regular occurrences in particular regions, but neither has yielded any considerable advantages against the EID crisis (Lakoff, 2017; Morens and Fauci, 2020). The reason is that all of the initiatives listed above require prior knowledge of the emergent pathogen. Public health needs to "know what they're looking for" to detect it and alert health systems at an early stage of the outbreak. Otherwise, the only clue of a recent emergence is the sudden spike of patients producing similar symptoms with unknown aetiology, as has been the case for SARS-CoV-2 (WHO, 2020a), the 2015 Zika epidemic (Schuler-Faccini et al., 2016), and

even the currently ongoing outbreaks of hepatitis among children (WHO, 2022a). Response also requires data on the clinical manifestations, morbidity, and mortality to adequately prepare health-care infrastructures, and focusing research requires an already identified and defined target pathogen or disease. In the case of a newly emerging disease, none of this descriptive information is available, so crisis response is therefore constantly lagging behind the spread of the epidemic. Taking into consideration the effects of globalized travel and trade (Findlater and Bogoch, 2018; Morens and Fauci, 2020), preparatory efforts will have little success in halting an epidemic in fulfilling its pandemic potentials, and crisis response is, by definition, a reactive measure.

To successfully address the emergence of novel diseases, a new paradigm has to be introduced into global health security, shifting our main focus from preparedness to prevention, and moving our intervention further back on the infection timeline. However, in order to change our approach, we must understand the gaps that have thus far allowed EIDs to ravage our societies.

What Are We Missing?

Health security measures are developed by close collaboration networks between public health and fundamental science. With the constant advancements in both technology and research, numerous defense strategies have been improved. Nevertheless, the EID crisis represents a completely novel challenge, which requires understanding the limitations of our current approaches, particularly about the predictability and the scope of EIDs.

Predictability

Despite the extent to which epidemics and pandemics damage a wide range of socioeconomic landscapes, concerningly few initiatives aim to prevent the large-scale effects of EIDs (Cazzolla Gatti et al., 2021; Vianna Franco et al., 2022). This lack comes from the prevailing evolutionary paradigm used in public health and research regarding the ability of pathogens to colonize new hosts, aka emerging as a novel disease. The traditional scientific paradigm states that strong selection is acting on parasite characteristics, which leads to extreme specialization in a narrow range, often to a single host species. Such specialized parasites are able to better exploit host resources but at the same time lose their ability to infect novel host organisms; therefore, any novel colonization must necessarily be preceded by the right mutation appearing at the right time (Parrish and Kawaoka, 2005). Because of the random and unpredictable nature of such genetic changes, host-switching events are assumed to be rare and unpredictable (Brooks et al., 2019; Molnár et al., 2022). However, this coevolutionary theory (CT) suffers from severe shortcomings when compared to empirical data: (1) CT's key assumption of parasites being tightly coadapted to a narrow range of hosts lacks empirical support, (2) CT's prediction regarding EIDs being rare occurrences is sharply contradicted by the accelerating EID crisis (WHO, 2007; de Vienne et al., 2013; Nylin et al., 2018), and (3) CT fails to connect such novel colonization events to environmental changes when there is evidence that emergences cluster around climate change perturbations (Brooks et al., 2015; Hoberg and Brooks, 2015).

This contradiction between the prevailing paradigm and empirical observations is referred to as the "parasite paradox" (Agosta et al., 2010), and it has significant consequences in how public health addresses EIDs. Public health deems emergence rare, and therefore of low global health concern, and at the same time unpredictable, thus judging prevention efforts to be impossible. These wrong predictions are the main reasons public efforts aiming to address the EID crisis have been futile, and precisely why we need a novel evolutionary paradigm to resolve this paradox.

The Stockholm Paradigm

The Stockholm Paradigm (SP) (Brooks et al., 2014, 2019; Hoberg and Brooks, 2015) relies on two Darwinian principles that lead to fundamentally different conclusions than those produced by the standard coevolutionary theory.

First, evolutionary outcomes are always local. Pathogens are genetically capable of infecting a certain range of hosts, translated as their "fundamental fitness space," but they infect only a subset of these that are available to them in their environment, creating their "realized fitness space." Selection acts only on traits within the realized fitness space and has no effect on other potential hosts in other environments. Pathogens that have a proportionally smaller realized fitness space therefore have a higher potential of colonizing a novel host, without the necessity of evolving new capacities. This potential is referred to as "ecological fitting" (Janzen, 1985). When viewed from a public health perspective, this means emergence is a built-in attribute of host-pathogen associations and is therefore expected to happen frequently, especially when environmental perturbations increase species encounters, which is what we are witnessing with the EID crisis.

Second, evolution is conservative. To use particular resources, pathogens will develop specialized traits. Since these traits are phylogenetically conservative, pathogens will be able to utilize distantly related, naive host species upon encounter, while the same host can serve as a resource for various pathogens (McCullough, 2014; Dicken et al., 2021; Lytras et al., 2021). The recently emerged SARS-CoV-2 uses the angiotensin-converting enzyme 2 (ACE2) as its main receptor, which is widely shared among phylogenetically disparate groups of mammals and is the primary reason the pathogen had established itself in mustelids, felids, and cervids, among other mammals (Damas et al., 2020; Hoberg et al., 2022). Translated to a practical view, conservative traits allow us to predict the risk an unknown pathogen poses to human populations without having to wait for an outbreak. Pathogenic microbes can therefore be sampled from reservoir species, and action can be taken not only to contain emergence but to prevent it altogether.

The SP therefore changes the theoretical foundation on which our global health security infrastructure is built. The bad news is that EIDs are indeed frequent and should only be expected to increase in occurrence with the intensifying globalization and climate change. The good news is that EIDs are predictable, and preventive action can and should be taken to avoid the next epidemic and pandemic.

Scope

With regard to EIDs, literature and policy refer almost exclusively to human pathogens (Jones et al., 2008; Findlater and Bogoch, 2018; Morens and Fauci, 2020). Preparatory efforts and early action plans exclusively target human diseases, which manifests in recommended actions for rapid response and focued research (Palagyi et al., 2019; Leach et al., 2021). But this human focus narrows our view to a small subset of potentially dangerous pathogens while we ignore those that affect crops and livestock. Infectious diseases that decimate agricultural production are dealt with by food safety, agri-food sciences, and agricultural policies and are barely put in the context of EIDs. Nevertheless, the loss of production and associated costs affect regions' economies just as much if not more than human diseases do. Coconut lethal yellowing disease destroyed 95% of the coconut palms in a region of Mexico; killed millions of trees in Nigeria, affecting the livelihood of 30,000 families; and ruined 72% to 99% of the trees in West Africa (Gurr et al., 2016; Datt et al., 2020). Wheat stem rust (*Puccinia graminis* f. sp.) was considered eradicated until 1998, when a new, highly virulent strain emerged in Uganda (Pretorius et al., 2000). Since then, it has spread throughout eastern and southern Africa, the Middle East, and western Europe and poses a threat to more than 80% of the world's wheat varieties (Saunders et al., 2019). From those affecting livestock, the 2014–15 avian influenza (AIV) epidemic led to the culling of 45 million birds in the US and export bans in 75 countries (Newton and Kuethe, 2015), and the ongoing H5N1 avian influenza outbreak has already led to the loss of 77 million birds (Miller, 2022). Within a few years, African swine fever (ASF) swept through Europe and Asia, destroying 20% of Vietnam's swine population and resulting in a US$141 billion economic loss for China, collapsing half the world's pork export market in a single year (FAO, 2019). Apart from the obvious socioeconomic effects of food shortage and skyrocketing food prices, policy interventions aimed at relieving damages of ASF were suggested to have led to the emergence of COVID-19 (Xia et al., 2021).

Although currently considered to be separate issues from human well-being, food security and global health security are threatened by the same thing: emerging infectious diseases. If we understand the dynamic that allows novel pathogens to explore and colonize new hosts, then we must also understand that this applies not only to humans invading natural habitats but to our crops and livestock being placed in close proximity to natural reservoirs (Brooks et al., 2022; Trivellone et al., 2022).

With EIDs being predictable but much more frequent and abundant than previously thought, health security measures have to incorporate this new paradigm and adopt appropriate and much-called-for prevention measures (Bernstein et al., 2022). We therefore describe a comprehensive four-step protocol based on the SP that leads all the way to policy implementations.

The DAMA Protocol

The DAMA protocol—Document, Assess, Monitor, Act—is a policy plan derived directly from the evolutionary framework of the SP, which aims to connect evolutionary science with applied health security. It focuses on preventing outbreaks and facilitating communication between private and public actors, knowledge institutions, and the communities that are directly affected (Figure 10.1).

The **documenting** of pathogens has to be extended from only those that are already causing diseases to those existing in wild animal and plant populations. Taking advantage of the evolutionary context provided by the SP, anticipatory research has to focus on potential reservoirs. Pathogens that cause disease in humans, crops, or livestock are all present in at least one other species that manifests no symptoms.

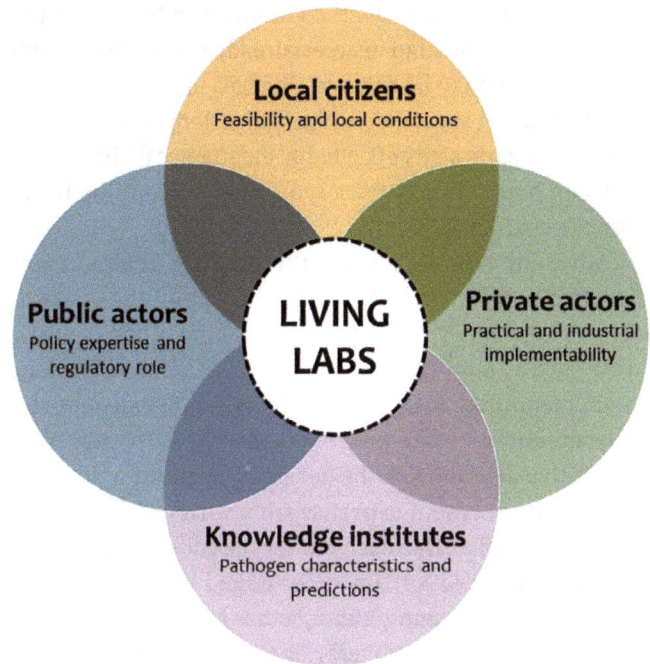

Figure 10.1. The stakeholder groups that construct Living Labs to target EIDs. Adapted and updated from Steen and van Bueren (2017).

Taxonomic inventories and virological and bacteriological studies have often revealed these pathogen-reservoir associations, which direct research focus on a subset of species within any given area. Pathogen transmission occurs at the interface between such reservoirs and human settlements, agricultural areas, and breeding facilities (Gallardo et al., 2015; Cyranoski, 2017; CDC, 2022). The primary step in establishing a preventive protocol is collecting all information into strategic inventories that feed into archives of host and pathogen specimens, modes of transmission, and potential vectors (Dunnum et al., 2018; Colella et al., 2021). Finally, inventories are also to include local and traditional knowledge on the distribution, behavior, and abundance of reservoirs, which calls for the establishment of robust science-society collaborative programs (Brooks et al., 2019; Földvári et al., 2022; Francisco et al., 2022).

Inventories then allow us to **assess** the risk posed by potential pathogens. A three-step process first separates our potential pathogens from those already known and those considered to be nonpathogenic through *phylogenetic triage*; then uses *phylogenetic assessment* to determine mode of transmission, reservoirs, and potential vectors; and finally maps population genetics and rare genotypes through *population modeling*.

Potential pathogens are then **monitored** to create a detailed distribution map in areas already confirmed as well as those deemed to be suitable. Changes in geographic distribution, host range, mode of transmission, or disease pathology are early signs of potential emergence on interfaces between populations of reservoirs and susceptible hosts (Brooks et al., 2022).

Adequate monitoring sets the stage for adequate **action** in policy making. Highly dependent on the context, such as legal environment, policy modifications concern areas such as food safety, wildlife management, veterinary medicine, public health, and education. Because of the large number of stakeholders affected by EID outbreaks, preventive action has to be designed by multi-actor task forces that represent expertise from various sectors and scales. In practice, this necessitates the collaborative work of scientists, actors from knowledge institutions (e.g., SMEs, industries) and public sectors (e.g., municipal administrations, policy-makers), and local experts. This collaboration can be realized by employing transdisciplinary approaches, which establish a common understanding and translation of expert knowledge (Antoine-Moussiaux et al., 2019). A transdisciplinary approach can be defined as "a critical and self-reflexive research approach that relates societal with scientific problems . . . [and] produces new knowledge by integrating different scientific and extra-scientific insights" (Jahn et al., 2012), and they are increasingly recognized for their potential to tackle complex real-life issues by integrating different kinds of knowledge (Haire-Joshu and McBride, 2013; Schäpke et al., 2018).

Unlike a pandemic or large epidemic, emergence always takes place on a small, local scale, which calls for the facilitation of bottom-up effects and the subsequent co-accommodation of grassroots and institutional settings. When establishing task forces to put science into action, initiators have to consider implementation strategies on various scales (local, regional, global) as well as policy environments (human, livestock, and crop health security).

Implementation Strategies on Different Scales

Global

Current global frameworks are all based on managing existing diseases and increasing palliation and preparedness for those that are newly emerging (FAO, OiE, WHO, 2006; FAO and WHO, 2020). Since they are all based on the assumption that EIDs are rare and unpredictable, plans to prevent outbreaks are slim to none. Nevertheless, most of the global frameworks in use name prevention of disease as their main aim, which refers to containing diseases at the level of outbreak, halting large-scale transmission and thereby avoiding the growth of an outbreak into epidemics. Although we can understand restricting pathogens from spreading beyond small, local outbreaks prevents epidemics, we argue that prevention should be used in the context of avoiding emergence in the first place. This shift in epistemics is also strongly supported by the grave predictions regarding the speed with which smaller outbreaks can spread in an increasingly globalized world (Khasnis and Nettleman, 2005; Findlater and Bogoch, 2018; Feronato et al., 2021), narrowing the time window available for containment measures.

On the one hand, global health security has to adopt a novel evolutionary paradigm to adjust risks and predictions regarding EIDs. On the other hand, the epistemology and definition of prevention needs to be unified across all global guidelines to focus efforts on both containing and preventing diseases in an evolutionary context. Therefore, current measures have to be evaluated to determine their applicability and limitations, and prevention has to be contextualized within global health security.

The Prevent-Prepare-Palliate (3P) framework offers a comprehensive, systemic characterization of existing health security initiatives and describes how prevention can be adopted into current infrastructures (Molnár et al., 2022). Implementing prevention into global health-care frameworks will help identify gaps that allow EIDs to emerge at an accelerating rate and would provide guidelines for health-care infrastructures to intervene on a regional level.

Regional

Managing diseases at a regional level faces the challenge of having to act in various policy and cultural environments. Ranging from upper regional levels such as international alliances (e.g., European Union) that operate within large-scale legal environments (such as EU regulations), through mid-regional levels concerning one or a few neighboring countries, to lower regional levels involving small municipalities that manage local communities, regional scales are the most diverse in terms of expertise, jurisdiction, and policies. Nevertheless, epidemics of national concern are dealt with on regional levels, involving municipalities directly affected as well as national health-care infrastructures and public health institutions (Lakoff, 2017). Therefore, implementing the DAMA protocol on a regional scale requires carefully selected methods that facilitate intersectoral collaboration and define outcomes to accommodate local policy environments.

Living Labs

From the toolkits of transdisciplinary methods, Living Labs (LLs) provide an opportunity to establish solid, well-thought-out task forces that bring together the skills and capacities required of different actors for addressing a particular issue (Romero Herrera, 2017). LLs can be defined as both "an arena (i.e., geographically or institutionally bounded spaces) and . . . an approach for intentional collaborative experimentation of researchers, citizens, companies and local governments" (Voytenko et al., 2016). It makes them suitable for dealing with health-care issues, as they are designed to foster co-innovation in terms of development and testing of new technologies, products, services, policy instruments, governance arrangements, ways of living (Voytenko et al., 2016). In doing so, the approach strengthens intersectoral collaboration and communication as well as increases the feasibility of intervention plans by fitting them to local policy environments and interests of affected stakeholders (Kim et al., 2020). If designed and implemented well, the LL approach can also help avoid stumbling blocks (e.g., disciplinary boundaries and silos between science, practice, and society; low feasibility in diverse policy environments; low levels of adaptability to local cultural, socio-economic, institutional, and environmental settings; decreasing trust in policy and politics; etc.) by involving diverse experts on legal limitations, local settings, and market conditions, and foster knowledge exchange and integration and widen professional networks. This potential makes LLs a "proliferating approach to working in a transdisciplinary fashion" (Schäpke et al., 2018)

Containing outbreaks or epidemics requires a collaboration between science and society, across sectors and between organizations, and prevention is no different. Current solutions are mostly characterized by hasty and temporary collaborations formed under the pressure of a health emergency. LLs are potentially a very impactful approach for dealing with EID crises. They have been proliferating in Europe since 2006, when the European Network of Living Labs (ENoLL) was founded as a platform for best practice exchange, and have since been successfully adopted in domains such as food bioeconomy, agriculture, environment, urban and rural development (Mirijamdotter et al., 2006; Voytenko et al., 2016; Menny et al., 2018). However, up to the current time, LLs have hardly ever been applied to the area of EIDs. Apart from the benefits of the LL approach just discussed, LLs can also enable and foster discussion between authorities, scientists, and the public, thereby addressing the dire consequences of public distrust in science and science-based policies, such as that revealed by the COVID-19 pandemic (Kreps and Kriner, 2020; Plohl and Musil, 2021).

With awareness of the various actors affected by infectious disease outbreaks, participants are selected based on their expertise and involvement in the context of EIDs, making them highly adaptable and specific to the issue investigated. Selection must also consider the highest-level decision makers needed for efficient intervention (municipality governance and policy makers, national government officials, regional public health authorities, etc.). LL members generally represent four larger sectors (Figure 10.1):

1. **Public actors** – Policy- and decision-makers, legal experts, and government officials; expertise in the legal environment and regulatory role in the long-term management of the outcome. Typical actors for disease management are public health, municipal governments, or food safety control authorities.

2. **Private actors** – Private institutions, organizations, and companies affected by the emergence; insights into practical and industrial implementability of intervention plans. Managing disease will be of interest to agricultural organizations and farmers' associations, livestock breeders, and food production companies, travel agencies, or pharmaceutical companies.

3. **Knowledge institutions** – Scientific expertise on the emerging pathogen generates predictions related to transmission, epidemic and pandemic potential, and risk assessment. Partners to consider in relation to EID are university research groups, independent research institutions, and scientific organizations (e.g., Chatham House, Milken Institute).

4. **Local citizens** – It is crucial to include members of the community directly affected by potential emergence. In addition to increasing feasibility of the intervention plans among local conditions, involvement raises awareness of health-care threats and provides the community with a sense of ownership over the situation. An emphasis must be placed on reaching out to local citizen science programs that have extensive experience in not only local settings but research processes.

LLs are fit to address the issue of feasibility and temporary collaborations through an inclusive, planned process in which solutions are planned in a precautionary manner. Participants representing diverse stakeholders and sectors jointly create an intervention plan, which aims to accommodate interests of all sides and respects limitations. Given the highly diverse legal environment on various regional levels, LLs consist of a select group of stakeholders with expertise relevant to the scientific, legal, geographical, economical, and social conditions of a well-defined pathogen system. This creates a highly flexible and adaptable tool that bases its operating on specific guidelines but is always adapted to the local environment.

LLs that address disease prevention on a regional level should furthermore always be in close collaboration with citizen science and community programs that are engaging members of the exposed, susceptible population. The following section describes the tools necessary for dealing with EIDs on a local scale.

Local

In case of a novel emergence, the initial absence of available information leaves an extremely narrow time window for reactive action. Additionally, because of the cultural and socioeconomic diversity of directly affected communities, disease management often suffers from low feasibility and inefficient implementation (Gebreyes et al., 2014; Benelli and Beier, 2017; Chen et al., 2021). Similar as described for the regional scale, the DAMA protocol facilitates bottom-up processes and the involvement of local actors in both data gathering and generating solutions. Building working relationships with members and local experts of exposed communities creates mutual benefits by increasing the efficiency of implementing disease control measures and building trust and collaboration between authorities, science, and society (Ajibaye et al., 2019).

Citizen/community science

Citizen or community science (CS) initiatives are founded to include locals as active participants in research projects that target their direct environment. With insights provided by patients, farmers, students, hunters, and hikers, CS has been invaluable in tracking insect and tick vectors (Palmer et al., 2017; Földvári et al., 2022), biodiversity in city parks (Marizzi et al., 2018), avian influenza in urban environments (Francisco et al., 2021), and wildlife health (Lawson et al., 2015).

Similar to preparation, disease prevention also relies heavily on completing scientific data with traditional insights and observations regarding reservoirs. CS programs provide an opportunity for outreach and contact with communities most exposed to emergence (e.g. urban minorities, underserved communities, students, rural farmers, etc.) (Den Broeder et al., 2018). Emphasis is placed on establishing communication channels and training programs

between susceptible communities, public health and scientific research, creating long-term science-practice collaborations which will serve as a foundation for continued monitoring and early warning systems (Alvarez-Hernandez et al., 2020). When mapping the distribution of a reservoir and/or vector of a suspected pathogenic microbe, local expertise and traditional insights on reservoir behavior and distribution will be fundamental for conducting efficient monitoring. A further benefit of such initiatives is building the science-society trust bridge, which faces a difficult test during a public health emergency. Finally, including locals in the process, including in LLs, grants ownership over outcomes and intervention plans to those whom policies act upon.

When targeting a community, it is just as important to make sure the members involved are relevant to the research and policy issue at hand as it is to contact groups that are likely to positively respond to the particular research collaboration. Local organizations not only have in-depth knowledge about their community members, they already have the relevant network and infrastructure for reaching out and advertising opportunities. Although each program should consider the local setting and structure of the community, good examples of groups to reach out to are (Figure 10.2):

- **Community-based organizations** – These organizations are founded and run by members of the community advocating for particular issues and rights and therefore collect proactive groups of locals that could be approached with a CS initiative. Typical examples are environmental protection groups, neighborhood associations, or volunteering clubs.

- **Educational institutions** – Gathering young members of the community who are currently or have been exposed to scientific knowledge makes educational institutions a prime target for CS programs. Often, cohorts of volunteers remain active in a research program even after leaving the institution. High schools, General Educational Development (GED) programs, or colleges are only a few examples.

When designing a CS program, we aim to establish a long-term, mutually beneficial relationship with key members of the community. This requires initial investment into recruiting data samplers and also running parallel with feedback and networking activities. The long-term goal of successful CS programs is not necessarily constant data

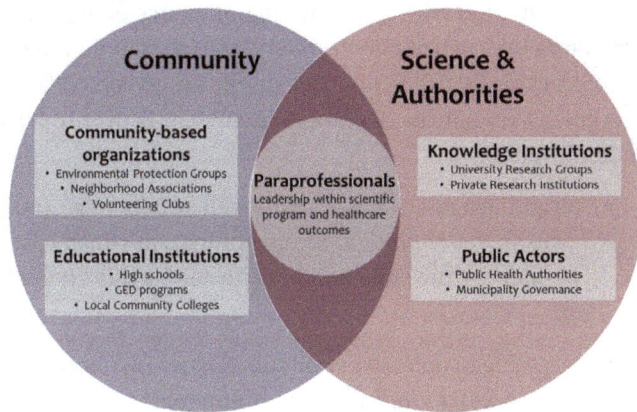

Figure 10.2. The stakeholders that participate in community science initiatives.

influx but building a dynamic network of locally trained experts and researchers who will remain in touch and can be mobilized or expanded. This permanent working relationship is further fostered by selecting reliable and committed participants for leadership positions within the CS program.

Paraprofessional networks

Although initiatives targeting health disasters have been widely introduced, including the integrated One Health surveillance, a common issue is barriers in the way of societal implementation, such as lack of efficient communication, fluctuating compliance and engagement, and territorial fragmentation (Uchtmann et al., 2015). Disease prevention relies on long-term monitoring of both pathogen and reservoir populations; it is therefore crucial to have a permanent program in place that engages citizens, collects data and knowledge, and feeds information back to the community.

After establishing a CS program, opportunities must be provided for consistently involved, engaged members to immerse within the project and gain agency and ownership over the issue at hand. Selecting candidates for leadership positions creates a network of so-called paraprofessionals who are then able to head particular tasks, assist with training, lead recruitment, and participate in feedback. A regular income assigned to such positions also improves livelihoods as well as trust and cooperation with authorities. Paraprofessionals have been indispensable in addressing livestock (Ilukor et al., 2015) and human diseases (Vollmer and Valadez, 1999), and should therefore be cornerstones of EID prevention efforts.

The strategies listed here provide guidance for implementing prevention measures at different scales of society. Nevertheless, efficient prevention requires the close collaboration of not only partners within a certain program and scale but also partners across diverse scales. Information and expertise coming from grassroots science must be used to design and implement intervention plans on regional levels, which will feed into global frameworks collecting exemplary cases and efficient methods that can be applied to other reservoir-pathogen systems in diverse policy environments.

Implementation Strategies in Different Policy Environments

When reviewing various EIDs, outbreaks and epidemics are often dealt with by different legal and economic frameworks depending not only on the scale they manifest on but also on the newly infected host. Pathogens that emerge in human communities, for instance, will come under the jurisdiction of public health and health-care institutions, a cumulativeness that has already resulted in the false interpretation that the EIDs are exclusively human diseases (Morens et al., 2004). Pathogens that damage livestock and crops are therefore seldom connected to those that create illness in humans, despite the anthropogenic drivers of their emergence (globalization, climate change, international travel and shipping, human intrusion, etc.) and the socioeconomic impacts being the same in both severity and magnitude (Burns et al., 2008; Gallardo et al., 2015; Cazzolla Gatti et al., 2021; Brooks et al., 2022). Policy silos can further be observed in livestock diseases being addressed by food safety regulations and production management, while crop diseases fall under the concern of agricultural policies. Nevertheless, the drivers of disease emergence the same, and pathogens colonize livestock and crops can increase the probability of emerging human diseases (Xia et al., 2021).

The One Health initiative considers all novel diseases to be a direct threat to human well-being and has been working to implement the One Health approach to medical, veterinary, and wildlife disease management, urging for large-scale, merged databases; expanding research focus to wild populations and reservoirs; and preparing for future emergences (Gebreyes et al., 2014; Kelly et al., 2017; Chatterjee et al., 2021). In line with these efforts, the DAMA protocol calls for preventive intervention against all pathogens with a potential to emerge in human, livestock, or crop populations. At the same time, this also means that prevention has to be planned and executed in three different policy environments. In the following sections, we present the main focus points, target stakeholders, and typical stumbling blocks of establishing LLs and CS programs in different policy infrastructures.

Human pathogens

Human diseases come under the deepest scrutiny and attract the most attention from authorities and the public alike. Nevertheless, there is major divergence between countries and regions in terms of health-care infrastructure, pathogen diversity, and sources of potential emergence. While temperate-zone regions are more exposed to air travel–related infections (Findlater and Bogoch, 2018), tropical and Mediterranean areas have a higher potential for wildlife-originated emergences (Wang et al., 2021). These patterns are then further complicated by climate change driving both species and human migration, providing opportunity for diseases to expand their geographic, vector, and host range.

Chikungunya is an arboviral infection spread by the yellow fever mosquito (*Aedes aegypti*) and was therefore referred to as a "tropical fever" because its distribution area was limited to that of its vector. However, 2010 saw the virus establish itself in the tiger mosquito (*Aedes albopictus*) and produce autochthonous cases in southern Europe (Franke et al., 2019), where it has since developed self-sustaining populations (Weaver and Lecuit, 2015). Also an arbovirus moving from the *Ae. aegypti* to the *Ae. albopictus*, the ZIKA virus caused its first local cases in Europe in 2019 (Brady and Hay, 2019) and is likely to threaten more than a billion people with its recent range expansion (Ryan et al., 2021). Finally, with the recent outbreaks of hepatitis of unknown etiology (WHO, 2022a) and the ongoing monkeypox outbreaks (WHO, 2022b), it is clear that preparation for the barrage of human EIDs is unsustainable. The focus needs to shift toward prevention by launching multi-actor task forces to handle emergence within local and regional settings.

Living Labs preventing human pathogens

Preventing EIDs that directly threaten human health will focus on the interfaces between human communities and identified reservoir populations from which pathogens are expected to switch over to their new, susceptible host. Exposure will often be increased by living and working in close contact with wildlife (e.g., rural farming and hunting communities) and/or having limited access to health-care services coupled with improper maintenance of hygienic standards (e.g., urban poverty, marginalized communities) (Brooks et al., 2019). Main focus points are to control and minimize the chances of pathogens switching over to humans, by either

- targeting a specific host-pathogen system (e.g., Zika virus in *Aedes albopictus* mosquitos), in which case we identify the stakeholders affected by this system, or

- targeting people whose circumstances (living conditions, occupation, habits, etc.) supposedly place them at higher exposure (e.g., having a job without a remote-work option during COVID-19), in which case we identify stakeholders connected to our target community.

From the groups outlined for LLs in general, the following actors should be considered relevant to preventing high risk human EIDs.

- **Public actors** – Government authorities who address public health–related matters, such as health services and public health authorities, national laboratories and epidemiological surveillance facilities, district health systems, and district public health authorities and food safety institutions.

- **Private actors** – Private companies and organizations affected by potential emergence or connected to endangered communities. Relevant examples include pharmaceutical companies producing treatment or vaccines against the potential threat, travel agencies mapping international routes of concern, or software development and data management companies offering digital tools for tracking and monitoring human-pathogen interfaces.

- **Knowledge institutions** – Scientific institutions focusing on human pathogens, such as epidemiology research laboratories, veterinary research groups working on reservoirs or vectors of the pathogen in question, and medical research institutions focusing on human diseases.

- **Local citizens** – Community members should be involved in multi-actor task forces to represent local interest and expertise. Priority should be given to those already participating in CS programs or civil organizations, with an emphasis on engaging students and early-career young adults.

LLs handling human diseases will rely heavily on personal data regarding local workforce, financial status, access to health services, medical history, and connectivity. It is therefore crucial to secure data protection and privacy. Furthermore, resources should be dedicated to communicating the process to the local community through paraprofessionals involved in the LL to establish solid working relationships and trust between the task force and individuals operating at local levels. Relying on a collaborative foundation will not only facilitate implementation but also foster long-term engagement for maintaining preventive monitoring and screening.

Community science programs preventing human pathogens

The focal point of disease prevention measures is engaging and working with communities directly exposed to the emergence of a novel pathogen. As described for LLs, the target population for CS programs is identified either through (1) their contact with a particular reservoir or (2) their circumstances making them susceptible to emergence of potential pathogens. Main focus points are to engage members of a community and initiate bidirectional communication channels between locals and researchers. On the one hand, prevention research relies heavily on knowledge of local habits and lifestyle, traditions, knowledge of reservoir behavior, and the interface between potential pathogens and community members. On the other hand, researchers can raise awareness of the lurking health-care threat, establish educational and training programs, involve locals in the project, and solidify collaboration by assigning leadership positions to paraprofessionals.

When planning recruitment within the community, factors to be considered include setting (e.g., rural vs. urban), occupation (local trade unions, commonalities between employment types, working conditions), socioeconomic status (access to health care, level of education, household income), and cultural background (ethnicity, language, cultural habits and traditions, religion, etc.). Recruiting and training programs should be designed to be accessible and comprehensible for the target population and should clearly explain to recruits why they have been selected as participants.

Attention should be given to providing regular and thorough feedback on the process to all members of the community. Activities should be planned to ensure bidirectional flow of information: benefiting from community engagement should always be coupled with feedback sessions planned around delivering preliminary results, reflecting on the experience of involved members (both academic and community), and discussing potential impacts. This bidirectional discussion builds the trust and engagement required to establish long-term programs and networks, and builds a reliable network of nonscientific, local experts. Feedback should be constant during the actual sampling and collecting to both give back to the community and collect reflections and observations that can improve methods and communication strategies. A public-facing website that is accessible for all stakeholders at all times is ideal, but regular newsletters or social media are also popular ways to communicate effectively.

Contrary to LLs, CS programs are widely used to target infectious disease threats by monitoring bacterial pathogens polluting water bodies (Agate et al., 2016), preventing Lyme disease (Seifert et al., 2016), or monitoring viruses in urban environments (Marizzi et al., 2018). Methods and practices developed in previous programs should be implemented into newly established initiatives that focus on prevention.

Livestock pathogens

Diseases emerging in livestock have been just as impactful as those affecting humans directly. The past decades have seen an increase in both frequency and magnitude, with pandemics plaguing livestock across regions and continents (Tomley and Shirley, 2009; Bett et al., 2017). However, studies often focus on zoonoses rather than diseases that affect livestock directly, creating a lack of available information on pathogens of domesticated species (Rajala et al., 2021). This bias in research is fueled by preferential funding for zoonotic diseases, which also manifests in lack of veterinary health-care infrastructure, low efficiency, or high-priced medications, lapses in vaccination programs, and knowledge discrepancies in breeders regarding diseases (Brooks-Pollock et al., 2015; Ashfaq et al., 2020). Adding to the effects of this asymmetry in knowledge and research is the management of diseases of domestic animals, which primarily aims to eliminate infected individuals from breeding stock, with extremely limited efforts dedicated to treatment development (te Beest et al., 2011; Nyerere et al., 2020). Although there have been suggestions to introduce preemptive hunting strategies to avoid livestock being contaminated from wild populations (Mysterud et al., 2020), prevention still has a lot of ground to cover regarding livestock disease. This is further certified by the major economic effects livestock pandemics have, which add to the costs and damages caused by human EIDs. Foot and mouth disease of cattle resulted in up to 88% market value losses, affecting all actors along the cattle marketing chain in Uganda (Baluka, 2016), while leading to the culling of 3.4 million animals during the UK epidemic (Blake et al., 2003). African swine fever has led to major economic losses in Southeast Asia and has triggered policy modifications linked to the emergence of SARS-CoV-2 (Gallardo et al., 2015; Xia et al., 2021). Avian influenza has led not only to dire losses in poultry production (Burns et al., 2008) but also to pandemic potential in humans (Watanabe et al., 2014). Therefore, in line with the One Health approach that calls for integrated investigation of livestock, wildlife, and human systems (Elmberg et al., 2017; Mohamed, 2020), the DAMA protocol calls for precautionary and preventive policies addressing livestock diseases.

Living Labs preventing livestock pathogens
Livestock disease will be of concern to a different set of stakeholders than those for human pathogens, although a considerable overlap is to be expected. The drivers behind any preventive or management intervention are mostly to maintain production and the livelihood of breeders and production plant operators. Exposure will increase in free-range breeding stocks and those housed partly in external enclosures, while outbreaks will be more likely to occur among high-density stocks (Meadows et al., 2018). Main focus points are to control and minimize the chances of pathogens switching over to livestock, by either

- targeting a specific host-pathogen system (e.g., ASF in wild boar populations), in which case we identify the stakeholders affected by this system, or

- targeting breeding facilities and game populations whose circumstances (housing conditions/distribution area, species, immediate surroundings, etc.) supposedly place them at higher exposure (e.g., frequent encounters with [other] wildlife, limited access to veterinary/wildlife services, lack of knowledge regarding livestock/wildlife diseases, etc.), in which case we identify stakeholders connected to target facilities.

From the groups outlined for LLs in general, the following actors should be considered relevant to preventing high-risk livestock EIDs.

- **Public actors** – Government agencies involved in food safety, including national-level institutions (e.g., Food Safety and Inspection Service [FSIS; US], Federal Institute of Risk Assessment [BIR; Germany], Austrian Agency for Health and Food Safety [AGES; Austria],

and fish and wildlife departments) as well as municipal-level departments of public health, agriculture, hunting, and food safety.

- **Private actors** – Private companies and enterprises whose main activity is related to the livestock and/or game exposed to emergence. A few examples include farms, processing plants, hunting associations, and the suppliers and veterinary institutions that provide vaccinations and medication. In case they are active in the area of potential emergence, companies offering digital tracking services that record movement, development, and other data on individual animals will also have valuable expertise in identifying interfaces and location of possible intervention to reduce encounters between livestock and reservoirs.

- **Knowledge institutions** – Research groups that target the livestock and game pathogen under investigation as well as those that conduct research in livestock and wildlife vaccination, treatment, methods to increase production, and reduce environmental effects on stock yield. Veterinary science is also a key stakeholder, contributing to knowledge on transmission, morbidity, and mortality and to potential direction of treatment and vaccine development.

- **Local citizens** – Required expertise will be found among individual farmers and workers at breeding and processing facilities as well as hunters, who not only hold valuable insights regarding animal behavior and diseases but are also directly exposed to any emerging pathogen because of their constant close contact with the breeding stock and wildlife.

LLs handling livestock and game diseases must always consider that, contrary to those dealing with human pathogens, they will have dual priorities of preventing emergence and maintaining or even increasing production. As the livelihoods of most stakeholders are closely connected to yield of breeding stocks and game populations, tools such as culling or restricting stock size, increasing hunting bag size, or applying targeted hunting should be used with extraordinary caution to establish long-term feasibility of prevention methods.

Community Science programs preventing livestock pathogens
CS programs that target livestock diseases are far less common than those that address human pathogens

because the community affected by them is much smaller and consists almost exclusively of citizens working in livestock breeding or processing and hunters. Whether a CS initiative is designed to target (1) a particular pathogen and the livestock or game exposed to it or (2) breeding stocks and game populations whose circumstances make them susceptible to emergence of potential pathogens, the initiative will be of interest to a narrower community of local experts.

The main focus points are to engage breeding experts and individual hunters who work in close contact with animals and are aware of the day-to-day issues and conditions of a breeding or processing facility or a particular hunting area. Livestock experts will be able to identify interfaces between the stock and wildlife accurately, while hunters will be familiar with movement and behavioral patterns of game and potential reservoirs as well as population sizes and demography.

When planning recruitment strategies, a close collaboration is required with the management of the breeding facility (or facilities) for efficient study design and institutional encouragement to participate. It is also necessary to align interests of larger breeding enterprises and small-scale, local farmers to ensure the homogeneity of data collected. An additional opportunity lies in designing studies for the general audiences, targeting those that are active outdoors and therefore have occasional encounters with wildlife. Recruiting and training programs should be designed to be accessible and comprehensible for the target population and should clearly explain to recruits why they have been selected as participants.

To avoid unnecessary investment, planning must always consider existing data collected by breeders and hunters, as both institutions collect particular types of data continually and permanently. This data is available either from government institutions that oversee wildlife management or private breeders who keep their own records, both subject to restricted access. Similarly, feedback sessions and reports have to be targeted to both citizen participants and the institutional board overseeing the stock in question, which can alter the format of the feedback.

CS programs have been introduced into research focusing on wildlife health surveillance (Lawson et al., 2015) as well as monitoring invasive vector species (Földvári et al., 2022). Studies have also used CS methods to target diseases plaguing wildlife and livestock simultaneously (Perrin, 2017) and to identify shortcomings of policies addressing foot and mouth disease (Kim, 2011).

Crop pathogens

Crop pathogens are commonly the most neglected EIDs since they pose no immediate health risk to humans and therefore mostly manifest in indirect effects caused by decreased production. Crop pathogens have typically been addressed by palliative efforts that eliminate them from the cultivated plant stock (Schulthess, 1761; Ayesha et al., 2021) or by the later application of defense priming against known crop diseases (Conrath et al., 2015). Macroscopic pests of crops have a longer history of defense strategies, as microscopic pathogens have been discovered to coincide with plant diseases only in the late 19th century and named as a cause decades later (Russell, 2006). Initial research focus gradually shifted from epidemiology toward control and founded commercial disease control with a wide range of bactericide, fungicide, and virucide treatments as well as extensive gene-modification research to breed resistant crops lineages (Russell, 2006). Although without such protection measures, losses in crop production could increase five-fold in Europe (Oerke et al., 2012), it has now become clear that global demand as well as changing climate and globalized trade have subjected crops to EIDs unmanageable by current measures. In addition to the coconut lethal yellowing disease and wheat stem rust described earlier, tomatoes are plagued by rapidly spreading, diverse viral diseases (Hanssen et al., 2010), grapevine downey mildew has spread from Europe and now threatens vineyards worldwide (Fontaine et al., 2021), and the *Fusarium incarnatum-equiseti* species complex had invaded leafy vegetable crops in novel European areas (Matić et al., 2020). Although still treated as an agricultural and production issue, plant diseases now have more studies connecting them to the larger context of EIDs (Vurro et al., 2010; Fones et al., 2020; Yadav et al., 2020). In line with this, evidence shows that plant pathogens follow similar evolutionary trajectories to those described in the SP; for instance phytoplasmas use common receptors distributed across several insects that serve as vectors to infect plants (Galetto et al., 2011; Trivellone et al., 2019).

Although the overlap between plant and human pathogens is presumably negligible, the effect of emerging plant pathogens on global food security is devastating, which justifies their inclusion within the preventive measures of the DAMA protocol.

Living Labs preventing crop pathogens

Crop pathogens will be of interest to stakeholders quite different from those described for human and livestock diseases, with smaller overlaps. However, some similarities will exist between motivation for preventing crop and livestock EIDs, namely the drive to maintain production and yield of crops. Also, exposure will increase in those planted in the vicinity of wild areas, with large-scale monocultural fields being at elevated risk for epidemics and outbreaks. Additionally, growing similar species in close spatial or temporal proximity may further increase the chances of transferring pathogens from one to the other (Bakker et al., 2016). Main focus points are to control and minimize the chances of pathogens switching over to crop plants, by either

- targeting a specific host-pathogen system (e.g., phytoplasma in their vector insects), in which case we identify the stakeholders affected by this system, or

- targeting areas or particular crops whose circumstances (distribution area, species, immediate surroundings, etc.) supposedly place them at higher exposure (e.g., large areas bordering natural habitats, limited access to agricultural and control services, lack of knowledge regarding crop diseases, etc.), in which case we identify stakeholders connected to our target areas or species.

From the groups outlined for LLs in general, the following actors should be considered relevant to preventing high-risk crop EIDs.

- **Public actors** – Government ministries involved in agricultural services, including national-level institutions (e.g., National Institute of Food and Agriculture [NIFA; US], Federal Ministry of Food and Agriculture [Germany], Federal Ministry of Agriculture [Austria] as well as municipal-level departments of public health, agriculture, and food safety.

- **Private actors** – Private companies and enterprises whose main activity is related to the crop and/or area exposed to emergence. A few examples include farms, plantations, suppliers, and agricultural institutions that provide protection methods. In case they are active in the area of potential emergence, companies offering digital mapping services that record distribution, density, species composition, and other data in high resolution will also have valuable expertise in identifying interfaces and location of possible intervention.

- **Knowledge institutions** – Research groups involved in agri-food sciences relating to the emergent threat, working on control measures such as resistant

lineages, pesticides, defense-priming techniques, and ways of increasing production as well as those studying the distribution and genetic mapping of the pathogen in question.

- **Local citizens** – Required expertise will be found among individual farmers and those working with investigated crops or in relevant areas, along with the general public visiting natural areas in the vicinity of the cultivated plants. Both will have direct insights into the manifestation and the distribution of the disease and will be able to point out significant interfaces between crops and wild reservoirs or hosts.

A further similarity to LLs handling livestock and game diseases, those addressing crop diseases must also aim to prevent emergence and maintain or increase production at the same time. Additionally, since current control measures hold off substantial losses in production, prevention measures must accommodate ongoing treatment protocols. Finally, different regions will often have very different infrastructure on cultivated areas, which will have a significant influence on the potential prevention plans and their feasibility.

Although particular plant pathogens have been addressed by multi-actor approaches that target pathogens such as cassava viruses (About CVAP, n.d.), the LL approach is still to be utilized to its full potential in preventing and controlling emerging crop diseases.

Community Science programs preventing crop pathogens

Unlike in livestock diseases, CS programs tend to have a more thorough representation in crop disease studies. This difference is mainly due to the economic drivers of controlling crop pests as well as the larger community of farmers and general public that is able to contribute. Whether a CS initiative is designed to target (1) a particular pathogen and the cultivated species exposed to it or (2) crops whose circumstances make them susceptible to emergence of potential pathogens, the initiative will be of interest to a wider audience than in the case of livestock diseases.

The main focus points are to engage farmers, cultivation experts, and individuals who live in or frequent endangered areas, all of which will possess the knowledge on crop and reservoir species as well as specifics on the area of cultivation. Training programs should primarily focus on developing skills to identify particular wild plant species and recognize signs of infection, which will also be

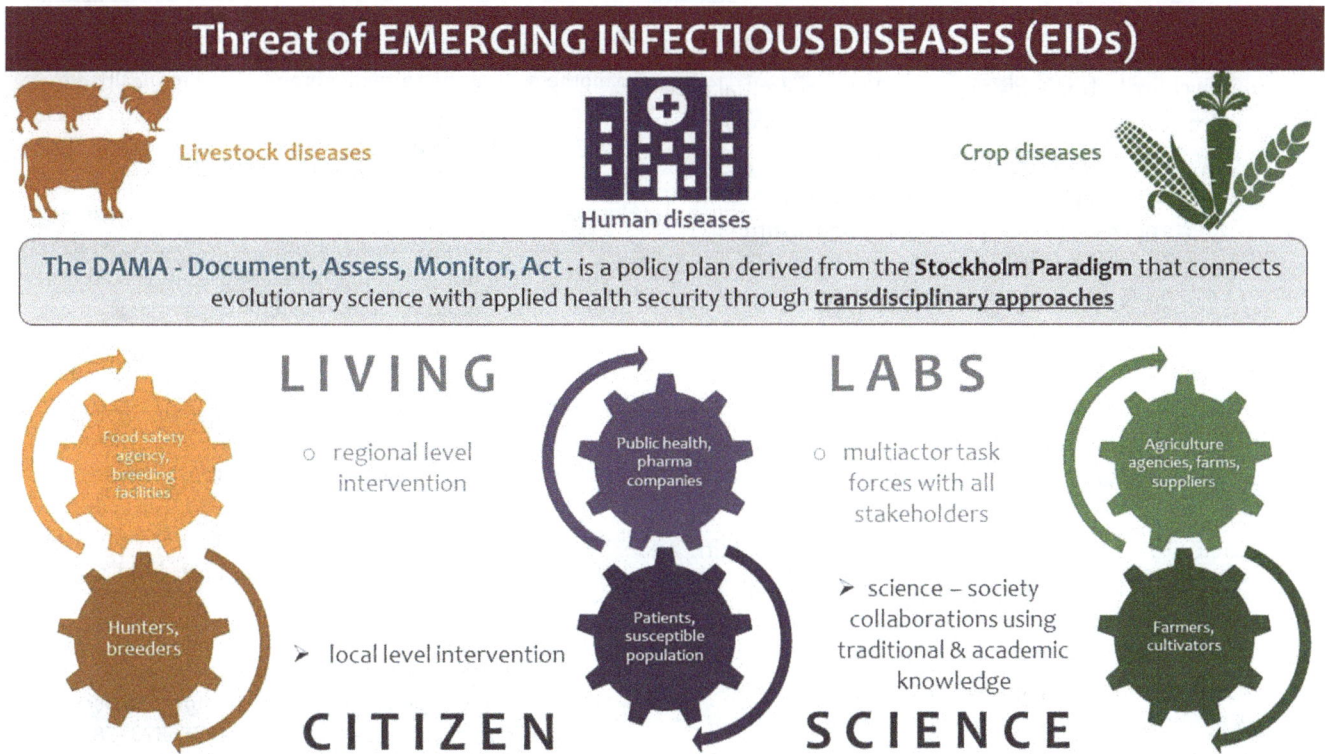

Figure 10.3. Visual representation of the interdependent processes between local and regional initiatives that address emerging infectious disease prevention.

useful in tracking invasive species in the future. Recruiting and training programs should be designed to be accessible and comprehensible for the target population, with an additional educational role in raising awareness about food security issues and conscientious consumer behavior. Depending on the setting of the study, a long-term return can be encouraging participants to grow produce at home, thereby increasing green areas and increasing self-sustaining households.

The benefit of CS programs in crop health has been established regarding potato diseases (Lidwell-Durnin, 2020) and identifying main threats of maize and soybean in the Amazon region (Hampf et al., 2021). Additionally, data collected by a relatively small number of expert citizens has been demonstrated to be highly accurate (Steinke et al., 2017), which makes CS programs very promising for implementing the DAMA protocol.

Conclusions

The EID crisis represents one of the largest threats to our modern lifestyle in history, endangering human health,

food security, and economic and societal systems. Isolated institutions dealing with various manifestations of EIDs have thus far been unsuccessful in stopping the wave of newly emergent pathogens. The SP provides a comprehensive evolutionary framework, which replaces current, false characterization of EIDs with clear predictions. The DAMA protocol provides a general step-by-step plan for constructing preventive interventions that target emergent pathogens before the onset of an outbreak. This paper focuses on the final step of implementing evolutionary theory into preventive policies that consider scales and policy environments.

Global, regional, and local scales require precise conceptualization and the introduction of adequate transdisciplinary methods to gather all relevant knowledge and expertise, and create feasible, cost-efficient intervention plans. Global integration of the DAMA protocol into existing frameworks is crucial to provide useful guidelines to regional and national institutions; this is described in the Prevent-Prepare-Palliate (3P) framework. Regional scales addressing EID threats are to introduce the widely tested approach of Living Labs, which can be seen as multi-actor platforms delivering solutions co-created by various stakeholders. Their application to infectious disease threats will be a unique contribution which has significant potential of dealing with diverging interests. Finally, local scales would benefit from a wide-range of community science initiatives that target affected populations directly and the assistance of local experts on various host-pathogen systems. Although each method is most suitable for its particular scale, it is crucial that all of them operate in close collaboration with each other, circulating knowledge from the grassroots toward institutions. The key to disease prevention is ongoing monitoring that engages local experts and citizens as well as relevant decision makers in bidirectional communication

Another important step toward more effectively controlling the EID crisis is elimination of the barriers among human health care, wildlife health care, livestock health care, and crop health care. The current lack of a unifying scientific understanding of health issues results in divergent policies providing palliative and perhaps preparatory solutions, none of which is efficient or sustainable in the face of accelerating EIDs. By understanding the common underlying evolutionary drivers, predictions can be adjusted appropriately across the board for human, livestock, and crop diseases, and prevention can be implemented in existing infrastructures and legal environments (Figure 10.3).

Our advancements in technology have brought with them novel threats in the shape of EIDs. Climate change and globalization have changed the evolutionary trajectory of diseases as we know them; it is therefore inevitable to change our approach to global health security and shift our focus from reactive approaches to those moving up the infection timeline toward prevention.

Literature Cited

About CVAP. n.d. Cassava Virus Action Project [website]. Accessed 29 September 2022. https://cassavavirusactionproject.com/about/

Agate, L.; Beam, D.; Bucci, C.; Dukashin, Y.; Jo'Beh, R; O'Brien, K.; Jude, B.A. 2016. The search for violacein-producing microbes to combat *Batrachochytrium dendrobatidis*: a collaborative research project between secondary school and college research students. Journal of Microbiology and Biology Education 17: 70–73. https://doi.org/10.1128/jmbe.v17i1.1002

Agosta, S.J.; Janz, N.; Brooks, D.R. 2010. How specialists can be generalists: resolving the "parasite paradox" and implications for emerging infectious disease. Zoologia (Curitiba) 27: 151–162. https://doi.org/10.1590/S1984-46702010000200001

Ajibaye, O.; Balogun, E.O.; Olukosi, Y.A.; Orok, B.A.; Oyebola, K.M.; Iwalokun, B.A.; et al. 2019. Impact of training of mothers, drug shop attendants and voluntary health workers on effective diagnosis and treatment of malaria in Lagos, Nigeria. Tropical Parasitology 9: 36–44. https://doi.org/10.4103/tp.TP_36_18

Alvarez-Hernandez, G.; Drexler, N.; Paddock, C.D.; Licona-Enriquez, J.D.; Delgado-de la Mora, J.; Straily, A., et al. 2020. Community-based prevention of epidemic Rocky Mountain spotted fever among minority populations in Sonora, Mexico, using a One Health approach. Transactions of the Royal Society of Tropical Medicine and Hygiene 114: 293–300. https://doi.org/10.1093/trstmh/trz114

Antoine-Moussiaux, N.; Janssens de Bisthoven, L.; Leyens, S.; Assmuth, T.; Keune, H.; Jakob, Z.; et al. 2019. The good, the bad and the ugly: framing debates on nature in a One Health community. Sustainability Science 14: 1729–1738. https://doi.org/10.1007/s11625-019-00674-z

Ashfaq, M.; Kousar, R.; Makhdum, M.S.A.; Naqvi, S.A.A.; Razzaq, A. 2020. Farmers' perception and awareness regarding constraints and strategies to control livestock diseases. Pakistan Journal of Agricultural Sciences 57: 573–583. https://doi.org/10.21162/PAKJAS/20.8346

Ayesha, M.S.; Suryanarayanan, T.S.; Nataraja, K.N.; Prasad, S.R.; Shaanker, R.U. 2021. Seed treatment with systemic fungicides: time for review. Frontiers in Plant Science 12: 654512. https://doi.org/10.3389/fpls.2021.654512

Bakker, M.G.; Acharya, J.; Moorman, T.B.; Robertson, A.E.; Kaspar, T.C. 2016. The potential for cereal rye cover crops to host corn seedling pathogens. Phytopathology 106: 591–601. https://doi.org/10.1094/PHYTO-09-15-0214-R

Baluka, S.A. 2016. Economic effects of foot and mouth disease outbreaks along the cattle marketing chain in Uganda. Veterinary World 9: 544–553. https://doi.org/10.14202/vetworld.2016.544-553

Benelli, G.; Beier, J.C. 2017. Current vector control challenges in the fight against malaria. Acta Tropica 174: 91–96. https://doi.org/10.1016/j.actatropica.2017.06.028

Bernstein, A.S.; Ando, A.W.; Loch-Temzelides, T.; Vale, M.M.; Li, B.V.; Busch, J.; et al. 2022. The costs and benefits of primary prevention of zoonotic pandemics. Science Advances 8: eabl4183. https://doi.org/10.1126/sciadv.abl4183

Bett, B.; Kiunga, P.; Gachohi, J.; Sindato, C.; Mbotha, D.; Robinson, T.; et al. 2017. Effects of climate change on the occurrence and distribution of livestock diseases. Preventive Veterinary Medicine 137: 119–129. https://doi.org/10.1016/j.prevetmed.2016.11.019

Blake, A.; Sinclair, M.T.; Sugiyarto, G. 2003. Quantifying the impact of foot and mouth disease on tourism and the UK economy. Tourism Economics 9: 449–465. https://doi.org/10.5367/000000003322663221

Blake, P.; Wadhwa, D. 2020. 2020 Year in Review: The impact of COVID-19 in 12 charts. World Bank Blogs. Accessed 29 September 2022. https://blogs.worldbank.org/voices/2020-year-review-impact-covid-19-12-charts

Brady, O.J.; Hay, S.I. 2019. The first local cases of Zika virus in Europe. Lancet 394: 1991–1992. https://doi.org/10.1016/S0140-6736(19)32790-4

Brooks, D.R.; Hoberg, E.P.; Boeger, W.A. 2015. In the eye of the Cyclops: the classic case of cospeciation and why paradigms are important. Comparative Parasitology 82: 1–8. https://doi.org/10.1654/4724C.1

Brooks, D.R.; Hoberg, E.P.; Boeger, W.A. 2019. The Stockholm Paradigm: Climate Change and Emerging Disease. Chicago University Press, Chicago.

Brooks, D.R.; Hoberg, E.P.; Boeger, W.A.; Gardner, S.L.; Galbreath, K.E.; Herczeg, D.; et al. 2014. Finding them before they find us: informatics, parasites, and environments in accelerating climate change. Comparative Parasitology 81: 155–164. https://doi.org/10.1654/4724b.1

Brooks, D.R.; Hoberg, E.P.; Boeger, W.A.; Trivellone, V. 2022. Emerging infectious disease: an underappreciated area of strategic concern for food security. Transboundary and Emerging Diseases 69: 254–267. https://doi.org/10.1111/tbed.14009

Brooks-Pollock, E.; de Jong, M.C.M.; Keeling, M.J.; Klinkenberg, D.; Wood, J.L.N. 2015. Eight challenges in modelling infectious livestock diseases. Epidemics 10: 1–5. https://doi.org/10.1016/j.epidem.2014.08.005

Burns, A.; van der Mensbrugghe, D.; Timmer, H. 2008. Evaluating the Economic Consequences of Avian Influenza. The World Bank [working paper]. Accessed 3 October 2022. http://documents.worldbank.org/curated/en/977141468158986545/Evaluating-the-economic-consequences-of-avian-influenza

Cameron, E.E.; Nuzzo, J.B.; Bell, J.A.; Nalabandian, M.; O'Brien, J.; League, A.; et al. 2019. Global Health Security Index: Building Collective Action and Accountability. Nuclear Threat Initiative and Johns Hopkins Bloomberg School of Public Health, Center for Health Security. 316 p. https://www.ghsindex.org/wp-content/uploads/2020/04/2019-Global-Health-Security-Index.pdf

Cazzolla Gatti, R.; Menéndez, L.P.; Laciny, A.; Bobadilla Rodríguez, H.; Bravo Morante, G.; Carment, E.; et al. 2021. Diversity lost: COVID-19 as a phenomenon of the total environment. Science of the Total Environment 756: 144014. https://doi.org/10.1016/j.scitotenv.2020.144014

CDC [Centers for Disease Control and Prevention]. 2022. Avian Influenza in Birds [website]. Accessed 29 September 2022. https://www.cdc.gov/flu/avianflu/avian-in-birds.htm

Chatterjee, P.; Nair, P.; Chersich, M.; Terefe, Y.; Chauhan, A.; et al. 2021. One Health, "Disease X" & the challenge of "unknown" unknowns. Indian Journal of Medical Research 153: 264–271. https://doi.org/10.4103/ijmr.ijmr_601_21

Chen, Y.; Wang, Y.; Robertson, I.D.; Hu, C.; Chen, H.; Guo, A. 2021. Key issues affecting the current status of infectious diseases in Chinese cattle farms and their control through vaccination. Vaccine 39: 4184–4189. https://doi.org/10.1016/j.vaccine.2021.05.078

Cheng, V.C.C.; Chan, J.F.W.; To, K.K.W.; Yuen, K.Y. 2013. Clinical management and infection control of SARS: lessons learned. Antiviral Research 100: 407–419. https://doi.org/10.1016/j.antiviral.2013.08.016

Colella, J.P.; Bates, J.; Burneo, S.F.; Camacho, M.A.; Bonilla, C.C.; Constable, I.; et al. 2021. Leveraging natural history biorepositories as a global, decentralized, pathogen surveillance network. PLOS Pathogens 17: e1009583. https://doi.org/10.1371/journal.ppat.1009583

Conrath, U.; Beckers, G.J.M.; Langenbach, C.J.G.; Jaskiewicz, M.R. 2015. Priming for enhanced defense. Annual Review of Phytopathology 53: 97–119. https://doi.org/10.1146/annurev-phyto-080614-120132

COVID-19 to Plunge Global Economy into Worst Recession since World War II. 2020. The World Bank [press release]. Accessed 29 September 2022. https://www.worldbank.org/en/news/press-release/2020/06/08/covid-19-to-plunge-global-economy-into-worst-recession-since-world-war-ii

Cyranoski, D. 2017. Bat cave solves mystery of deadly SARS virus—and suggests new outbreak could occur. Nature 552: 15–16. https://doi.org/10.1038/d41586-017-07766-9

Damas, J.; Hughes, G.M.; Keough, K.C.; Painter, C.A.; Persky, N.S.; Corbo, M.; et al. 2020. Broad host range of SARS-CoV-2 predicted by comparative and structural analysis of ACE2 in vertebrates. Proceedings of the National Academy of Sciences USA 117: 22311–22322. https://doi.org/10.1073/pnas.2010146117

Datt, N.; Gosai, R.C.; Ravuiwasa, K.; Timote, V. 2020. Key transboundary plant pests of Coconut [Cocos nucifera] in the Pacific Island Countries—a biosecurity perspective. Plant Pathology and Quarantine 10: 152–171. https://doi.org/10.5943/ppq/10/1/17

Den Broeder, L.; Devilee, J.; Van Oers, H.; Schuit, A.J.; Wagemakers, A. 2018. Citizen Science for public health. Health Promotion International 33: 505–514. https://doi.org/10.1093/heapro/daw086

DeSalvo, K.; Hughes, B.; Bassett, M.; Benjamin, G.; Fraser, M.; Galea, S.; et al. 2021. Public Health COVID-19 Impact Assessment: Lessons Learned and Compelling Needs. NAM Perspectives [discussion paper]. National Academy of Medicine, Washington, DC. https://doi.org/10.31478/202104c

de Vienne, D.M.; Refrégier, G.; López-Villavicencio, M.; Tellier, A.; Hood, M.E.; Giraud, T. 2013. Cospeciation vs host-shift speciation: methods for testing, evidence from natural associations and relation to coevolution. New Phytologist 198: 347–385. https://doi.org/10.1111/nph.12150

Dicken, S.J.; Murray, M.J.; Thorne, L.G.; Reuschl, A.-K.; Forrest, C.; Ganeshalingham, M.; et al. 2021. Characterisation of B.1.1.7 and Pangolin coronavirus spike provides insights on the evolutionary trajectory of SARS-CoV-2. bioRxiv preprint. https://doi.org/10.1101/2021.03.22.436468

Dunnum, J.L.; McLean, B.S.; Dowler, R.C.; Alvarez-Castañeda, S.T.; Bradley, J.E.; et al. 2018. Mammal collections of the western hemisphere: a survey and directory of collections. Journal of Mammalogy 99: 1307–1322. https://doi.org/10.1093/jmammal/gyy151

Elmberg, J.; Berg, C.; Lerner, H.; Waldenström, J.; Hessel, R. 2017. Potential disease transmission from wild geese and swans to livestock, poultry and humans: a review of the scientific literature from a One Health perspective. Infection Ecology and Epidemiology 7: 1300450. https://doi.org/10.1080/20008686.2017.1300450

FAO [Food and Agriculture Organization of the United Nations]. 2019. Food Outlook—Biannual Report on Global Food Markets. Rome.

FAO and WHO [Food and Agriculture Organization of the United Nations and World Health Organization]. 2020. Codex and the SDGs: How Participation in Codex Alimentarius Supports the 2030 Agenda for Sustainable Development. Rome. 60 p. https://doi.org/10.4060/CB0222EN

FAO, OiE, WHO [Food and Agriculture Organization of the United Nations, World Organisation for Animal Health (Office International des Epizooties), World Health Organization]. 2006. Global Early Warning and Response System for Major Animal Diseases, including Zoonoses (GLEWS). 26 p. https://www.glews.net/wp-content/uploads/2011/11/agre_glews_en.pdf

Feronato, S.G.; Araujo, S.; Boeger, W.A. 2021. "Accidents waiting to happen"—insights from a simple model on the emergence of infectious agents in new hosts. Transboundary and Emerging Diseases 69: 1727–1738. https://doi.org/10.1111/tbed.14146

Findlater, A.; Bogoch, I.I. 2018. Human mobility and the global spread of infectious diseases: a focus on air travel. Trends in Parasitology 34: 772–783. https://doi.org/10.1016/j.pt.2018.07.004

Földvári, G.; Szabó, É.; Tóth, G.E.; Lanszki, Z.; Zana, B.; Varga, Z.; Kemensi, G. 2022. Emergence of *Hyalomma marginatum* and *Hyalomma rufipes* adults revealed by citizen science tick monitoring in Hungary. Transboundary and Emerging Diseases 69: e2240–e2248. https://doi.org/10.1111/tbed.14563

Fones, H.N.; Bebber, D.P.; Chaloner, T.M.; Kay, W.;T.; Steinberg, G.; Gurr, S.J.; et al. 2020. Threats to global food security from emerging fungal and oomycete crop pathogens. Nature Food 1: 332–342. https://doi.org/10.1038/s43016-020-0075-0

Fontaine, M.C.; Labbé, F.; Dussert, Y.; Delière, L.; Richart-Cervera, S.; Giraud, T.; Delmotte, F. 2021. Europe as a bridgehead in the worldwide invasion history of grapevine downy mildew, *Plasmopara viticola*. Current Biology 31: 2155–2166.E4. https://doi.org/10.1016/j.cub.2021.03.009

Francisco, I.; Bailey, S.; Bautista, T.; Diallo, D.; Gonzalez, J.; Gonzalez, J.; Roubidoux, E.K.; et al. 2022. Detection of velogenic avian paramyxoviruses in rock doves in New York City, New York. Microbiology Spectrum 10: e02061-21. https://doi.org/10.1128/spectrum.02061-21

Franke, F.; Giron, S.; Cochet, A.; Jeannin, C.; Leparc-Goffart, I.; de Valk, H.; et al. 2019. Autochthonous chikungunya and dengue fever outbreak in Mainland France, 2010–2018. European Journal of Public Health 29 (Supplement 4): ckz186.628. https://doi.org/10.1093/eurpub/ckz186.628

Funding boom or bust? [editorial]. 2009. Nature Cell Biology 11: 227. https://doi.org/10.1038/ncb0309-227

Galetto, L.; Bosco, D.; Balestrini, R.; Genre, A.; Fletcher, J.; Marzzchi, C. 2011. The major antigenic membrane protein of "*Candidatus phytoplasma asteris*" selectively interacts with ATP synthase and actin of leafhopper vectors. PLOS ONE 6: e22571. https://doi.org/10.1371/journal.pone.0022571

Gallardo, M.C.; Reoyo, A.D.; Fernández-Pinero, J.; Iglesias, I.; Muñoz, M.J.; Arias, M.L. 2015. African swine fever: a global view of the current challenge. Porcine Health Management 1: 21. https://doi.org/10.1186/s40813-015-0013-y

Gebreyes, W.A.; Dupouy-Camet, J.; Newport, M.J.; Oliveira, C.J.B.; Schlesinger, L.S.; Saif, Y.M.; et al. 2014. The global One Health paradigm: challenges and opportunities for tackling infectious diseases at the human, animal, and environment interface in low-resource settings. PLOS Neglected Tropical Diseases 8: e3257. https://doi.org/10.1371/journal.pntd.0003257

Gurr, G.M.; Johnson, A.C.; Ash, G.J.; Wilson, B.;A.L.; Ero, M.M.; Pilotti, C.A.; et al. 2016. Coconut lethal yellowing diseases: a phytoplasma threat to palms of global economic and social significance. Frontiers in Plant Science 7: 1521. https://doi.org/10.3389/fpls.2016.01521

Haire-Joshu, D.; McBride, T.D. (eds.). 2013. Transdisciplinary Public Health: Research, Education, and Practice. Wiley, NY. 432 p.

Hampf, A.C.; Nendel, C.; Strey, S.; Strey, R. 2021. Biotic yield losses in the southern Amazon, Brazil: making use of smartphone-assisted plant disease diagnosis data. Frontiers in Plant Science 12: 621168. https://doi.org/10.3389/fpls.2021.621168

Hanssen, I.M.; Lapidot, M.; Thomma, B.P.H.J. 2010. Emerging viral diseases of tomato crops. Molecular Plant-Microbe Interactions 23: 539–548. https://doi.org/10.1094/MPMI-23-5-0539

Hoberg, E.P.; Boeger, W.A.; Brooks, D.R.; Trivellone, V.; Agosta, S.J. 2022. Stepping-stones and mediators of pandemic expansion: a context for humans as ecological super-spreaders. MANTER: Journal of Parasite Biodiversity 18. https://doi.org/10.32873/unl.dc.manter18

Hoberg, E.P.; Brooks, D.R. 2015. Evolution in action: climate change, biodiversity dynamics and emerging infectious disease. Philosophical Transactions of the Royal Society B, Biological Sciences 370: 20130553. https://doi.org/10.1098/rstb.2013.0553

Ilukor, J.; Birner, R.; Rwamigisa, P.B.; Nantima, N. 2015. The provision of veterinary services: who are the influential actors and what are the governance challenges? A case study of Uganda. Experimental Agriculture 51: 408–434. https://doi.org/10.1017/S0014479714000398

Jahn, T.; Bergmann, M.; Keil, F. 2012. Transdisciplinarity: between mainstreaming and marginalization. Ecological Economics 79: 1–10. https://doi.org/10.1016/j.ecolecon.2012.04.017

Janzen, D.H. 1985. On ecological fitting. Oikos 45: 308–310.

Jones, K.E.; Patel, N.G.; Levy, M.A.; Storeygard, A.; Balk, D.; Gittleman, J.L.; Daszak, P. 2008. Global trends in emerging infectious diseases. Nature 451: 990–993. https://doi.org/10.1038/nature06536

Kading, R.C.; Cohnstaedt, L.W.; Fall, K.; Hamer, G.L. 2020. Emergence of arboviruses in the United States: the boom and bust of funding, innovation, and capacity. Tropical Medicine and Infectious Disease 5: 96. https://doi.org/10.3390/tropicalmed5020096

Kelly, T.R.; Karesh, W.B.; Johnson, C.K.; Gilardi, K.V.K.; Anthony, S.J.; Goldstein, T.; et al. 2017. One Health proof of concept: bringing a transdisciplinary approach to surveillance for zoonotic viruses at the human–wild animal interface. Preventive Veterinary Medicine 137 (Part B): 112–118. https://doi.org/10.1016/j.prevetmed.2016.11.023

Khasnis, A.A.; Nettleman, M.D. 2005. Global warming and infectious disease. Archives of Medical Research 36: 689–696. https://doi.org/10.1016/j.arcmed.2005.03.041

Kim, J.; Kim, Y.L.; Jang, H.; Cho, M.; Lee, M.; Kim, J.; Lee, H. 2020. Living labs for health: An integrative literature review. European Journal of Public Health 30: 55–63. https://doi.org/10.1093/eurpub/ckz105

Kim, J.-S. 2011. Environmental problem and citizens science owing to the failure of foot and mouth disease (FMD) policy. The Korean Association for Environmental Sociology 15: 85–119.

Kreps, S.E.; Kriner, D.L. 2020. Model uncertainty, political contestation, and public trust in science: evidence from the COVID-19 pandemic. Science Advances 6: eabd4563. https://doi.org/10.1126/sciadv.abd4563

Lakoff, A. 2017. Unprepared: Global Health in a Time of Emergency. University of California Press, Oakland.

Lawson, B.; Petrovan, S.O.; Cunningham, A.A. 2015. Citizen science and wildlife disease surveillance. EcoHealth 12: 693–702. https://doi.org/10.1007/s10393-015-1054-z

Leach, M.; MacGregor, H.; Ripoll, S.; Scoones, I.; Wilkinson, A. 2021. Rethinking disease preparedness: incertitude and the politics of knowledge. Critical Public Health 32: 82–96. https://doi.org/10.1080/09581596.2021.1885628

Lidwell-Durnin, J. 2020. Cultivating famine: data, experimentation and food security, 1795–1848. British Journal for the History of Science 53: 159–181. https://doi.org/10.1017/S0007087420000199

Lytras, S.; Xia, W.; Hughes, J.; Jiang, X.; Robertson, D.L. 2021. The animal origin of SARS-CoV-2. Science 373: 968–970. https://doi.org/10.1126/science.abh0117

Marizzi, C.; Florio, A.; Lee, M.; Khalfan, M.; Ghiban, C.; Nash, B.; et al. 2018. DNA barcoding Brooklyn (New York): a first assessment of biodiversity in Marine Park by citizen scientists. PLOS ONE 13: e0199015. https://doi.org/10.1371/journal.pone.0199015

Matić, S.; Tabone, G.; Guarnaccia, V.; Gullino, M.L.; Garibaldi, A.. 2020. Emerging leafy vegetable crop diseases caused by the *Fusarium incarnatum-equiseti* species complex. Phytopathologia Mediterranea 59: 303–317. https://doi.org/10.14601/Phyto-10883

McCullough, J. 2014. RBCs as targets of infection. Hematology, American Society of Hematology (ASH) Education Program 2014: 404–409. https://doi.org/10.1182/asheducation-2014.1.404

Meadows, A.J.; Mundt, C.C.; Keeling, M.J.; Tildesley, M.J. 2018. Disentangling the influence of livestock vs. farm density on livestock disease epidemics. Ecosphere 9: e02294. https://doi.org/10.1002/ecs2.2294

Menny, M.; Palgan, Y.V.; McCormick, K. 2018. Urban Living Labs and the role of users in co-creation. GAIA–Ecological Perspectives for Science and Society 27: 68–77. https://doi.org/10.14512/gaia.27.s1.14

Miller, B.J. 2022. Why unprecedented bird flu outbreaks sweeping the world are concerning scientists. Nature 606: 18–19. https://doi.org/10.1038/d41586-022-01338-2

Mirijamdotter, A.; Ståhlbröst, A.; Sällström, A.; Niitamo, V.-P.; Kulkki, S. 2006. The European Network of Living Labs for CWE—user-centric co-creation and innovation. In: Exploiting the Knowledge Economy: Issues, Applications, and Case Studies. P. Cunningham; M. Cunningham (eds.). IOS Press, Barcelona.

Mohamed, A. 2020. Bovine tuberculosis at the human-livestock-wildlife interface and its control through One Health approach in the Ethiopian Somali pastoralists: a review. One Health 9: 100113. https://doi.org/10.1016/j.onehlt.2019.100113

Molnár, O.; Hoberg, E.; Trivellone, V.; Földvári, G.; Brooks, D.R. 2022. The 3P Framework—a comprehensive approach to coping with the emerging infectious disease crisis. Authorea preprint. https://doi.org/10.22541/au.166176189.90109497/v1

Morens, D.M.; Fauci, A.S. 2020. Emerging pandemic diseases: how we got to COVID-19. Cell 182: 1077–1092. https://doi.org/10.1016/j.cell.2020.08.021

Morens, D.M.; Folkers, G.K.; Fauci, A.S. 2004. The challenge of emerging and re-emerging infectious diseases. Nature 430: 242–249. https://doi.org/10.1038/nature02759

Mysterud, A.; Hopp, P.; Alveseike, K.R.; Benestad, S.L.; Nilsen, E.B.; Rolandsen, C.M.; et al. 2020. Hunting strategies to increase detection of chronic wasting disease in cervids. Nature Communications 11: 4392. https://doi.org/10.1038/s41467-020-18229-7

Newton, J.; Kuethe, T. 2015. Economic Implications of the 2014–2015 Bird Flu. farmdoc daily 5: 104. Department of Agricultural and Consumer Economics, University of Illinois at Urbana-Champaign. https://farmdocdaily.illinois.edu/2015/06/economic-implications-of-the-2014-2015-bird-flu.html

Nyerere, N.; Luboobi, L.S.; Mpeshe, S.C.; Shirima, G.M. 2020. Optimal control strategies for the infectiology of brucellosis. International Journal of Mathematics and Mathematical Sciences 2020: 1214391. https://doi.org/10.1155/2020/1214391

Nylin, S.; Agosta, S.; Bensch, S.; Boeger, W.A.; Braga, M.P.; Brooks, D.R.; et al. 2018. Embracing colonizations: a new paradigm for species association dynamics. Trends in Ecology and Evolution 33: 4–14. https://doi.org/10.1016/j.tree.2017.10.005

Oerke, E.-C.; Dehne, H.-W.; Schönbeck, F.; Weber, A. 2012. Crop Production and Crop Protection: Estimated Losses in Major Food and Cash Crops. Elsevier, Amsterdam. 829 p.

Palagyi, A.; Marais, B.J.; Abimbola, S.; Topp, S.M.; McBryde, E.S.; Negin, J. 2019. Health system preparedness for emerging infectious diseases: a synthesis of the literature. Global Public Health 14: 1847–1868. https://doi.org/10.1080/17441692.2019.1614645

Palmer, J.R.B.; Oltra, A.; Collantes, F.; Delgado, J.A.; Lucientes, J.; Delacour, S.; et al. 2017. Citizen science provides a reliable and scalable tool to track disease-carrying mosquitoes. Nature Communications 8: 916. https://doi.org/10.1038/s41467-017-00914-9

Parrish, C.R.; Kawaoka, Y. 2005. The origins of new pandemic viruses: the acquisition of new host ranges by canine parvovirus and influenza A viruses. Annual Review of Microbiology 59: 553–586. https://doi.org/10.1146/annurev.micro.59.030804.121059

Perrin, L.D. 2017. Exploration of the spatial epidemiology of tick borne pathogens of livestock in southern Cumbria. PhD diss., University of Salford, Manchester, United Kingdom. http://usir.salford.ac.uk/id/eprint/44647/

Plohl, N.; Musil, B. 2021. Modeling compliance with COVID-19 prevention guidelines: the critical role of trust in science. Psychology, Health and Medicine 26: 1–12. https://doi.org/10.1080/13548506.2020.1772988

Pretorius, Z.A.; Singh, R.P.; Wagoire, W.W.; Payne, T.S. 2000. Detection of virulence to wheat stem rust resistance gene *Sr31* in *Puccinia graminis* f. sp. *tritici* in Uganda. Plant Disease 84: 203. https://doi.org/10.1094/PDIS.2000.84.2.203B

Quaglio, G.L.; Goerens, C.; Putoto, G.; Rübig, P.; Lafaye, P.; Karapiperis, T.; et al. 2016. Ebola: lessons learned and future challenges for Europe. Lancet Infectious Diseases 16: 259–263. https://doi.org/10.1016/S1473-3099(15)00361-8

Rajala, E.; Lee, H.S.; Nam, N.H.; Huong, C.T.T.; Son, H.M.; Wieland, B.; Magnusson, U. 2021. Skewness in the literature on infectious livestock diseases in an emerging economy—the case of Vietnam. Animal Health Research Reviews 22: 1–13. https://doi.org/10.1017/S1466252321000013

Romero Herrera, N. 2017. The emergence of Living Lab methods. In: Living Labs: Design and Assessment of Sustainable Living. D. Keyson; O. Guerra-Santin; D. Lockton (eds.). Springer, Cham, Switzerland. 14 p. https://doi.org/10.1007/978-3-319-33527-8_2

Russell, P.E. 2006. The development of commercial disease control. Plant Pathology 55: 585–594. https://doi.org/10.1111/j.1365-3059.2006.01440.x

Ryan, S.J.; Carlson, C.J.; Tesla, B.; Bonds, M.H.; Ngonghala, C.N.; Mordecai, E.A.; et al. 2021. Warming temperatures could expose more than 1.3 billion new people to Zika virus risk by 2050. Global Change Biology 27: 84–93. https://doi.org/10.1111/gcb.15384

Saunders, D.G.O.; Pretorius, Z.A.; Hovmøller, M.S. 2019. Tackling the re-emergence of wheat stem rust in Western Europe. Communications Biology 2: 51. https://doi.org/10.1038/s42003-019-0294-9

Schäpke, N.; Stelzer, F.; Caniglia, G.; Bergmann, M.; Wanner, M.; Singer-Brodowski, M.; et al. 2018. Jointly experimenting for transformation? Shaping real-world laboratories by comparing them. GAIA–Ecological Perspectives for Science and Society 27: 85–96. https://doi.org/10.14512/gaia.27.S1.16

Schuler-Faccini, L.; Ribeiro, E.M.; Feitosa, I.M.L.; Horovitz, D.D.G.; Cavalcanti, D.P.; Pessoa, A.; et al. 2016. Possible association between Zika virus infection and microcephaly—Brazil, 2015. Centers for Disease Control and Prevention, Morbidity and Mortality Weekly Report (MMWR) 65: 59–62. http://dx.doi.org/10.15585/mmwr.mm6503e2

Schulthess, H. 1761. Vorschlag einiger durch die Erfahrung bewährter Hilfsmittel gegen den Brand im Korn. Abhandlungen Der Naturforschenden Gesellschaft in Zurich, 1, 498–506.

Seifert, V.A.; Wilson, S.; Toivonen, S.; Clarke, B.; Prunuske, A. 2016. Community partnership designed to promote Lyme disease prevention and engagement in citizen science. Journal of Microbiology and Biology Education 17: 63–69. https://doi.org/10.1128/jmbe.v17i1.1014

Singh, R.K.; Rani, M.; Bhagavathula, A.S.; Sah, R.; Rodriguez-Morales, A.J.; Kalita, H.; et al. 2020. Prediction of the

COVID-19 pandemic for the top 15 affected countries: advanced autoregressive integrated moving average (ARIMA) model. JMIR Public Health and Surveillance 6: e19115. https://doi.org/10.2196/19115

Steen, K.; van Bueren, E. 2017. Urban Living Labs: A Living Lab Way of Working. Amsterdam Institute for Advanced Metropolitan Solutions, Delft University of Technology, Amsterdam. https://www.ams-institute.org/documents/28/AMS_Living_Lab_Way_of_Working-ed4.pdf

Steinke, J.; van Etten, J.; Zelan, P.M. 2017. The accuracy of farmer-generated data in an agricultural citizen science methodology. Agronomy for Sustainable Development 37: 32. https://doi.org/10.1007/s13593-017-0441-y

te Beest, D.E.; Hagenaars, T.J.; Stegeman, J.A.; Koopmans, M.P.G.; van Boven, M. 2011. Risk based culling for highly infectious diseases of livestock. Veterinary Research 42: 81. https://doi.org/10.1186/1297-9716-42-81

Tomley, F.M.; Shirley, M.W. 2009. Livestock infectious diseases and zoonoses. Philosophical Transactions of the Royal Society B: Biological Sciences 364: 2637–2642. https://doi.org/10.1098/rstb.2009.0133

Trivellone, V.; Hoberg, E.P.; Boeger, W.A.; Brooks, D.R. 2022. Food security and emerging infectious disease: risk assessment and risk management. Royal Society Open Science 9: 211687. https://doi.org/10.1098/rsos.211687

Trivellone, V.; Ripamonti, M.; Angelini, E.; Filippin, L.; Rossi, M., Marzachí, C.; Galetto, L. 2019. Evidence suggesting interactions between immunodominant membrane protein Imp of Flavescence dorée phytoplasma and protein extracts from distantly related insect species. Journal of Applied Microbiology 127: 1801–1813. https://doi.org/10.1111/jam.14445

Uchtmann, N.; Herrmann, J.A.R.; Hahn, E.C.; Beasley, V.R. 2015. Barriers to, efforts in, and optimization of integrated One Health surveillance: a review and synthesis. EcoHealth 12: 368–384. https://doi.org/10.1007/s10393-015-1022-7

Vianna Franco, M.P.; Molnár, O., Dorninger, C.; Laciny, A.; Treven, M.; Weger, J.; et al. 2022. Diversity regained: precautionary approaches to COVID-19 as a phenomenon of the total environment. Science of the Total Environment 825: 154029. https://doi.org/10.1016/j.scitotenv.2022.154029

Vollmer, N.A.; Valadez, J.J. 1999. A psychological epidemiology of people seeking HIV/AIDS counselling in Kenya: an approach for improving counsellor training. AIDS 13: 1557–1567. https://doi.org/10.1097/00002030-199908200-00017

Voytenko, Y.; McCormick, K.; Evans, J.; Schliwa, G. 2016. Urban living labs for sustainability and low carbon cities in Europe: towards a research agenda. Journal of Cleaner Production 123: 45–54. https://doi.org/10.1016/j.jclepro.2015.08.053

Vurro, M.; Bonciani, B.; Vannacci, G. 2010. Emerging infectious diseases of crop plants in developing countries: impact on agriculture and socio-economic consequences. Food Security 2: 113–132. https://doi.org/10.1007/s12571-010-0062-7

Wang, Y.X.G.; Matson, K.D.; Santini, L.; Visconti, P.; Hilbers, J.P.; Huijbregts, M.A.; et al. 2021. Mammal assemblage composition predicts global patterns in emerging infectious disease risk. Global Change Biology 27:4995–5007. https://doi.org/10.1111/gcb.15784

Watanabe, T.; Watanabe, S.; Maher, E.A.; Neumann, G.; Kawaoka, Y. 2014. Pandemic potential of avian influenza A (H7N9) viruses. Trends in Microbiology 22: 623–631. https://doi.org/10.1016/j.tim.2014.08.008

Weaver, S.C.; Lecuit, M. 2015. Chikungunya virus and the global spread of a mosquito-borne disease. New England Journal of Medicine 372: 1231–1239. https://doi.org/10.1056/NEJMra1406035

WHO [World Health Organization]. 2007. The World Health Report 2007: A Safer Future: Global Public Health Security in the 21st Century. Accessed 3 October 2022. https://apps.who.int/iris/handle/10665/43713

WHO [World Health Organization]. 2020a. COVID-19—China [Pneumonia of unknown cause—China]. Disease Outbreak News. Accessed 3 October 2022. https://www.who.int/emergencies/disease-outbreak-news/item/2020-DON229

WHO [World Health Organization]. 2020b. The COVID-19 pandemic: lessons learned for the WHO European Region: a living document, 15 September 2020. Accessed 3 October 2022. https://apps.who.int/iris/handle/10665/334385

WHO [World Health Organization]. 2022a. Multi-Country—Acute, severe hepatitis of unknown origin in children. Disease Outbreak News. Accessed 3 October 2022. https://www.who.int/emergencies/disease-outbreak-news/item/2022-DON376

WHO [World Health Organization]. 2022b. WHO working closely with countries responding to monkeypox [press release]. Accessed 3 October 2022. https://www.who.int/news/item/20-05-2022-who-working-closely-with-countries-responding-to-monkeypox

Xia, W.; Hughes, J.; Robertson, D.; Jiang, X. 2021. How one pandemic led to another: was African swine fever virus (ASFV) the disruption contributing to severe acture respiratory syndrome coronavirus 2 (SARS-CoV-2) emergence? Preprints preprint. https://doi.org/10.20944/preprints202102.0590.v2

Yadav, S.; Gettu, N.; Swain, B.; Kumari, K.; Ojha, N.; Gunthe, S.S. 2020. Bioaerosol impact on crop health over India due to emerging fungal diseases (EFDs): an important missing link. Environmental Science and Pollution Research 27: 12802–12829. https://doi.org/10.1007/s11356-020-08059-x

11

Citizen Science Can Help Prevent Emerging Infectious Diseases

Éva Szabó and Gábor Földvári

Abstract

Citizen science has emerged as a popular research approach in recent times, owing to its cost effectiveness and ability to gather large amounts of data across vast areas, frequently yielding outcomes beyond the initial research goals. However, like any research methodology, citizen science has certain drawbacks. For instance, the data obtained through this method can be challenging to analyze, and special attention needs to be paid toward data protection. In addition, community involvement into research may not be easily implemented in underdeveloped regions. Despite these limitations, an increasing number of researchers have successfully used data collected via citizen science in various infectious disease–related projects. Here we review published examples from the scientific literature and highlight the fact that citizen science is essential in implementing the DAMA (Document, Assess, Monitor, Act) protocol. We conclude that research projects involving volunteers not only educate the public about the scientific process and increase awareness but can greatly improve our efforts to anticipate and mitigate the risks of emerging infectious diseases.

Keywords: citizen science, community science, crowd science, crowd-sourced science, civic science, participatory monitoring, volunteer monitoring, emerging infectious diseases, prevention, DAMA protocol

Introduction

Climate change, destruction of natural habitats, and globalization bring several challenges for humanity, and one of the most significant is the threat of emerging infectious diseases (EIDs). The Stockholm Paradigm has recognized that the risk for EIDs is far larger than previously thought, and it also explains how climate change and anthropogenic impacts play a key role in setting the stage for emergence, through reducing ecological isolation and providing the opportunity for EID (Brooks et al., 2019; Brooks et al., 2022; Hoberg et al., 2022).

We as humans and the rest of our coinhabitants on Earth are interconnected to a previously unimaginable extent, which brings new pathogens into new areas, resulting in increased opportunity for worldwide difficulties (Daszak, 2012). Fortunately, in a way, one of the causes of this global problem includes a possibility for treating it. Globalization connects humans to domesticated animals and wildlife, and densely populated cities to isolated regions, but along with emerging diseases an interconnected global humanity also brings us new, faster ways of sharing knowledge and novel channels of communication. These opportunities can be the foundation of the DAMA (Document, Assess, Monitor, Act) protocol, which is a preventive procedure for detecting and preventing future infectious diseases. The DAMA protocol was developed as a framework for coping with EIDs, where **documentation** of potential pathogens is followed by an **assessment** of the risks they pose. High-risk pathogens, their vectors, and their reservoir hosts are then **monitored** to reveal spatial and temporal dynamics. With these data, an **action** plan can then be developed to avoid infections of humans and/or economically significant species (livestock, crops, etc.). As a comprehensive protocol, the DAMA's final action phase requires the efficient and fluent collaboration of multiple diverse fields, such as science, society, health care, economics, politics, and policy making. In practice, this can be done in several ways, and one of them is citizen science (Brooks et al., 2020; Molnár, Hoberg, et al., 2023, this volume; Molnár, Knickel, et al., 2023, this volume).

Citizen science is a method of collecting and analyzing data produced by the public; it builds on the strength of communities. This approach has a history of roughly 120 years, but thanks to the new and fast ways of sharing information through the internet and other media, it is becoming more and more widespread. Several possibilities exist for gathering information this way, from surveys, such as those used to track symptoms during the COVID-19 pandemic, to monitoring known disease vectors based on public reports (Eisen and Eisen, 2021). Citizen science fits perfectly in all four stages of the DAMA protocol, but its strengths might be most useful in the Document and the Monitor phases.

Advantages of Applying Citizen Science

Cost effectiveness

Citizen science has many benefits, one of them being cost effectiveness (Figure 11.1, Table 11.1). For example, programs for monitoring certain species of ticks or mosquitoes in Belgium, the Netherlands, Czech Republic, and Spain use this method for gathering data. These programs usually cover the entire country, and the researchers can receive hundreds or thousands of inputs each year. Producing that amount of data from across the country without the help of the public would be way too expensive for any government, research fund, or institution. With citizen involvement,

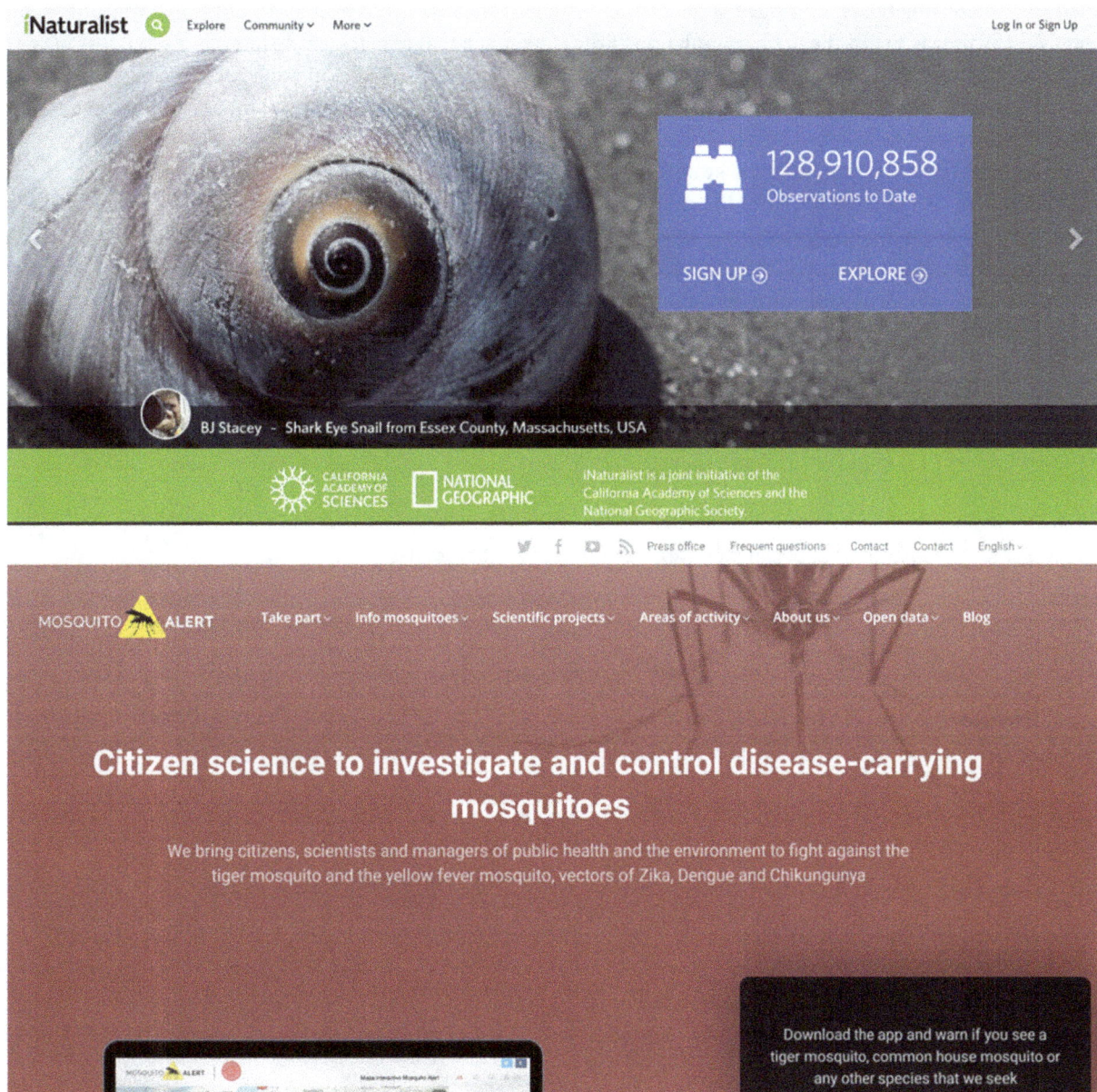

Figure 11.1. The home page of iNaturalist and the Mosquito Alert website.

scientists can document the emergence of new vector species in the country without an extensive budget or a large team (Bartumeus et al., 2018; Lernout et al., 2019; Uiterwijk et al., 2021; Daněk et al., 2022). The global cost of an infectious disease cannot be easily converted to numbers. We have only estimates of the price of the direct and indirect consequences. Even well before the COVID-19 pandemic, the expenses caused by zoonotic diseases were in the hundreds of billions of US dollars per decade (Narrod et al, 2012). It is clear that preventing the emergence of an infectious disease is more cost effective than managing an ongoing pandemic with a rapidly changing nature and experimenting with different responses.

A smaller but good example is the mass testing that occurred in Slovakia during the second wave of the COVID-19 pandemic. The plan was to test the entire population of the country with relatively cheap antigen tests every weekend and quarantine everyone who did not have a negative test result. On the first weekend, 3,625,332 individuals were tested, which corresponds to 66% of the state's population. A campaign of this volume required the training and transporting of the participating health-care workers and Slovak armed forces and volunteers, the preparation and maintenance of the hundreds of testing sites, the elimination of the hazardous waste, and mainly the purchasing of 5,276,832 rapid antigen tests. Although the campaign had a visible effect on the infection prevalence, the mass testing of the entire country lasted only a few weeks. The program was deemed too expensive and too difficult to organize; it simply was not sustainable for more than a short period of time. After that, the number of positive cases started to rise again, despite other interventions (Hledík et al., 2021; Pavelka et al., 2021).

It is easy to see that preventing the emergence of such diseases is the cheaper and easier way, but it still requires trained professionals and various expenses. Citizen science can help to achieve this outcome by reducing the cost of surveillance programs and outsourcing the tasks to the volunteering participants. The drop in expenses can be substantial—a campaign focused on mosquito detection in southern Australia demonstrated that the cost of their citizen science program is only 20% that of a similar program executed by professionals (Braz Sousa et al., 2020).

In relation to prevention, there is the One Health approach that recognizes the interconnectedness of the health of humans, animals, and the environment (Poh et al., 2022). In addition to focusing on humans and pathogens that are directly dangerous to humans, it is also important to organize citizen science projects that focus on other organisms, such as migratory and urban birds. Many already-existing programs have successfully applied the tools of citizen science, such as the Taiwanese Avian Influenza Virus program, which was based on data from the eBird citizen science database; the New York City–based project that focused on velogenic avian paramyxoviruses; and a survey that examined haemosporidian infections in wild passerine birds (Himmel et al., 2021; Francisco et al., 2022; Wu et al., 2023).

Table 11.1. Some of the advantages of community research and related examples

Advantages	Example	A brief summary	Reference
Time efficiency	The Dutch Great Influenza Survey	Loyal participants report their symptoms every week	Land-Zandstra et al., 2016
Cost efficiency	The citizen science mosquito surveillance program in southern Australia	Use of citizen science is estimated to have reduced expenses to 20%	Braz Sousa et al., 2020
Wide geographic coverage	The COVID-19 pandemic online survey	The University of Maryland Global COVID-19 Trends and Impact Survey collected 31,142,582 responses from 114 countries with the help of Facebook	Astley et al., 2021
Valuable additional results (bycatch)	The Czech Najdi pijáka [Find a drinker] program	The program focused on *Dermacentor* ticks, but the participants additionally found 12 nonindigenous *Hyalomma* ticks	Daněk et al., 2022
Strengthening local communities	The program for malaria control in Rwanda	The program showed that citizen science can raise curiosity and a sense of responsibility among participants	Asingizwe et al., 2020

Wider geographical coverage and larger sample size

Collecting a large amount of data over a short period of time can be the key part of research, especially at the beginning of a pandemic. A good example for this is the COVID-19 pandemic online survey, which collected data from Facebook users. The program started as early as 23 April 2020 and by 20 December 2020 had collected information from 31,142,582 participants from 114 countries. Another notable example is the the Dutch Great Influenza Survey, which collects data from its loyal participants on a weekly basis regarding their flu-like symptoms. Likewise, a 2021 research project on the distribution of vectors was successfully implemented through a citizen science initiative that involved collecting *Ixodes* ticks. Within a span of only four years, researchers were able to receive submissions of tick records from various regions of the United States and thereby analyze the patterns of distribution for both the ticks and the pathogens they carry. These studies showed that findings can at least complement the work of healthcare professionals (Land-Zandstra et al., 2016; Astley et al., 2021; Porter et al., 2021).

There are examples for using volunteer-classified images to train machine-learning algorithms to identify species. In terms of the amount of data produced and geographical spread, iNaturalist is one of the most significant citizen science platforms currently in use. The program, which started in 2008 and was organized by the University of California, Berkeley and the California Academy of Sciences was aimed at providing a common platform on which participants can upload photos and observations of plants, animals, and other organisms they encounter (Figure 11.2). These records are then identified and verified by other users, including experts in the field, and later can be the basis of scientific research and conservation programs (Mesaglio et al., 2021).

A 2021 research program focused on the extent to which the submissions from the application can be used later in scientific studies. The objective of this study was to analyze data from the iNaturalist online community to identify medically significant tick species across a vast region of North America. The first submission of tick data in this region dates back to 1997, and the quantity of submissions of records of ticks taken by citizens has grown steadily since then, with 39% of the total submissions occurring in 2021. The study excluded reports that were not clearly identifiable to the level of species of tick or had been misidentified as a tick, resulting in 17,172 valid hits. Incredibly, using this filter, only 3% of the submissions turned out to be incorrect, with the remaining 97% deemed usable. The study

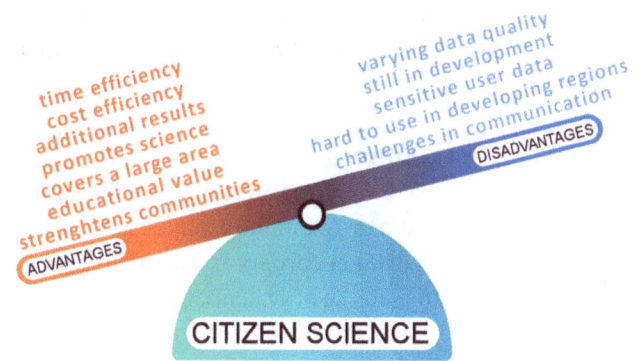

Figure 11.2. Some advantages and disadvantages of citizen science.

found that the outcomes related to the geographic distribution and seasonal patterns of the tick species were consistent with current knowledge. Moreover, the analysis identified a limited number of new county occurrence records for both *Amblyomma americanum* and *Ixodes scapularis*, which indicates that this data source could be a valuable passive surveillance method to supplement other vector surveillance strategies. This method could help confirm the distribution data and recognize regions where tick populations are emerging, thus posing a risk for disease (Cull, 2022).

Another small but captivating example of the program's role in biodiversity science is the rediscovery of the orange gull butterfly, *Cepora iudith malaya*, in Singapore. This species had not been reliably recorded since 1975 and was deemed locally extinct, but in recent years it was rediscovered with the help of uploaded photos on iNaturalist (Jain et al., 2019).

In addition, citizen science programs can also predict the spread of vectors of emerging infectious diseases, as shown by the Mosquito Alert program. Mosquito Alert is a citizen science project focused on monitoring the distribution of mosquitoes that are potential vectors of EIDs. The program was launched in 2014 in Spain and originally concentrated on the prevalence and distribution of *Aedes albopictus*. The program was successful, so starting in 2016 it was extended to several species and several countries. The mobile phone application that gives the basis of the research is now available worldwide, and participants can choose from 18 languages, find out about the dangers posed by mosquitoes, and in addition to identified specimens they can also report bites and breeding sites (Figure 11.2). Thanks to this project, a lot of useful data can be collected from a large area at relatively low cost. The program has shown it can play a major role in predicting the new appearance of invasive species, for example, the first detection of *Aedes japonicus* in Spain (Carney et al., 2022).

Valuable additional results

Sometimes citizen science projects can get additional results beyond expectations. A notable example is the Czech Najdi pijáka [Find a drinker] program. The main goal for this research was to map the prevalence of the nonindigenous species of *Dermacentor* ticks in the country. The project was a success, with 558 photo reports of ticks and 250 specimens submitted, but on top of that, participants unexpectedly reported 12 adult *Hyalomma* ticks, which are less widespread in Central Europe but are the main vectors of the Crimean-Congo hemorrhagic fever virus (Daněk et al., 2022). Additional case studies have been included by Ortiz and Juarrero (2023, this volume) on Zika, mosquito-borne diseases in Costa Rica, and rodent outbreaks in Minneapolis.

Strengthening local communities

At this point, we wish to shift the focus from the researchers' benefits to the benefits that participants accrue from this kind of participation. For example, citizen science projects can increase the enthusiasm in a community toward the examined topic. The participants in a program for malaria control in Rwanda reported that in the first months of the project they were so excited to find out in the morning whether their handmade mosquito trap managed to catch anything that they could not sleep properly. This shows a great involvement; the participants can feel more like a part of the solution, not just the subject of it, which can be beneficial in the prevention or the control of emerging infectious diseases (Asingizwe et al., 2020).

Another good example of community-strengthening methods within citizen science are bioblitz programs. Bioblitzes are short-term surveying projects conducted by scientists, citizen scientists, and other volunteers. The purpose of a bioblitz is to gather as much information as possible about the biodiversity of a given location in a short amount of time, usually in less than 24 hours. Such an event can be focused on identifying and recording different species of plants, animals, fungi, and other organisms. Bioblitzes can be conducted in a variety of environments, such as lakes, rivers, forests, cities, national parks, local parks, and more. These programs have many advantages, including strengthening communities, giving the participants the opportunity to learn about their environment, and building trust between experts and residents. Because

of the joint work done together, these programs may be one of the best opportunities for professionals and residents to get to know each other's perspectives on a given topic. In addition to these advantages, bioblitz projects are an efficient way to gather a vast amount of valuable data in a short period of time (Meeus et al., 2023).

A 2017 study, held in the Brocéliande Forest of Brittany, France, built on this advantage. The examined wildland field had undergone substantial ecological change during the last 60 years. In the past, it was used as an agricultural area, but later it gradually regained its natural state. The total of 209 participating citizen scientists discovered 660 species during the 24 hours of the program. These results, supplemented by the photos and additional notes of the participants, helped a great deal in understanding the changes that had occurred over the past 60 years. For example, in addition to the historical data collected prior to the bioblitz program, citizen scientists discovered 5 bird, 7 mammal, and 52 vascular plant species that had never been recorded in this region (Nicolai et al., 2020). This kind of added value of citizen science has many unexplored future possibilities for EID prevention as well.

Educational value

Citizen science can also be a useful tool in education. During research between 2017 and 2019, 400 middle school students between the ages of 11 and 14 were able to learn about ticks, diseases caused by ticks, and Lyme disease. Then, in groups of 20 to 25, they could talk with experts about the change in the prevalence of Lyme disease. In addition, the participants were able to try DNA isolation themselves. After the session, the students reported on their experiences in a questionnaire. One of the goals of the program was to promote a career in science, and this goal can be considered successfully achieved since, based on the questionnaire, 76.1% of the students indicated they would like to participate in similar community science projects, and 45% stated they were now considering a career in science. In addition to the students' learning about diseases caused by ticks and trying out participation in scientific research, the program also had other results. The DNA isolated by the participants gave reliable results in 66% of the cases during the subsequent polymerase chain reaction (PCR) test. Around 85% of the students successfully executed the isolation protocol without encountering any complications, one of them saying "you just needed to follow the directions" (Prunuske et al., 2021).

Citizen science is a useful method for educating secondary school children as well. In the Nuestra Señora del Puy secondary school in Spain, biology and geography courses are regularly held in locations where students can actively participate in various programs related to conservation biology and biodiversity. In 2016 and 2018 the school decided to base that year's project on iNaturalist. In the nearby Basaula Reserve, the students identified trees and shrubs and created a virtual herbarium from them with the application. At the end of the program, the students were able to express their opinion on the inclusion of the application in their studies. Here the students marked the usage of technology, interactivity, and ease of use as positive aspects, while the lack of a stable internet connection was listed as the main shortcoming. It is worth mentioning that the students participating in the second year of the study were able to see how their older classmates' observations were already in the iNaturalist database, which helped them understand how their small contributions can help track changes over time in the environment (Echeverria et al., 2021).

Challenges and Possible Solutions

As with perhaps every method of research, citizen science has its downsides too. Some of its flaws come from this approach still being in its development phase, and so are its tools. Smartphones and other similar devices, which are frequently used in projects based on citizen science, have emerged and spread widely in the last decade, leaving little time for researchers to catch up and catch on. Nowadays, whole professional sectors are specialized in managing a brand's social media presence or finding new approaches in design to catch the attention of new users. In this environment, scientists leading citizen science programs can find themselves on social media platforms pondering which eye-catching font to use in a new post about the dangers of a mosquito-borne disease or plastic pollution. Some of these new challenges can appear to be unnecessary and filled with absurd details, but often among these tasks we find the foundation of a successful project.

In 2020 the participants in the Citizen Science Conference categorized their own failures in hope of future improvements. In 55% of the programs, one of the biggest missteps was the inadequate communication of the research aims and the tasks of the participants (Westreicher et al., 2021). Designing a research program using citizen science can sometimes feel like a crusade against emerging diseases or climate change, but to be successful, it must be realized that most of the participants do not have the time or interest to read long and detailed paragraphs about the goals and importance of the program. In short, these

projects should be based on clear, straightforward messages and achievable tasks that can mostly be inserted into the daily life of the participating community. The tone of the communication is also important—fearful news, especially regarding emerging diseases, might travel faster and reach more people, but magnifying the threats is unethical and also an error because it shifts the participants' focus from the original question.

Another disadvantage can be the varying quality of the incoming data. Our citizen science–based tick monitoring project (https://kullancsfigyelo.hu/) is focused on the presence of nonindigenous *Hyalomma* ticks in Hungary (Földvári et al., 2022). The program started in 2021, and it immediately showed us that the inputs from the participants often can be either lacking important elements or, in contrast, be full of irrelevant information. Moreover, these additional details sometimes contain highly sensitive data regarding, for example, the contributor's health insurance status. We can see that the handling of personal data can be a delicate issue, even if it is not our intention to collect these types of information. From this viewpoint, it is even more important to use the received data carefully when the research is primarily focused on monitoring, for example, the prevalence of a disease based on the volunteer's symptoms.

Data privacy is an important topic even when the participants share only the most essential data about themselves. For example, almost all citizen science projects require some information about the volunteer's location. These questions—the "where?" and "when?"—would be hard, if not impossible, to replace with something else if we want our research to make sense. And still, these types of information can raise very important questions regarding personal privacy. From a participant's location, we can easily guess their hometown or even their address; moreover, citizen science programs that require regular submission can record a lot about a volunteer's traveling habits and daily routes. These concerns could affect not only the participants but also the subjects of the studies, which could include threatened and endangered species. In the case of projects focused on emerging diseases, this aspect can be a minor concern, but the picture changes if we consider the hosts of our studied parasites. For example, the submitted location of a tick found on a deer, if the data get into the wrong hands, can lead poachers to the exact spot (Bowser et al., 2017).

Data protection issues have to be considered when launching a citizen science project because most participants share sensitive data about themselves without concern regarding its subsequent use. While we collected the submissions for our tick monitoring project, we never directly received any questions from participants about their personal data safety. Volunteers often submit information about, for example, the route they take while walking their dog, the destination and duration of their vacation, the location of a playground they frequently visit with their child, or the exact bench they sit on during their daily afternoon walk. The most concerned volunteer feared that the news of a nonindigenous and dangerous tick he found could cast a bad light on his hometown's good reputation, so he would reveal his location only if the tick was indeed a *Hyalomma*. Apart from this, this person shared other sensitive information, including daily habits.

Converting the participants' reports and stories into plain data storable in a spreadsheet can be more time consuming than researchers might think during the planning phase of the project, but valuable time can be lost in other areas too. Projects similar to ours often work closely with people in the media. A good connection with journalists and media outlets is crucial if we want future participants to know about the existence of the research and thus make it a functional project. Timing of media appearances might also heavily influence success. For example, we experienced an unexpectedly warm spring in 2021, which was suitable for the tick vectors we were searching for; however, a delay of our first media appearances because of administrative reasons led to a loss of data during the early tick season (Földvári et al., 2022).

High-volume citizen science projects often heavily rely on internet access and the use of smartphones. These tools, while being undeniably helpful in the process of collecting a vast amount of data, can be a privilege of the more developed regions and wealthiest participants, thereby automatically excluding potential citizen scientists living in developing countries or in areas of modest income. The leaders of a 2020 citizen science project focused on dengue prevention in Paraguay decided to gather data by using both smartphones and paper questionnaires. Each possibility had its own challenges. Smartphone users often had low-end devices unable to capture quality images with high enough resolution, or they would not have enough storage for the data. Volunteers who used the paper questionnaires created the challenge of collecting, replacing, and digitizing the datasheets. But even with these obstacles, the program benefited from this hybrid nature of data collection (Parra et al., 2020).

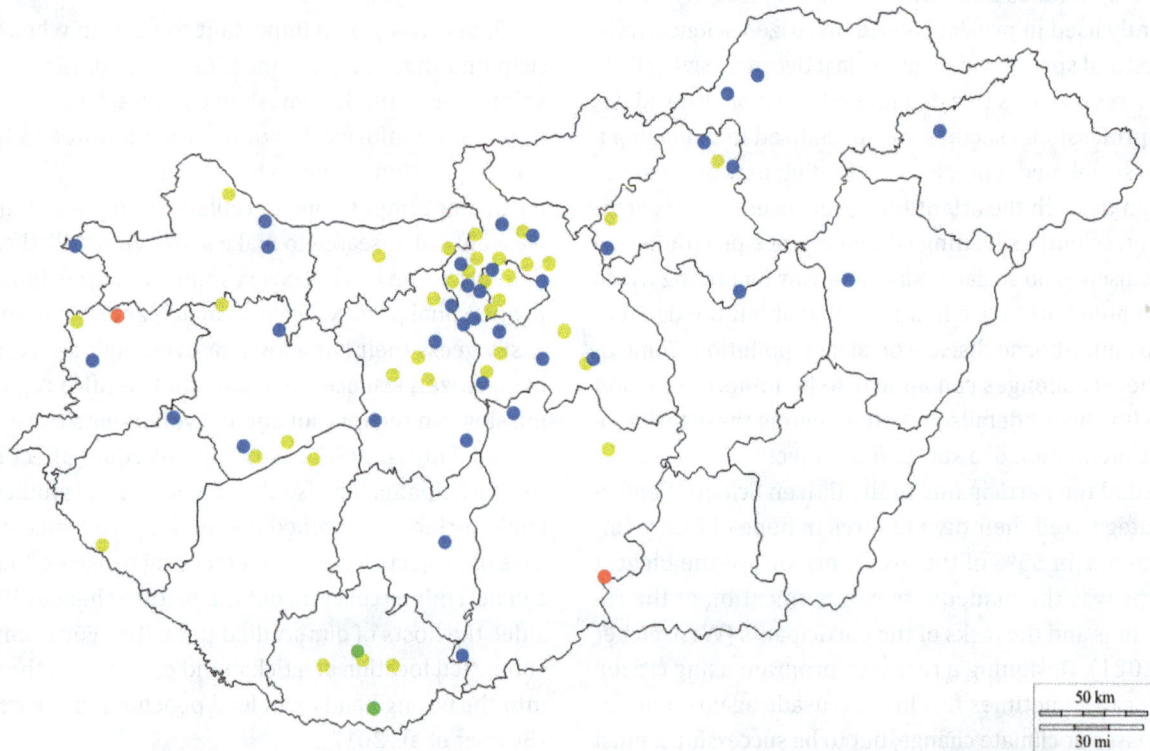

Figure 11.3. Map of distribution of the received ticks in the first year of the Hungarian tick monitoring program. The colored dots represent submitted species. It is visible that ticks were sent in from mostly from around the north-central capital.

Sometimes the challenge is caused not by the participants' inability to purchase an adequate quality of device for sample submission but by the lifestyle of people who rely less on the use of smartphones and the internet. In this case, we often do not get to discuss camera quality and data storage because the potential participants are not even aware of the existence of the study. Since the start of our tick monitoring project (Földvári et al., 2022), we have received a great quantity of data submissions by dog owners who live in the city center of the capital—and besides the fact that every participant contribution is valuable, the ticks in our focus study mostly prefer larger mammals, like cattle and horses, which mainly live in rural areas (Figure 11.3). It is easy to see how online media platforms, which tend to reach the most people in a short time, may easily leave out target groups that may not have quick access to social media, in this case horse stable owners or farmers. A participant, who successfully identified *Hyalomma* ticks on his horse, pointed out that in his experience, most of those owners who have time to regularly check social media and online news platforms probably will not be thorough enough to discover unusual parasites on their horses. His view might be polarized and pessimistic, but considering the dominance of urban participants in our study, relying not exclusively on online platforms in each stage of the program might pay off.

In addition, the problem may not be only where the population encounters the program but also where it encounters the ticks themselves. Citizen science tick collections typically measure the exposure of humans to ticks, while active surveillance involves collecting ticks in various settings, including places that are not frequently visited by the public, to assess actual tick density. Identifying these systematic biases can lead to more reliable and useful data (Tran et al., 2021).

Conclusion

With every challenge and concern, citizen science has the potential to develop into a fast and cost-effective way of gathering and evaluating data regarding emerging diseases. Based on the experiences of vector monitoring and other emerging pathogen–related citizen science projects, it is clear that participants are often enthusiastic, curious, and willing to contribute. With the development of smartphones, expanding internet coverage, and new ways of reaching possible participants, in the next few years we can hope it will be even easier to participate in such projects.

If we manage to overcome the current difficulties that humanity faces, citizen science has the potential to establish trust and facilitate collective decision-making among communities, researchers, and policy makers by enabling knowledge sharing and two-way communication (Tan et al., 2022).

Literature Cited

Asingizwe, D.; Poortvliet, P.M.; Koenraadt, C.J.M.; van Vliet, A.J.H.; Ingabire, C.M.; Mutesa, L.; Leeuwis, C. 2020. Why (not) participate in citizen science? Motivational factors and barriers to participate in a citizen science program for malaria control in Rwanda. PLOS ONE 15: e0237396. https://doi.org/10.1371/journal.pone.0237396

Astley, C.M.; Tuli, G.; McCord, K.A.; Cohn, E.L.; Rader, B.; Varrelman, T.J.; et al. 2021. Global monitoring of the impact of the COVID-19 pandemic through online surveys sampled from the Facebook user base. Proceedings of the National Academy of Sciences 118: e2111455118. https://doi.org/10.1073/pnas.2111455118

Bartumeus, F.; Oltra, A.; Palmer, J.R.B. 2018. Citizen science: a gateway for innovation in disease-carrying mosquito management? Trends in Parasitology 34: 727–729. https://doi.org/10.1016/j.pt.2018.04.010

Bowser, A.; Shilton, K.; Preece, J.; Warrick, E. 2017. Accounting for privacy in citizen science: ethical research in a context of openness. Proceedings of the 2017 ACM Conference on Computer-Supported Cooperative Work and Social Computing (CSCW 2017), Portland, OR, February 25–March 1, 2017. Pp. 2124–2136. https://doi.org/10.1145/2998181.2998305

Braz Sousa, L.; Fricker, S.R.; Doherty, S.S.; Webb, C.E.; Baldock, K.L.; Williams, C.R. 2020. Citizen science and smartphone e-entomology enables low-cost upscaling of mosquito surveillance. Science of the Total Environment 704: 135349. https://doi.org/10.1016/j.scitotenv.2019.135349

Brooks, D.R.; Boeger, W.A.; Hoberg, E.P. 2022. The Stockholm Paradigm: lessons for the emerging infectious disease crisis. MANTER: Journal of Parasite Biodiversity 22. https://digitalcommons.unl.edu/manter/22

Brooks, D.R.; Hoberg, E.P.; Boeger, W.A. 2019. The Stockholm Paradigm: Climate Change and Emerging Disease. University of Chicago Press, Chicago.

Brooks, D.R.; Hoberg, E.P.; Boeger, W.A.; Gardner, S.L.; Araujo, S.B.L.; Bajer, K.; et al. 2020. Before the pandemic ends: making sure this never happens again. World Complexity Science Academy Journal 1: 8. https://doi.org/10.46473/WCSAJ27240606/15-05-2020-0002/full/html

Carney, R.M.; Mapes, C.; Low, R.D.; Long, A.; Bowser, A.; Durieux, D.; et al. 2022. Integrating global citizen science platforms to enable next-generation surveillance of invasive and vector mosquitoes. Insects 13: 8. https://doi.org/10.3390/insects13080675

Cull, B. 2022. Monitoring trends in fistribution and deasonality of medically important ticks in North America using online crowdsourced records from iNaturalist. Insects 13: 404. https://doi.org/10.3390/insects13050404

Daněk, O.; Hrazdilová, K.; Kozderková, D.; Jirků, D.; Modrý, D. 2022. The distribution of *Dermacentor reticulatus* in the Czech Republic re-assessed: citizen science approach to understanding the current distribution of the *Babesia canis* vector. Parasites & Vectors 15: 132. https://doi.org/10.1186/s13071-022-05242-6

Daszak, P. 2012. Anatomy of a pandemic. Lancet 380: 1883–1884. https://doi.org/10.1016/S0140-6736(12)61887-X

Echeverria, A.; Ariz, I.; Moreno, J.; Peralta, J.; Gonzalez, E.M. 2021. Learning plant biodiversity in nature: the use of the citizen-science platform iNaturalist as a collaborative tool in secondary education. Sustainability 13: 735. https://doi.org/10.3390/su13020735

Eisen, L.; Eisen, R.J. 2021. Benefits and drawbacks of citizen science to complement traditional data gathering approaches for medically important hard ticks (Acari: Ixodidae) in the United States. Journal of Medical Entomology 58: 1–9. https://doi.org/10.1093/jme/tjaa165

Földvári, G.; Szabó, É.; Tóth, G.E.; Lanszki, Z.; Zana, B.; Varga, Z.; Kemenesi, G. 2022. Emergence of *Hyalomma marginatum* and *Hyalomma rufipes* adults revealed by citizen science tick monitoring in Hungary. Transboundary and Emerging Diseases 69: e2240–e2248. https://doi.org/10.1111/tbed.14563

Francisco, I.; Bailey, S.; Bautista, T.; Diallo, D.; Gonzalez, J.; Gonzalez, J., et al. 2022. Detection of velogenic avian paramyxoviruses in rock doves in New York City, New York. Microbiology Spectrum 10: e02061–21. https://doi.org/10.1128/spectrum.02061-21

Himmel, T.; Harl, J.; Matt, J.; Weissenböck, H. 2021. A citizen science–based survey of avian mortality focusing on haemosporidian infections in wild passerine birds. Malaria Journal 20: 417. https://doi.org/10.1186/s12936-021-03949-y

Hledík, M.; Polechová, J.; Beiglböck, M.; Herdina, A.N.; Strassl, R.; Posch, M. 2021. Analysis of the specificity of a COVID-19 antigen test in the Slovak mass testing program. PLOS ONE 16: e0255267. https://doi.org/10.1371/journal.pone.0255267

Hoberg, E.; Boeger, W.; Molnár, O.; Földvári, G.; Gardner, S.; Juarrero, A.; et al. 2022. The DAMA protocol, an introduction: finding pathogens before they find us. MANTER: Journal of Parasite Biodiversity 21. https://doi.org/10.32873/unl.dc.manter21

Jain, A.; Chan, S.K.M.; Soh, M.; Chow, L. 2019. Rediscovery of the orange gull butterfly, *Cepora iudith malaya*, in Singapore. Singapore Biodiversity Records 2019: 22–23.

Land-Zandstra, A.M.; van Beusekom, M.; Koppeschaar, C.; van den Broek, J. 2016. Motivation and learning impact of Dutch flu-trackers. Journal of Science Communication 15: A04. https://doi.org/10.22323/2.15010204

Lernout, T.; De Regge, N.; Tersago, K.; Fonville, M.; Suin, V.; Sprong, H. 2019. Prevalence of pathogens in ticks collected from humans through citizen science in Belgium. Parasites & Vectors 12: 550. https://doi.org/10.1186/s13071-019-3806-z

Meeus, S.; Silva-Rocha, I.; Adriaens, T.; Brown, P.M.J.; Chartosia, N.; Claramunt-López, B.; et al. 2023. More than a bit of fun: the multiple outcomes of a bioblitz. BioScience 73: 168–181. https://doi.org/10.1093/biosci/biac100

Mesaglio, T.; Callaghan, C.T. 2021. An overview of the history, current contributions and future outlook of iNaturalist in Australia. Wildlife Research 48: 289–303. https://doi.org/10.1071/WR20154

Molnár, O.; Hoberg, E.P.; Trivellone, V.; Földvári, G.; Brooks, D.R. 2023. Prevent-Prepare-Palliate: the 3P framework—integrating the DAMA protocol into global public health systems. In: An Evolutionary Pathway for Coping with Emerging Infectious Disease. S.L. Gardner, D.R. Brooks, W.A. Boeger, E.P. Hoberg (eds.). Zea Books, Lincoln, NE.

Molnár, O.; Knickel, M.; Marizzi, C. 2023. All hands on deck: turning evolutionary theory into preventive policies. In: An Evolutionary Pathway for Coping with Emerging Infectious Disease. S.L. Gardner, D.R. Brooks, W.A. Boeger, E.P. Hoberg (eds.). Zea Books, Lincoln, NE.

Narrod, C.; Zinsstag, J.; Tiongco, M. 2012. A One Health framework for estimating the economic costs of zoonotic diseases on society. EcoHealth 9: 150–162. https://doi.org/10.1007/s10393-012-0747-9

Nicolai, A.; Guernion, M.; Guillocheau, S.; Hoeffner, K.; Le Gouar, P.; Ménard, N.; et al. 2020. Transdisciplinary bioblitz: rapid biotic and abiotic inventory allows studying environmental changes over 60 years at the Biological Field Station of Paimpont (Brittany, France) and opens new interdisciplinary research opportunities. Biodiversity Data Journal 8: e50451. https://doi.org/10.3897/BDJ.8.e50451

Ortiz, E.; Juarerro, A. 2023. Monitoring: an emerging infectious disease surveillance platform for the 21st century. In: An Evolutionary Pathway for Coping with Emerging Infectious Disease. S.L. Gardner, D.R. Brooks, W.A. Boeger, E.P. Hoberg (eds.). Zea Books, Lincoln, NE.

Parra, C.; Cernuzzi, L.; Rojas, R.; Denis, D.; Rivas, S.; Paciello, J.; et al. 2020. Synergies between technology, participation, and citizen science in a community-based dengue prevention program. American Behavioral Scientist 64: 1850–1870. https://doi.org/10.1177/0002764220952113

Pavelka, M.; Van-Zandvoort, K.; Abbott, S.; Sherratt, K.; Majdan, M.; CMMID COVID-19 Working Group; et al. 2021. The impact of population-wide rapid antigen testing on SARS-CoV-2 prevalence in Slovakia. Science 372: 635–641. https://doi.org/10.1126/science.abf9648

Poh, K.C.; Evans, J.R.; Skvarla, M.J.; Machtinger, E.T. 2022. All for One Health and One Health for all: considerations for successful citizen science projects conducting vector surveillance from animal hosts. Insects 13: 492. https://doi.org/10.3390/insects13060492

Porter, W.T.; Wachara, J.; Barrand, Z.A.; Nieto, N.C.; Salkeld, D.J. 2021. Citizen science provides an efficient method for broad-scale tick-borne pathogen surveillance of *Ixodes pacificus* and *Ixodes scapularis* across the United States. MSphere 6: e00682-21. https://doi.org/10.1128/mSphere.00682-21

Prunuske, A.; Fisher, C.; Molden, J.; Brar, A.; Ragland, R.; vanWestrienen, J. 2021. Middle-school student engagement in a tick testing community science project. Insects 12: 1136. https://doi.org/10.3390/insects12121136

Tan, Y.-R.; Agrawal, A.; Matsoso, M.P.; Katz, R.; Davis, S.L.M.; Winkler, A.S.; et al. 2022. A call for citizen science in pandemic preparedness and response: beyond data collection. BMJ Global Health 7: e009389. https://doi.org/10.1136/bmjgh-2022-009389

Tran, T.; Porter, W.T.; Salkeld, D.J.; Prusinski, M.A.; Jensen, S.T.; Brisson, D. 2021. Estimating disease vector population size from citizen science data. Journal of the Royal Society Interface 18: 20210610. https://doi.org/10.1098/rsif.2021.0610

Uiterwijk, M.; Ibáñez-Justicia, A.; van de Vossenberg, B.; Jacobs, F.; Overgaauw, P.; Nijsse, R.; et al. 2021. Imported *Hyalomma* ticks in the Netherlands 2018–2020. Parasites & Vectors 14: 244. https://doi.org/10.1186/s13071-021-04738-x

Westreicher, F.; Cieslinski, M.; Ernst, M.; Frigerio, D.; Heinisch, B.; Hübner, T.; Rüdisser, J. 2021. Recognizing failures in citizen science projects: lessons learned. Proceedings of Austrian Citizen Science Conference 2020, Vienna, Austria, September 14–16, 2020. PoS(ACSC2020)007. https://doi.org/10.22323/1.393.0007

Wu, H.-D. I.; Lin, R.-S.; Hwang, W.-H.; Huang, M.-L.; Chen, B.-J.; Yen, T.-C.; Chao, D.-Y. 2023. Integrating citizen scientist data into the surveillance system for avian influenza virus, Taiwan. Emerging Infectious Diseases 29: 45–53. https://doi.org/10.3201/eid2901.220659

Conclusion

Emerging Diseases and Institutional Stumbling Blocks, or Why We Cannot Afford Business as Usual

Daniel R. Brooks and Salvatore J. Agosta

Abstract

The emerging infectious disease (EID) crisis is accelerating toward a largely unprepared humanity. The contributions to this volume show that we can change this situation by following principles and policies rooted in evolutionary biology and the natural history of disease-causing organisms and their hosts. The lessons from nature for how to cope with EID are clear, and the applied framework to implement them is in place, so why has so little progress been made? We believe the answer is that our socioeconomic institutions have become stumbling blocks, large bureaucracies that favor preservation of the status quo to their own benefit, adverse to new ideas and technologies that threaten business as usual. The contributions to this volume illuminate an optimistic, evolutionarily informed pathway into the future of EID prevention and mitigation, recognizing that the most fundamental stumbling block to success is not technological innovation or scientific know-how but human behavioral change.

Keywords: institutional stumbling blocks, DAMA protocol, disease prevention, proactive management, Stockholm Paradigm, crisis response, myth of control, COVID-19

With respect to the emerging infectious disease (EID) crisis accelerating toward humanity, the time is short, the danger is great, and we are largely unprepared. As the contributions in this volume show, however, we do not have to remain unprepared. We know the scope of what we are facing, we know why it is happening, and we know what to do to mitigate the impacts. We have a robust conceptual framework supported by substantial empirical data and a useful modeling platform. This has led to a more accurate assessment of the risk space and the risk potential than was previously known. And that knowledge has led to a comprehensive plan for coping effectively with EID. So, what are we waiting for? In short, a sea change in institutional behavior.

Socioeconomic institutions become *stumbling blocks* when actions that seem adequate and reasonable in view of the problems to be addressed cannot be executed effectively for reasons that are not clear, inhibiting the institutions own stated functions (Vasbinder and Sim, 2022; Brooks and Agosta, forthcoming 2024). Five categories of social institutions have become stumbling blocks that are impeding effective action on EID.

Finances (the myth of control)

> *Many . . . have grown fat by taking few risks. One cannot blame them for this; one can only despise them.*
> —Frank Herbert (*Dune*, 1965)

> *Technology, in common with many other activities, tends toward avoidance of risk by investors. Uncertainty is ruled out if possible. Capital investment follows this rule, since people generally prefer the predictable. Few recognize how destructive this can be, how it imposes severe limits on variability and this makes whole populations fatally vulnerable to the shocking ways the universe can throw the dice.*
> —Frank Herbert (*Heretics of Dune*, 1984)

Disease is a fact of life and always has been. Half the species on this planet are pathogens of some sort and the rest are their hosts. Treatment costs and production losses are always associated with disease. We factor them into our taxes and our food and water prices and pest-control programs. This is business as usual. Only a small allocation is made for crisis response, in the belief that EIDs will be rare, and if there is an outbreak, it will be short-lived. The total commitment to business as usual and crisis response means no funds are available for even pilot studies in proactive measures, much less long-term and broad commitments to them. This is true despite the fact that the costs of crisis response are much higher than the costs of prevention. The usual response to a request for funding for novel approaches is that funding is a zero-sum game, but this means we have to set a ceiling on our response, even if the urgency is growing.

The Stockholm Paradigm shows that humanity has been poor at assessing the magnitude of the risk space for EID as well as the existential nature of the threat in the context of global climate change and globalized trade and travel by humans, the world's primary ecological super-spreaders of disease. Global estimates suggest that the annual cost of EIDs before the COVID-19 pandemic was about US$1.3 trillion, more than the gross domestic product (GDP) of all but fifteen of the world's countries. In the United States, if business as usual continues, treatment expenses and production losses caused by EIDs that affect crops and livestock will exceed the national GDP within eighty years (Brooks et al., 2022; Trivellone et al., 2022). *This is not survivable.* The global financial system must fund proposals calling for private and public investment in effective *prevention* to decrease the chances that another pandemic will occur; that is, making investments in the DAMA (Document, Assess, Monitor, Act) protocol (e.g., Hoberg, Boeger, Molnar, et al., 2022). Every funded DAMA project would be an investment in "an ounce of prevention is worth a pound of cure." Increased spending on prevention when there is no ongoing pandemic to prevent catastrophic losses later would be true investment in the future.

Vested interest (maximizing exploitation at the expense of potential for exploration)

> *Quite naturally, holders of power wish to suppress wild research. Unrestricted questing after knowledge has a long history of producing unwanted competition. The powerful want a "safe line of investigation," which will develop only those products and ideas that can be controlled and, most important, that will allow the larger part of the benefits to be captured by inside investors. Unfortunately, a random universe full of relative variables does not ensure such a "safe line of investigation."*
> —Frank Herbert (*Heretics of Dune*, 1984)

Social institutions are meant to act in the public good, but some use those institutions to amplify personal power and shield themselves from threats to their vested interests, placing their job security above all other considerations. In the arena of emerging disease, the two most powerful vested interests are clinical practice and clinical science. For more than a century and a half, clinical practice has followed a formula of medicating the ill, vaccinating those at risk, and eradicating portions of biodiversity associated with disease transmission. It is evident that this traditional triad is not working in terms of coping with the emerging disease crisis, but it is making a lot of money for people and keeping many clinicians and diagnostic labs operating. Clinical research has become largely domesticated within molecular laboratories in universities, research institutes, and pharmaceutical companies. Driven by both the desire to maintain public and philanthropic financial vested interest, clinical research is also driven by a massive profit motive, which attracts investment funding. This has created a massive incentive for lying about the efficacy of results, obscuring, for example, the reality that the most productive 20-year period for vaccine production was 1880 to 1900 (see Brooks et al., 2019) and that antibiotic resistance has become a marketing tool. Neither of these categories of vested interest will willingly share their resources; in fact, under current conditions they cannot share without risking extinction. Thus, if new resources are suggested, these vested interests argue they should get them. Clinical practice and research are comfortable with their way of operating, even if it is not working. Simply put, there is too much profit in peoples' misery and not enough in their well-being.

If implemented, the DAMA protocol would alleviate pressure on clinicians through prevention, for example, but would require the services of evolutionary biologists. The COVID-19 pandemic showed that clinicians and technicians are not yet willing to break from their vested interests. Like clinical research, scientific research about EIDs is done within molecular laboratories in universities, research institutes, and pharmaceutical companies. Yet researchers have little direct contact with either clinicians or patients. The need to maintain public and philanthropic financial backing drives researchers to behave as if clinical practice was merely a by-product of their work. The vested interests of this group also preclude discussions of prevention;

prevention would divert funds from vaccine and medication production. Neither group can share resources for the common good without one or the other risking the loss of support to do their work. Neither group will cooperate so long as each sees the other as an impediment to its vested interests. Both sides feel weak and vulnerable because what they are doing is not working. Whenever new resources become available, both groups engage in turf wars, arguing about who gets them. It is no wonder that there is considerable resistance to the entire outlook presented in this volume, but that does not alter the facts. Unless the vested interests are able to replace those who see their positions threatened by efforts to change institutions with those who believe that "my vested interests are served by supporting your vested interests," efforts to cope effectively with the EID crisis will be nonexistent.

Political will and support (collective action trap)

> At the quantum level our universe can be seen as an indeterminate place, predictable in a statistical way only when you employ large enough numbers. Between that universe and a relatively predictable one where the passage of a single planet can be timed to a picosecond, other forces come into play. For the in-between universe where we find our daily lives, that which you believe is a dominant force. Your beliefs order the unfolding of daily events. If enough of us believe, a new thing can be made to exist. Belief structure creates a filter through which chaos is sifted into order.
> —Frank Herbert (*Heretics of Dune*, 1984)

Without disasters, there is little to no public pressure for longer-term solutions. No immediate reward! No possibility of appearing heroic! This is the reason there has been no effort to take proper advantage of any of the myriad pandemics of the past decade to use them as a catalyst and rationale for redirecting funds and changing policies toward proactive measures. Elected politicians have restricted event horizons; totalitarians are preoccupied with survival. The DAMA bet is that within 7 years of introducing a DAMA program, a country will save the cost of the DAMA program in reduced crisis-response direct and indirect costs. Seven years is beyond the event horizon for most politicians, and given planning cycles, we are looking at more like 10 years to get a DAMA program underway, so 17 years before the true economic benefits are realized. This leads politicians

to listen to the vested interests that prefer their own financially stable status quo. And since none of the pandemics of the past 20 years has catalyzed a sense of immediate peril and a need for prevention, it is possible that the magnitude of a pandemic necessary to move political will to action will be so great that it will come too late.

Contemporary political will is increasingly characterized by political weakness. Politicians and the systems of government in which they operate cannot think and act proactively. Poor understanding of the complex causes for emerging problems leads to the *collective action trap*; members of a collective pursue individual profit or satisfaction rather than behave in the group's best long-term interest. This makes it impossible to act on knowledge that consequences of disasters can be mitigated if proactive planning is applied. Those in charge increasingly think only of disbursing money when dealing with problems in an effort to appear heroic. And that does not enhance the long-term survival of the people whose best interests they are supposed to serve.

The world needs politicians who can resist the impulse to wait until it is too late and then make a heroic gesture. They would be people who put public service ahead of career advancement, being stewards of the public good. How can humanity find such ethical leaders in this time of fear and uncertainty?

Bureaucracy (institutional resistance)

> Bureaucracy destroys initiative. There is little that bureaucracies hate more than innovation, especially innovation that produces better results than the old routines. Improvements always make those at the top look inept. Who enjoys looking inept?
> —Frank Herbert (*Heretics of Dune*, 1984)

All organizations are bureaucratized, be they corporations, political entities, nongovernmental organizations (NGOs), or religions. Bureaucracies are the way in which different vested interests try to preserve their financial base. But there is a cost. By their nature, bureaucracies are rigid and responsive only to questions of "what" or "how." Bureaucracies can be efficient and rigid (what), or they can be flexible and inefficient (how). We have tended toward efficient and rigid, which progressively makes bureaucracies less capable of coping with changing conditions. Bureaucracies are never good at "why." The assumption is that the entities that created the bureaucracy know "why," and bureaucracies are there only to implement "what" and "how." They are not flexible enough to cope with change, and they cannot initiate new policies. Nor can they cooperate with each other. Concerned with their own job security, bureaucratic leaders become masters of deflecting responsibility. As the world grows more complex, bureaucracies subdivide, increasing organizational complexity and inflexibility. They spend less time delivering services they were mandated to provide.

Disasters never unfold in accordance with organizational charts. Bureaucrats stubbornly went into crisis-response mode during the COVID-19 pandemic, hoping to contain the problem until it fixed itself and disappeared. They spoke past the crisis about returning to normal, exacerbating the financial, vested interest, and political stumbling blocks. Early in the pandemic, bureaucrats altered policies so rapidly that the institutions they represent lost public credibility, opening the doors to a wide range of conspiracy theories. Bureaucratic institutions have the potential to be helpful if they embed lessons learned from previous emergencies in preparing for those to come. But no institution associated with the pandemic has changed any policies, so public health institutions have been hit unexpectedly by the monkeypox outbreak and recurrence of the poliovirus. Bureaucracies must be led by people who appreciate the value of being as efficient as possible without losing the flexibility to cope with change by changing.

Talents (getting the right people)

> There is another form of temptation, even more fraught with danger. This is the disease of curiosity. . . . It is this which drives us to try and discover the secrets of nature, those secrets that are beyond our understanding, which can avail us nothing, and which man should not wish to learn.
>
> —St. Augustine

When too few people are adequately trained or temperamentally suited for essential jobs in a time of crisis, the lack of talent becomes a stumbling block. The people who are employed to do the work of the institutions too often have been chosen for their lack of curiosity. Those few who show assertive initiative and innovation are too often ostracized or fired. That there is no general outcry against such treatment is a testament to the efficiency with which we have selected our talent.

Lack of talent exposes weaknesses in educational systems and hiring practices. Maintaining business as usual leads to a preference for people who lack curiosity or see institutions as a means of amplifying personal power. Anticipating and mitigating disasters requires understanding the nature of the complex processes that lead to them. Education becomes an obstacle. Even if we have developed proactive action plans to deal with potential disasters, who will be able to carry them out when they are needed? The capacity to understand and cope with complexity needs to be a foundational element of all educational systems, but few educators can teach a next generation about complex systems.

We have failed to train, hire, retain, and reward innovators in areas critical to finding an effective way to cope with emerging disease. We have failed to train, hire, retain, and reward a new generation of technologically adept field biologists knowledgeable about the natural history and evolutionary biology of pathogens and pathogen-host systems in wildlands and across habitat interfaces. These are the people who understand why medication, vaccination, and eradication are working more poorly with each passing year and why accelerating climate change creates a heightened sense of urgency. *And most importantly, we have failed to train, hire, and reward scientists who see the emerging disease crisis as a public service with possible career opportunities rather than simply as a pathway to career advancement.*

A lack of talent was the greatest stumbling block to coping with the COVID-19 pandemic. Humanity had failed to train, hire, and reward the critical innovators to cope with the problem. Instead, we witnessed denial of the problem and dithering. Then enormous amounts of public and private funding went into crisis response rather than prevention. As a result, there are currently more variants of the virus in nonhuman hosts than in humans because none of the entrenched institutional health personnel could possibly imagine that (1) the outbreak would become a pandemic, so air travel between Wuhan and the rest of the world was not stopped early enough to prevent the virus from becoming global; (2) having come from wildlife to humans, the virus could then move from humans back to wildlife in areas where infected humans had been allowed to travel; and (3) this could result in new human-infecting variants (e.g., Omicron) emerging from the newly infected nonhuman hosts (Boeger et al., 2022; Hoberg, Boeger, Brooks, et al., 2022). Training and research in infectious disease still has little input from evolution. The world does not train, hire, or reward technologically adept field biologists knowledgeable about the natural history and evolutionary biology of pathogens and pathogen-host systems to collaborate with clinicians and clinical researchers. Educational institutions still do not provide students with a substantial understanding of complex systems, how they function, and how to survive in a world that itself is a complex system.

Institutions must hire people who question the status quo and offer novel approaches within the context of working collaboratively. Managers must resist the urge to hire people who remind them of younger versions of themselves. They need to hire more people who see their job as a public service with career opportunities rather than simply as a career opportunity. Managers must then understand they are not behavior problems but instead potential allies whose new ideas can effectively meld with successful elements of traditional approaches.

We Are Out of Time

> It turned out that we didn't have much time
> after all. . . . But little as it was, we threw it
> away unused. . . . What do we do now?
> —Isaac Asimov (*Foundation*, 1951)

The danger from EID during this period of global climate change and in conjunction with all other threat multipliers is existential. Everything we do from now on will be triage. We can see a window of near time narrowing almost daily toward a human population bottleneck predicted to occur about 2050. The job of those concerned with EID is to try to save as much as we can—people, crops, livestock, wildlife, and the infrastructure of public, wildlife, veterinary, and crop health. We need to implement DAMA programs as widely as possible as soon as possible to buy us time to try to prevent some outbreaks and mitigate the damage from outbreaks we cannot prevent. Even a few years would be welcome.

In his classic book, Elton (1958) cautioned that the major impacts of global climate change would be migration and conflict, an assessment with which Audy (1958) would undoubtedly have agreed, given his understanding of pathogen movements and the long historical association between conflict, migration, and disease (de Kruif, 1926; Zinnser, 1935; Garrett, 1994, 2001; Diamond, 1997). Audy (1958) further provided the underlying framework for the truth of what we don't see can hurt us: The primary threat of disease emergence and reemergence is in the reservoir hosts and habitat interfaces, where the threat of recurring outbreaks takes on an aura

of "stop me before I kill again," leading to an economic impact akin to the "death of a thousand cuts." The resulting *pathogen pollution* means that the pathogen is always there not only before it announces itself but long after, and its true cost begins only after the first acute outbreak is over. Once a pathogen has broken out, therefore, efforts to prevent additional outbreaks must be in place the day the outbreak is declared over. Prevention in the form of DAMA operations is not a luxury, but a necessity.

The migrants we need to worry about most with respect to EID are not people. Pathogens that cause diseases of wildlife, crops, and livestock far outnumber pathogens that cause disease in humans. The second greatest threat is zoonoses from global trade—again, we are doing a better job, but it is still not enough. There are not enough people to inspect everything coming into each point of entry for imported plants and animals, and demand for fresh agricultural products puts increasing pressure on ever more rapid movement of goods from point of entry to market. The lack of inspectors at the entry points places increasing pressure on inspection at the point of origin, but those places tend to be where funds for inspection are limited and pressure to move the goods out is highest. The third greatest threat is tourism and global travel—40% of the world's 1.24 billion tourists in 2016 visited the European Union (EU); additionally, the EU received 640 million foreign tourists (as well as 1.3 billion within-EU trips) in 2017 along with 2.4 million refugees and migrants. Only at the fourth level of threat do we find a public health risk from refugees and migrants displaced by conflict. In the near future, the increasing number of internally displaced climate refugees will become a significant issue as well (Brown and Brooks, 2021).

Summary

Institutional stumbling blocks are an enormous impediment to coping effectively with the EID crisis. Humans have a strong need for drama and heroism, a strong attraction to magic, and a strong aversion to bad news, especially news that involves taking personal responsibility. Knowing this makes us generally more hopeful than optimistic that humanity will change course and adopt a more effective approach to coping with the EID crisis. The contributions in this volume illuminate an optimistic pathway into the future. An effort to resolve the parasite paradox led us to a renewed appreciation for Darwin's fundamental views about evolution, and that led us to the Stockholm Paradigm. Understanding the Stockholm Paradigm in the context of EID led to the DAMA protocol. Finding a place for the DAMA protocol in contemporary responses to EID led us to a coherent public-policy platform. All that has led us to understand the fundamental stumbling blocks to achieving success and to recognize that the greatest need is not technological innovation but human behavioral change. We have the technology to cope effectively with the problem; do we have the fortitude to change a losing game?

Literature Cited

Asimov, I. 1951. Foundation. Gnome Press, New York.

Audy, J.R. 1958. The localization of disease with special reference to the zoonoses. Transactions of the Royal Society of Tropical Medicine and Hygiene 52: 308–334. https://doi.org/10.1016/0035-9203(58)90045-2

Boeger, W.A.; Brooks, D.R.; Trivellone, V.; Agosta, S.J.; Hoberg, E.P. 2022. Ecological super-spreaders drive host-range oscillations: Omicron and risk space for emerging infectious disease. Transboundary and Emerging Diseases 69: e1280–e1288. https://doi.org/10.1111/tbed.14557

Brooks, D.R.; Agosta, S.J. Forthcoming 2024. A Darwinian Survival Guide: Hope for the Twenty-First Century. MIT Press, Cambridge, MA.

Brooks, D.R.; Hoberg, E.P.; Boeger, W.A. 2019. The Stockholm Paradigm: Climate Change and Emerging Disease. University of Chicago Press, Chicago.

Brooks, D.R.; Hoberg, E.P.; Boeger W.A.; Trivellone, V. 2022. Emerging infectious disease: an underappreciated area of strategic concern for food security. Transboundary and Emerging Diseases 69: 254–267. https://doi.org/10.1111/tbed.14009

Brown, H.A.; Brooks, D.R. 2021. How 'managed retreat' from climate change could revitalize rural America: revisiting the Homestead Act. The Conversation. https://theconversation.com/how-managed-retreat-from-climate-change-could-revitalize-rural-america-revisiting-the-homestead-act-169007

de Kruif, P. 1926. Microbe Hunters. Houghton Mifflin Harcourt, New York.

Diamond, J. 1997. Guns, Germs, and Steel: The Fates of Human Societies. W.W. Norton, New York.

Elton, C.S. 1958. The Ecology of Invasions by Animals and Plants. Methuen, London.

Garrett, L. 1994. The Coming Plague: Newly Emerging Diseases in a World Out of Balance. Farrar, Straus and Giroux, New York.

Garrett, L. 2001. Betrayal of Trust: The Collapse of Global Public Health. Oxford University Press, Oxford.

Herbert, F. 1965. Dune. Chilton Books, Philadelphia.

Herbert, F. 1984. Heretics of Dune. Putnam, New York.

Hoberg, E.P.; Boeger, W.A.; Brooks, D.R.; Trivellone, V.; Agosta, S.J. 2022. Stepping-stones and mediators of pandemic expansion—a context for humans as ecological super-spreaders. MANTER: Journal of Parasite Biodiversity 18. https://doi.org/10.32873/unl.dc.manter18

Hoberg, E.P.; Boeger, W.A.; Molnár, O.; Földvári, G.; Gardner, S.; Juarrero, A.; et al. 2022. The DAMA protocol, an introduction: finding pathogens before they find us. MANTER: Journal of Parasite Diversity 21. https://doi.org/10.32873/unl.dc.manter21

Trivellone, V.; Hoberg, E.P.; Boeger, W.A.; Brooks, D.R. 2022. Food security and emerging infectious disease: risk assessment and risk management. Royal Society Open Science 9: 211687. https://doi.org/10.1098/rsos.211687

Vasbinder, J.W.; Sim, J.Y.H. (eds.). 2022. Buying Time for Climate Action: Exploring Ways around Stumbling Blocks. Exploring Complexity series, vol. 8. World Scientific, Singapore.

Zinsser, H. 1935. Rats, Lice and History. Little, Brown and Company, Boston.